Chance in Evolution

Chance in Evolution

Edited by Grant Ramsey and
Charles H. Pence

The University of Chicago Press CHICAGO & LONDON

The University of Chicago Press, Chicago 60637
The University of Chicago Press, Ltd., London
© 2016 by The University of Chicago
All rights reserved. Published 2016.
Printed in the United States of America

25 24 23 22 21 20 19 18 17 16   1 2 3 4 5

ISBN-13: 978-0-226-40174-4 (cloth)
ISBN-13: 978-0-226-40188-1 (paper)
ISBN-13: 978-0-226-40191-1 (e-book)
DOI: 10.7208/chicago/9780226401911.001.0001

Library of Congress Cataloging-in-Publication Data

Name: Ramsey, Grant, 1972– editor. | Pence, Charles H., editor.
Title: Chance in evolution / edited by Grant Ramsey and Charles H. Pence.
Description: Chicago ; London : University of Chicago Press, 2016. | © 2016 |
    Includes bibliographical references and index.
Identifiers: LCCN 2016009805 | ISBN 9780226401744 (cloth : alkaline paper)
    | ISBN 9780226401881 (paperback : alkaline paper) | ISBN 9780226401911
    (e-book)
Subjects: LCSH: Evolution (Biology)—Philosophy. | Chance.
Classification: LCC QH360.5 .C474 2016 | DDC 567. 801—dc23 LC record
    available at http://lccn.loc.gov/2016009805

♾ This paper meets the requirements of ANSI/NISO Z39.48-1992
(Permanence of Paper).

CONTENTS

ACKNOWLEDGMENTS

The editors offer manifold thanks to all the contributors for their lucid, accessible, and groundbreaking contributions; to the University of Chicago Press, especially to Christopher Chung and Evan White, for their extensive patience as we put this volume together during what turned out to be a far longer process than we (and they) had expected; and to Christie Henry, who stepped in just as we were finishing the volume to guide it to completion. This book was completed while one of us (Ramsey) was on a National Endowment for the Humanities–supported fellowship at the National Humanities Center. We thank the NEH and the NHC for their support. Any views, findings, conclusions, or recommendations expressed in this book do not necessarily reflect those of the National Endowment for the Humanities.

Chance in Evolution from Darwin to Contemporary Biology

Grant Ramsey and Charles H. Pence

The roles that the concept of chance plays in understanding the world around us have changed dramatically over the centuries. For many Enlightenment thinkers, for instance, chance was at best an admission of ignorance. Consider Voltaire, who quite forcefully said, "Chance is a word void of sense; nothing can exist without a cause."[1] In Darwin's *On the Origin of Species*, we still see traces of this sentiment in statements like this: "I have hitherto sometimes spoken as if the variations . . . had been due to chance. This, of course, is a wholly incorrect expression, but it serves to acknowledge plainly our ignorance of the cause of each particular variation" (C. Darwin 1859, 131). But this quote stands against dozens of other uses of chance in the *Origin* that seem to imply the illuminating power of chance. Darwin uses the term *chance* sixty-seven times in the *Origin*, and most of these instances are in the context of either the origin of variation ("the chance of their appearance will be much increased by a large number of individuals being kept" [41]) or survival and reproduction ("the best chance of surviving and of procreating their kind" [81]). And in some places Darwin seems to endow chance with causal powers of its own ("Mere chance, as we may call it, might cause one variety to differ in some character from its parents" [111]); he elsewhere steadfastly denied such a causal version of chance.

We thus see in Darwin a tension—one that derives from his unique position as one of the last great scientific thinkers who worked exclusively prior to a profound shift in our understanding and use of chance, probability, and (later) statistics. What occurred in the subsequent decades in evolutionary biology (and other sciences as well, including physics and the fledgling social sciences) is a fascinating transition from chance as a placeholder for unknown mechanisms, something to be replaced with deeper understanding, to chance taking a central explanatory role (Hacking 1990; Krüger, Daston, and Heidelberger 1987; Krüger, Gigerenzer, and Morgan 1987). This "probabilistic revolution" or "taming of chance," as it has variously been described by commentators, came a bit too late for Darwin, who owned, but never used, the seminal works of Adolphe Quetelet, the first person to extensively develop the use of statistics in science. Statistics were seized upon, however, around the turn of the twentieth century by a group of biologists keen to form the life sciences in a mathematical image (Pence 2015). By the time they were done and their work was taken up by the modern synthesis, biology had forever become a statistical science.

This often-told story of the development of chance in evolutionary theory, as our contributors make clear in this volume's chapters, is at best overly simplified. For one, the history of chance in evolution does not begin with Darwin, nor does it end with the modern synthesis. For the idea that the natural world was chancy by no means spontaneously generated in the late nineteenth century. It may be contextualized, rather, within a debate that spans many of the most important figures in the history of philosophy, from Democritus through Aristotle, Aquinas, and Kant, down to the present day.

And as the following chapters on contemporary biology make clear, the "chancy" nature of evolution by no means forms a settled, static picture. All agree that it is now nearly impossible to discuss contemporary evolutionary theory in any depth at all without making reference to at least some concept of "chance" or "randomness": many processes are described as chancy, outcomes are characterized as random, and many evolutionary phenomena are thought to be best described by stochastic or probabilistic models. Chance is taken by a variety of authors to be central to the understanding of fitness, genetic drift, macroevolution, mutation, foraging theory, and environmental variation, to mention but a few examples. The full impact of chance, however, as many of our contributors point out, is only beginning to be fully understood—whether in the realm of paleontology, mutation, or experimental evolution. This is a history to which more is added every day.

Beyond the question of whether particular biological phenomena should be considered chancy, there is a major challenge in understanding what is meant by chance: what conceptually, as well as empirically, does it mean to ascribe chance to a process or an outcome? We may intend a wide variety of different things when we use terms like *chance, randomness*, or *stochasticity*. To get a handle on this diversity of meanings, let us begin by drawing a few broad distinctions. First, we might be making reference to a particular sequence of outcomes, claiming that this sequence is random, in the sense often deployed by mathematicians. Randomness in this sense is a property of outcomes, such as a large number of flips of a fair coin or rolls of a fair die. Second, we might be speaking about a subjective sense of unpredictability, describing the inability of a particular agent to make a certain prediction within the context of a particular theory (Plutynski et al., in chapter 3, illuminatingly describe this sense of chance as a "proxy for probability"). Third, we could consider events to be chancy in the sense of being unusual or unlucky for a particular kind of organism within a particular environment or selective regime (as, say, being squashed by a meteor is for earthly organisms). Finally, we might be speaking either of a variety of indeterminism that arises, sui generis, at the level of evolutionary theory, or of indeterminism in evolutionary theory connected to or based on indeterminism in other areas of science, such as fundamental physics. (It is telling that this last sense appears so little throughout this volume—its most prominent use to date has been in the work of Fisher, as Plutynski et al. detail.)

These, of course, by no means exhaust the philosophical distinctions that can be drawn among concepts of chance. And for each of these notions, there are yet more stories to tell. Each weaves itself into the various branches of evolutionary theory in myriad different ways, with a wide variety of effects on the history and current state of life on earth. Each is grounded in a particular trajectory in the history of philosophy and the history of biology and has inspired a variety of responses throughout science and culture.

This dizzying multiplicity of ways of understanding chance inspired us to bring together the contributors to this book. We begin with chapters on the historical development of chance in evolutionary theory. David J. Depew chronicles the history of the idea of contingency in biology. Contingency, he deftly argues, has been a hallmark of biology since the time of Aristotle. There were, to be sure, some countervailing influences arising under the rubric that Mayr later called "essentialism" or "Platonism." These come, however, not from Plato, but from the seventeenth-century botanical community, and they must be weighed against a long-standing focus in the life sciences on contingency and the similar-but-not-identical relationship between parents and offspring. Genuine indeterminacy, too, is an old idea, looming in the background of debates over the life sciences in the figure of Epicurus. The contemporary prominence of chance should thus be seen less as an historical aberration than as an attempt to understand, formalize, and quantify chance in its many manifestations, dating back to the earliest speculation on the sciences of life.

Darwin's own views of chance did, however, provide a truly novel approach. While chance is certainly a factor in Darwin's picture of evolution, this is chance harnessed as a positive force for the generation of adaptation. This picture of chance, nascent and not fully elaborated in Darwin's own work, sets the context for its later combination with the tools of statistics and probability at the end of the nineteenth and beginning of the twentieth centuries, as well as discoveries in genetics from the 1920s to the present, that allows biological science to recapture what Depew calls "a sense of wonder about biological purposiveness, untethered from cosmic teleology."

Darwin's views, in great and illuminating detail, provide the material for Jonathan Hodge's contribution, chapter 2. Hodge's analysis penetrates deeply into Darwin's work in his early notebooks, arguing that his views on the nature of chance—despite occasional claims to the contrary—remained largely constant and were not greatly altered by his first formulation of the theory of natural selection in 1838. Darwin is wholly committed, Hodge argues, to the metaphysical position that we would now call determinism, and to the understanding of chance as nothing more than our ignorance of the precise details of biological systems. The central focus of these invocations of chance is, and remains throughout Darwin's published works, the reference to natural selection as a process that is, while chancy, not what Hodge calls "fortuitous"—for it is for a very well-determined *reason* that the fitter organisms survive while the less fit perish.

In a provocative conclusion, Hodge expands this claim about Darwin's works to the realm of contemporary philosophy of biology. For just as a causal process of nonfortuitous differential reproduction was central to Darwin's picture of natural selection, Hodge argues, we have received no evidence in the intervening years since 1838 that would lead us to interpret natural selection, and the chance that we find therein, in any other way.

Moving forward from Darwin, the thinkers of the modern synthesis are the subject of chapter 3, by Anya Plutynski, Kenneth Blake Vernon, Lucas John Matthews, and Daniel Molter. Despite a widespread acknowledgment that the question of the role of chance in evolution shifted dramatically during the period that we now describe as the modern synthesis, during which the Darwinian understanding of evolution by natural selection was integrated with the Mendelian or genetic picture of heredity and breeding, no comprehensive study of the ways the key authors from that period understand chance has yet been prepared, and this is the task that Plutynski and colleagues set for themselves. Examining the central works of R. A. Fisher, Sewall Wright, Theodosius Dobzhansky, Ernst Mayr, G. G. Simpson, and Ledyard Stebbins, the authors describe and track five separate notions of chance through a pivotal period in the history of biology, spanning some two decades from 1930 to 1950.

They conclude that, while these authors certainly found much about which to disagree, a core set of five meanings of chance is by and large held in common. These are expressed in a fairly widely shared set of evolutionary processes, including mutation, meiosis, geographic isolation, inbreeding, genetic drift, and even microscale indeterminism and macroscale contingency. The changes in the modern synthesis, then, do not represent a conceptual shift in how chance is understood, or even a transformation in views about where chance appears in evolution; rather, they have to do with the relative *importance* attributed—as an empirical matter of fact—to each of the various notions of chance in evolutionary processes.

In chapter 4, J. Matthew Ashley considers the impact of evolution's "chanciness" on its reception by two quite different theological communities: one is typified by the Princeton theologian Charles Hodge in the late nineteenth century, and the other is the contemporary Roman Catholic community typified by Cardinal Christoph Schönborn. In both cases, Ashley compellingly demonstrates, chance was instrumental to the way these thinkers understood the impact of evolution on their theological commitments. And to the extent that both are relatively hostile to Darwinian evolution, this hostility can profitably be understood by pointing to the impact of evolution's chanciness on certain core theological commitments shared by both authors.

Ashley then turns his focus to the resolution of these debates, arguing that much of the acrimony could be resolved by a combination of two factors: first, a clearer understanding of just what chance in evolution amounts to, with insights garnered from contemporary biological and philosophical understanding, and second, a less constrained theological picture, drawing on inspiration from both biblical and early-Church sources, which can make room for the impact of chance in a way that could still fully satisfy the demands of theology.

Further pursuing the question of how we should understand ourselves in light of chance and the Darwinian revolution, part 1 of the volume closes with chapter 5, by Michael Ruse, examining the perennial question of the impact of chance on the question of human evolution. Ruse begins the tale with pre-Darwinian, theistic pictures of the development of mankind, passing through the proto-evolutionary theories of Erasmus Darwin to Darwin's nineteenth-century predecessors, Jean-Baptiste Lamarck and Robert Chambers. Throughout all, he argues, we find a progressive development in the natural world, with mankind at its peak. Darwin changed all that. The struggle for existence as described in the *Origin* leaves no room for progress of this kind, and that book devotes only a single sentence to the question of the evolution of humans.

Despite the near silence regarding human evolution in the *Origin*, Darwin had been privately applying evolutionary theory to humans and their characteristics in his notebooks for some time. It was Darwin's subsequent book, *Descent of Man*, that made the human evolution connection explicit. But despite what we might expect, Darwin still seems to be a firm believer in the progressive nature of evolution. Humans are still, indeed, special. But it was Darwin's genius, Ruse argues, to show that it was *possible* to believe in evolution without the baggage of a progressive picture of the development of life. Stephen Jay Gould, in the aftermath of the sociobiology controversy, was perhaps the most vocal person to take him up on the bargain, arguing powerfully against the idea of biological progress. But the issue is by no means settled—Ruse offers a variety of evidence from biologists as diverse as Julian Huxley, Richard Dawkins, and Simon Conway Morris who have, in turn, taken the other side, offering support for at least some tempered idea of biological progression. While none would now argue that human beings were a necessary outcome, Ruse concludes that it is still an open question whether the kind of complexity and intelligence exhibited by our species might be a more likely outcome of evolution than we might otherwise think.

In part 2, we turn from historical and contextual issues to the conceptual understanding of chance in evolution and the role of chance in the processes of evolution. We begin with a contribution from Michael Strevens (chapter 6), who applies his extensive body of work on the understanding and interpretation of evidence in and for chancy processes to the understanding of what might be evolution's most quintessentially chancy process: genetic drift. Strevens shows that a problem that is fundamental to the philosophy of probability, the reference class problem, requires a solution in order to disentangle genetic drift from natural selection.

Understanding drift, in Strevens's picture, amounts to seeing how the reference class problem can be solved for the probabilities associated with evolutionary processes. He offers a solution to the problem of the reference class for evolutionary biology and thereby provides a way of distinguishing selection and drift, providing a deeper understanding of the character of the chanciness of drift.

Francesca Merlin, in chapter 7, considers another classic source of chance in evolution: mutations. We know that mutations are taken to be "random" not only with respect to their impact on organismic fitness, but also with respect to their phenotypic effects. Philosophically, the question has long been one of whether these are "genuinely" chancy mutations. That is, are they caused by processes such as those arising in quantum mechanics (making chance an irreducible element of their description), or are they chancy only insofar as we lack precise knowledge of the biochemical and physical influences on the genome?

Merlin challenges this dichotomy by presenting a common biological model of mutation, the Lea-Coulson model, developed in the 1940s. Analyzing this model, and attempting to classify just what uses of chance are found within it, Merlin argues that the best understanding of mutation follows from considering mutations to be what she calls "weakly random" events. This work, she claims, can even impact the future direction of biological models used to determine mutation rates.

Part 2 closes with chapter 8, by Thomas Lenormand, Luis-Miguel Chevin, and Thomas Bataillon, evaluating the relationship between the role of chance in evolution and the evidence provided by repeated instances of, and experiments in, parallel evolution. Two views of parallel evolution, they claim, have traditionally been offered in the literature, and these correspond, in turn, to two differing broader ways of understanding evolutionary processes. On the "selectionist" view, natural selection is the primary driver of evolutionary change, with variation as only a minimal limiting factor. The "mutationist" view, by contrast, takes the availability of variation to be the primary limit on evolutionary change, with selective pressures by far insufficient to understand the trajectory of life over time. They carefully and lucidly detail the varieties and causes of parallelism and argue that the causes that parallelism might be traced to can all be interpreted in line with both views by shifting emphasis between phenotypic parallelism (selectionist) and genotypic parallelism (mutationist).

Further, they work to advance the debate over parallel evolution by providing a model for the prediction of the probability of parallelism. Their model offers helpful insights that can, they hope, clarify and guide future research into parallel evolution. The chapter concludes with a series of open conceptual and empirical questions that, if resolved, hold out the hope of offering us much more clarity on just what the evidence of parallel evolution should tell us.

In part 3 we turn to the history of life and consider the question of how chance, especially in the sense of contingency, has played a role in large-scale evolutionary outcomes. Eric Desjardins, in chapter 9, considers the relationship between chance and the supposedly nonchancy influence of convergent evolution. The debate between Stephen Jay Gould and Simon Conway Morris involved in no small part the question of whether widespread evolutionary convergence was (as Conway Morris argued) evidence of a broad-scale predictability and inevitability in evolution, or (as Gould argued) nonetheless generated by a deeply historically contingent process.

Building on his analysis of the concept of path dependence in evolution, Desjardins offers us a formal analysis of evolutionary histories, detailing precisely the cases in which history can be said to have an impact on outcomes in addition to chance, and when chance instead obscures evidence of that impact. This is far from being an inaccessible question, Desjardins argues, we can describe the precise experimental conditions that would let us evaluate which of these two hypotheses obtains. Such work has, in fact, already been performed—the Long-term Experimental Evolution (LTEE) Project shepherded by Richard Lenski at Michigan State University has, according to Desjardins, exemplified precisely the sorts of experimental conditions that allow us to make such a determination. But we are not required to work at microevolutionary scales: Desjardins claims that work on macroevolutionary constraints (combined with philosophical work on the concept of generative entrenchment) can equally well be interpreted as examining the historicity of evolutionary trajectories.

Desjardins's chapter, then, provides a perfect segue to the final section of the book, which examines the influence of chance in evolution on particular episodes in the history of life. Zachary D. Blount offers, in chapter 10, a reconstruction of the Long-term Experimental Evolution Project, with a focus on what the experiments performed in this context can tell us about the role of chance in evolution. After beginning with a welcome primer on experimental evolution techniques in microbes and the structure of the LTEE experiments, Blount describes the most exciting results obtained thus far.

Those results have, perhaps unsurprisingly, been complex. The role of contingency in evolution appears to be fairly tightly constrained, and it is perhaps best understood in terms of contingency's effect on changes in evolvability. This makes history more than a merely destructive or confounding influence in evolution (though, as Blount notes, historicity can result in simultaneous convergence and divergence at differing levels); he shows that it is a profound creative force.

Blount's chapter closes with an impassioned appeal to increase collaboration in work on historical contingency, between microbiologists, paleontologists, and systems biologists, as well as philosophers of science. This appeal is one with which we as editors wholeheartedly agree.

In chapter 11, Betul Kacar considers the role of assumptions regarding chance in evolution in a novel class of studies merging ancestral protein reconstruction with experimental evolution. Ancestral protein reconstruction, which involves executing Gould's "replaying life's tape" experiment not at the organismic but at the biochemical level, recreates hypothesized ancestral forms of individual proteins in vitro. Kacar's new approach then endeavors to understand the function of these ancestral proteins not in a hypothetical ancestral cellular context, but within modern bacteria, bringing ancestral protein reconstruction in vivo. Seeing the ways these ancestral proteins function as elements of contemporary gene networks can be a vital ingredient in coming to understand how such complex networks might have evolved.

Selecting one particular protein, Kacar and colleagues were able to create a viable strain of *Escherichia coli* that expressed an ancestral protein as an element of a broad and highly interactive gene network—though the strain experiences dramatically reduced fitness. While this result is already of profound interest, the next step is to use this novel strain as the basis for experimental evolution. The evolutionary trajectory of the ancestral protein in the modern context provides a particularly unique way to understand the impact of chance on evolution. How much of the organism's change in fitness will be traceable to this particular reconstructed protein? How much of that fitness change will parallel the observed evolutionary track in contemporary *E. coli*? How much will be the result of interactions between the ancestral protein and other modern components in the cell, or the modern extracellular environment? Answers to these questions will provide an unparalleled look at the influence of chance on evolution at the biochemical level, an exciting and entirely new perspective.

The volume concludes with Douglas H. Erwin's masterful account, in chapter 12, of the historical development and current best understanding of the fossils of the Burgess Shale. It is these fossils that led Stephen Jay Gould, in his 1989 *Wonderful Life*, to argue for a powerful role for contingency in the history of life. But the years since the publication of Gould's work have not been silent. They have produced manifold fossil discoveries, revisions to a variety of phylogenetic relationships, and theoretical and conceptual models that have changed our understanding of macroevolution. Our appraisal of these important fossils must, then, change as well.

For his part, Gould famously argued that any "replay of the tape of life" would result in the evolution of dramatically different forms, making historical contingency one of the most powerful forces in evolution. Erwin reconsiders this argument in light of an extensive and expertly surveyed catalog of

Ediacaran-Cambrian Radiation research in the past two and a half decades. Enhanced methods have made the "strange" fauna of this period less so, elucidating their connections to known stem groups but confirming Gould's emphasis on the impressive changes in morphology that took place during this period. Tests of the contingency hypothesis itself, including the LTEE project and studies of convergent evolution, combined with a classification of five different types of contingency, allow Erwin to evaluate the status of Gould's argument. We have, in short, a mixed bag—the status of Gould's contingency hypothesis depends dramatically on the evidence at issue, the variety of contingency under discussion, and the level of evolutionary process of interest (whether molecular, developmental, phenotypic, or macroevolutionary). Erwin's discussion deftly explores this vast and complex conceptual and empirical landscape.

A few broad themes emerge from the book and can be discerned throughout the various contributions. First, *Chance in Evolution* knits together our knowledge of an incredibly wide array of information. The fossil record, mathematical theories of probability and random variables, human culture, microbial experiments, the popular and theological responses to evolution, the metaphysics of causation, and the interpretation of Aristotle's attacks on the pre-Socratics—all of these are important and relevant to our thorough understanding of chance's role in the history of life. If this book can help to reinforce among our readers the breadth of intellectual influences bearing on issues of chance in evolution, it will have been a resounding success.

Second, this is a vital and lively area of research: chance in evolution does not begin and end with Darwin. The Long-term Experimental Evolution Project, described and analyzed masterfully by Blount's and Desjardins's chapters, is ongoing; the work on models of mutation developed by Lenormand et al. and contextualized by Merlin continues as well. Kacar's work on ancestral protein reconstruction is part of a fast-moving area of biological and biochemical research. One of Erwin's main theses points to the rapidity of change in our interpretation of the Burgess Shale fossils. Much of the historical analysis of Darwin that grounds Hodge's contribution was only made available with the study of Darwin's notebooks in the past thirty years, while Plutynski et al. offer the first-ever systematic study of chance in the modern synthesis. Depew's chapter leverages (and contributes to) recent work on the species problem; Ashley's chapter amply shows that the theological reception of Darwin's theory is (if a bad pun is allowed) still evolving today. The methods of tackling objective probabilities deployed by Strevens have largely been developed within the past decade, and Ruse's chapter demonstrates the ongoing relevance of all these questions to our understanding of human beings' place in the world. In short, these chapters have their finger on the pulse of a cutting-edge research program that stretches across disciplinary boundaries. If they are any indication, the state of the field is strong, and will remain so, offering exciting and provocative research questions for a long time to come.

Finally, we hope to have offered an example of fruitful collaboration in action, between biologists, historians, theologians, and philosophers. While all our respective fields proudly trumpet the importance of interdisciplinary research, the barriers to the successful performance of such research are many and well known. We are hopeful, however, that we have harnessed here all its advantages while simultaneously avoiding its pitfalls. The chapters that follow speak to one another as an integrated whole, responding to (and in a few cases, profitably debating among) one another, in a way that belies the varied disciplinary backgrounds of their authors. Taking the liberty to speak for all the authors, we believe such collaboration has been both instructive and exciting, and we hope that it is so for our readers as well. Without it—to echo the sentiment with which Zachary Blount closes his chapter—a full understanding of chance in evolution may forever elude our grasp.

NOTE

1. "Le hazard est un mot vuide de sens; rien ne peut exister sans cause" (Voltaire 1772, 372).

* 1 *

The Historical Development and Implications of Chance in Evolution

CHAPTER 1

Contingency, Chance, and Randomness in Ancient, Medieval, and Modern Biology

David J. Depew

TOWARD A MORE COMPLICATED HISTORY OF BIOLOGY

In an oft-republished speech commemorating the centenary of Darwin's birth and the fiftieth anniversary of the publication of the *Origin of Species*, the philosopher John Dewey claimed that Darwin came close to completing the scientific revolution by undermining the "Greek" or "classic" assumption of Plato and Aristotle that species of animals and plants are immutably fixed natural kinds, each defined by an unchanging set of necessary and sufficient conditions for class membership (Dewey 1910, 1–4, 9–10). I say "came close" because Dewey saw evidence of the incompleteness of, and active resistance to, the Darwinian revolution in "the recrudescence of absolutistic philosophies" that pay lip service to Darwinian evolution by turning on its side the Great Chain of Being, with its hierarchically ordered ranks of essentially defined types ranging from monad to man, and watching it unfold in time.[1] This reactionary impulse can be pushed back, Dewey concluded, only if psychology and philosophy are quickly transformed into "Darwinian genetic and experimental logic" (18). He was alluding to his own "instrumentalist" view that ideas are tools for changing social and natural environments rather than pictures of an unchanging stability lying behind shifting appearances.

Fifty years later, the Nobel Prize–winning geneticist Hermann Muller took advantage of the *Origin*'s centenary and of Darwin's 150th birthday to proclaim, albeit without Dewey's plug for evolutionary psychology and pragmatic epistemology, that "One Hundred Years without Darwinism Are Enough" (Muller 1959). Two years later, as neocreationists were girding their loins to prevent school boards from taking Muller's advice and incorporating evolution by natural selection into high school biology classes, the paleontologist George Gaylord Simpson used Muller's title to say much the same thing (Simpson 1961). By then, the avian systematist Ernst Mayr was congratulating himself and other founding fathers of the modern evolutionary synthesis, including Simpson, for at last putting Darwinian natural selection onto genetic foundations solidly scientific enough to block regression from "population thinking" to Plato's "typological essentialism" (Mayr 1963). Not to be outdone, the philosopher of biology David Hull argued that the effect of Aristotelian essentialism on taxonomic theory and practice was "two thousand years of stasis" (Hull 1965). Aristotle was more at fault than Plato because, unlike his mentor, he was a biologist—and a bad one.

Binaries this categorical are usually poor guides to history. Their rhetorical function is to simplify the past in order to induce an intended audience to take this fork in the road rather than that fork at an uncertain moment. If one thinks, however, that nothing is better calculated to advance knowledge than accurate history, one will soon come to agree with Polly Winsor (2006), Ron Amundson (2005), John Wilkins (2009), Richard Richards (2010), and Phillip Sloan (2013) that the "essentialism story" told by Mayr, Hull, and their predecessors is a myth.[2] By the same token, we can also agree with Jonathan Hodge and Gregory Radick that "complications . . . make wholly unacceptable any version, no matter however qualified, of the Deweyan thesis about Darwin's place in the intellectual long run" (M. J. S. Hodge and Radick 2009, 247). Premodern biology, they report, was not nearly as typologically essentialist as Dewey alleged, and Darwin, who arose in the midst of a transition to evolution already under way, did not initiate as great a break with the biological philosophy of the past as Dewey believed.

For Hodge and Radick, greater continuity becomes visible when the influence on modern biology of ancient philosophers other than Plato and Aristotle, notably Epicurus and Lucretius, is taken into account, and when the subject of reproduction, which twentieth-century Darwinians screened off from evolutionary biology when they shifted to what Mayr called "population thinking," is restored to its crucial role in the history of evolutionary thought.[3] In this chapter, I will develop these points by tracing a series of shifting alignments since antiquity among the closely related, overlapping, but distinguishable concepts of contingency, chance, and randomness. These alignments affect how biological teleology—the ascription of functions, end-directedness, and adaptedness—is conceived and how it relates to natural selection. I will make six main points:

1. The history of biological contingency is much older than Darwin. There are elements of contingency even in Aristotle's theory of generation and, since contingency is the opposite of necessity, in his seemingly catachrestic concept of hypothetical necessity (Lennox 2000, 138–40). Although reproduction accurate enough to maintain species lineages is too regular and too end-directed to be due to chance, Aristotle held that species are reproductive lineages, not types. Although it is natural for each organism faithfully to reproduce the form of its male parent, he acknowledged that particular parent-offspring links in such chains vary and might not have occurred at all. "It is not necessary," Aristotle writes, "that if your father came to be you must have come to be. But if you came to be it is necessary that he did too" (*Generation and Corruption* [*GC*] II.11.338b10–11).[4] The element of contingency in Aristotle's theory of reproduction was defended by Galen in late antiquity (129–217 CE) and, through Galen's lasting influence on medical education, was revived in early modernity by William Harvey (1578–1657).[5] Harvey coined the term *epigenesis* to name the Aristotelian approach. After the embryological discoveries of Caspar Friedrich Wolff (1735–1794), epigenesis entered into the ferment surrounding the emergence of a unified discipline of biology a few decades before Darwin. Dewey and Co.'s contrast between ancient fixism and the contingency of species, I conclude, is overdrawn.

2. Species fixism of the sort that figures in Dewey's, Mayr's, and Hull's story did exist, but it isn't nearly as ancient as they imply. It arose in the late seventeenth century (Wilkins 2009, 96). Before then, naturalists, especially botanists, were comfortable with the mutability they observed in hybrids. What changed was the superimposition of the logical category of species onto biological classification and closely related efforts to involve God in fixing species boundaries (89). To maintain these boundaries, preformationists, who opposed epigenesis, hypothesized that God encapsulated into the germinal material of the first, directly created instance of each species the germs of all subsequent members of that kind. There ensued a long in-house debate among preformationists about whether he placed this information into the egg or the sperm. Strange as it may seem to us, anti-Aristotelian advocates of the mechanical philosophy were among the most ardent preformationists. They realized that the functional and goal-oriented characteristics of organisms are inconsistent on their face with modern physics, so they placed these teleological properties into the divinely created prototype of each species and let them mechanically come rolling out, true to type, generation after generation (Richards 1992; Arthur 2006, 172). A strong dose of necessitarianism, marked stress on the language of natural forces, and a correspondingly dimin-

ished role for contingency and chance characterized this period in biology's history, or, to be more accurate, prehistory. Species fixism and typological essentialism, I conclude, are comparatively recent phenomena.

3. The idea that chance is the origin of the natural world has been around since Empedocles (c. 490–c. 430 BCE), Epicurus (341–270 BCE), and the Roman scientist-poet Lucretius (c. 99–c. 55 BCE) (M. J. S. Hodge and Radick 2009, 248). Chance adds to contingency, whose opposite is necessity, an element of indeterminacy that, whether it is located in the nature of things or only in our ignorance, defies explanation, especially explanation in terms of purposes. Roman materialists, to relieve people's fear of death, promoted the idea that we are accidental confluences of atoms that will dissipate when we die. As late as the fourth century CE, St. Augustine acknowledged at an especially discouraged moment in his youthful quest for wisdom that he "would have awarded the prize to Epicurus" if he had been able to rid himself of his "belief that the soul lives on after death to account for its acts" (Augustine 1992, sec. VI.16). Since fear of death is for Christians the beginning of wisdom and is best cured by trusting in Providence, the idea that chance accounts for biological order served later thinkers mostly as the butt of reductio ad absurdum arguments in which natural theologians, preformationists, and epigeneticists alike reaffirmed the functional and goal-oriented characteristics of organisms that partisans of chance, supposed or real, denied.[6] Appeals to chance in biology are tolerant of species mutability. Accordingly, anyone who was perceived as advocating transmutation in the early nineteenth century was likely to be accused of being a materialist and a partisan of chance. That happened to Darwin.

4. There was genuine novelty in Darwin's anti-Lamarckian stipulation that the causes of variation are independent of its subsequent utility in coping with environments. When variants first arise, they only happen to be useful. But if they are heritable, they can gradually evolve into adaptations by the enhanced reproductive success of the organisms that have them. Darwin uses "chance" to signify the causal independence of variation's origin from subsequent biological adaptedness (Beatty 2013, 147). But he also used "chance" in other, closely related ways. There is *a* chance that variations that would prove useful in a certain environment will not be available. There is also *a* chance that useful variations, even if available, will not be passed on. There is *a* chance, too, that a useful and heritable variation might not spread through a population (C. Darwin 1859, 126–27; M. J. S. Hodge 1987; Hodge, this volume, chap. 2; Beatty 2013). It is no wonder, then, that Darwin had to fend off accusations that he was a friend of chance. Far from being a theory of chance in the way

Empedocles's, Epicurus's, or Lucretius's ideas were, however, Darwin's view was that natural selection working selectively on chance variation in the ways I have briefly described has the power, amid much failure, to evolve traits that are as genuinely functional as Aristotle's, Galen's, or the great comparative anatomist Georges Cuvier's. In Darwinism, contingency and chance combine with environmental determinism to evolve natural purposiveness. Adapted traits exist to perform certain functions. Accordingly, there is a certain unexpected affinity between Aristotelian functionalism and Darwinian adaptationism (Lennox 1993, 1994; Depew 2008, 2015). The price for embracing this affinity, however, is to admit that by making adaptedness depend on local environmental conditions and on the contingent availability of useful variations, Darwin further undermined the already vulnerable Great Chain of Being that placed humans at the top of the well-ordered, thoroughly purposeful biological continuum. It was an implication that made Darwin's mentor, friend, and supporter Charles Lyell queasy. It also troubled Asa Gray, a trusted correspondent with whom Darwin thought aloud about these issues. Darwin opposed both friends. Nonetheless, he assured himself that a temporalized version of systematics in which less advanced kinds are succeeded by more advanced followed as an ex post facto consequence of his theory.[7] In the long run, however, this may be among the least enduring elements of Darwin's response to the conceptual currents swirling around him.[8]

5. Darwin*ians* were able to undercut the long-standing dichotomy between chance and purposiveness because Darwin's theorizing stood on the edge of the "probability revolution" that, by mathematically "taming" them, gave chance and chances for the first time a key role in law-governed explanations of natural phenomena (Porter 1986; Krüger, Daston, and Heidelberger 1987; Krüger, Gigerenzer, and Morgan 1987; Gigerenzer et al. 1989; Hacking 1990; Gayon 1998). There are statistical regularities. In fact, the basic laws of physics are statistical. The rise of Darwinism as the dominant research tradition in evolutionary biology is inseparable from the fact that Darwin's successors not only took advantage of the explanatory resources of the probability revolution to develop what Mayr called "population thinking," but were also leading contributors to this revolution on its mathematical side. A factor in the continuing resistance of populist religious and political culture to Darwinism, I surmise, is the lagging uptake of the probability revolution into common sense.

6. The discovery that the ultimate source of heritable variation is spontaneous mutation in DNA has amply justified Darwinism's postulate that variation arises independently of its subsequent adaptive utility.[9] But this discovery has

also resulted in an assimilation of chance to randomness conceived as inherent unpredictability. Twentieth-century physics tells us that even if statistical regularities and patterns can be found at higher levels of resolution, there may be a pervasive indeterminacy at the level of individual events. It is even possible that indeterminacy of this sort plays a large role in evolutionary change (McShea and Brandon 2010). This possibility goes beyond Darwin's assumption that what we call chance events are as caused as any other occurrences, even if their causes are so tangled that they will forever elude our grasp (Beatty 2013, 147). But it also weakens the long-standing conception of chance as good (or bad) effects accidentally caused, a conception shared by figures as different as Empedocles, Aristotle, Epicurus, and Darwin. It is increasingly likely that randomness does affect the course of *evolution*, though just how much is a disputed question. For Stephen Jay Gould it dominates macroevolution, the larger history of life (Gould 1989; Blount, this volume, chap. 10; Erwin, this volume, chap. 12). In the form of genetic drift, it is a wild card in microevolution, evolution at or below the species level (Millstein, Skipper, and Dietrich 2009; Strevens, this volume, chap. 6). Nonetheless, any interpretation of *natural selection* in which sheer randomness is allowed to color what Darwin meant by the chance relationship between the adaptive utility of variation and the causal origin of that variation misconstrues his most basic and most fertile idea.

In this chapter, I will focus on the historically more remote of these theses. They set the context for a volume that contains detailed explorations of the more recent ones. If the history recounted in this chapter, sketchy as it is, lends some clarity to the sixth point, it will have proved its worth. For it is on the relationships among the concepts of randomness, chance, natural selection, and adaptation, I think, that most of the seemingly intractable deficiencies in contemporary public discourse about Darwinism turn.

CONTINGENCY, CHANCE, AND TELEOLOGY IN ARISTOTLE'S EPIGENETIC BIOLOGY

In a seminal chapter in his *Physics*, Aristotle divides chance into two sorts. The first is luck (*tuché*), a species of chance that because of its importance in our lives gives its name to the genus. Luck is chance in matters typically under the sway of choice, such as chancing to run into an acquaintance who owes you money you had been meaning to recover (*Physics* [*Ph.*] II.5.197a5–6). The other species is spontaneity (*to automatou*), the analogue of luck in matters under the sway of natural tendencies and impulses (*Ph.* II.6.197a36–b15).

In discussing the latter, Aristotle reports at length the views of Empedocles, who assigned greater objectivity to the functional and goal-directed appearances presented by living things than did Democritus, who was a more determinist and necessitarian materialist, at least according to Aristotle, and who in consequence believed that the appearance of purposiveness in organisms was just that: appearance (*Ph.* II.8.198a15–16). Genuinely good results, said Empedocles, happen in nature as they do in human affairs. Organic kinds originated in a way not unlike the way you happened to run across your debtor in the marketplace (*Ph.* II.8.198b15, 199a8).

Here is how Empedocles thinks it works. Biochemical compounding of the four elements, he admits, is necessitated. In an era when the cosmos was under the sway of Love, these compounds were more likely to form parts that self-assemble into organisms than we see today. Even during the reign of Love, most of these aggregations didn't last, because their components didn't reliably stick together. They stick together even less often in our own Strife-ridden times. Still, Aristotle reports Empedocles writing in his poem about cosmogony, "Whenever all the parts did coalesce together (*sumbanei*) *as if* they had come to be for the sake of something, these animals were preserved (*esothê*), having been aptly (*epitedeios*) constituted spontaneously (*ap tou automatous*). Those of which this is not the case perished and continue to perish, as Empedocles says of 'human-faced oxen'" (Aristotle, *Ph.* II.8.198b27–33, my italics; see also *Ph.* II.4.196a24).

For Empedocles, organisms persist because their parts happened to be physically stable in relation to each other even when, as in our own age, the cosmic cycle is too strife-ridden to form new combinations that endure. What sticks together and doesn't is necessitated. But insofar as the effects of such affinities are life preserving, and so, from the perspective of the living beings that survive, are good, their cause is nothing like that of a well-planned and well-executed series of intentional actions. What links the happy outcome with its causal antecedents are not means put to ends. It is spontaneity (*to automatou*), the natural counterpart of luck.

Empedocles realized that the continued existence of oxen or humans or any other stable kind requires a mechanism by which they replicate themselves. In an early but not unprecedented version of pangenesis—the theory of inheritance favored since antiquity by partisans of chance, including Darwin—the body parts of organisms throw off shoots, stirps, or seeds. These travel to the sexual parts, where they are recombined. Commingled shoots then reassemble as they grow into new copies of the species we observe around us. Since Aristotle tells us that, for Empedocles, organisms grow solely by the addition, inflation, or augmentation of seminal structures, it is possible that he viewed organisms as nothing but expanded seeds whose contracted form comprises structures that have been faithfully transmitted by the reproductive machinery ever since their initial coalescence (*GC* II.6.333a35–b4).

This is clever. Still, Aristotle contests this theory as vigorously as he ever contested anything (*Generation of Animals* [*GA*] I.17.721b9–18.724a11 and elsewhere). Unfair to the phenomena, he cries. By means of careful observation of the embryological process, he argues that seeds do not in fact return through the body to collect in the sexual organs. After initiating a new turn of the life cycle, sperm evaporates (*GA* I.18.722a1, 725a22–25). What drives the cycle of reproduction is the process of digesting food and converting it into what we call energy. For Aristotle, reproduction is the final phase of embryogenesis. The amount of energy expended to complete a life cycle that regularly culminates in the "production of another like itself" (*De Anima* [*De An.*] I.4.415a26–30) depends on the degree to which food is broken down, concocted, refined, or processed. The place of living kinds on the scale of life depends on how far concoction or the refinement of food goes, and thus on how much energy can be mustered to push this process of self-formation along. The process is most powerful, and gestation time longest, in humans. Aristotle does admit that kinds at the low end of the scale (insects, for example) are spontaneously generated in Empedocles's sense. But in Aristotle's analysis of spontaneous generation, the biochemical compounding that in higher kinds is a subordinate material cause of the end-oriented process of embryological differentiation is not set afoot by mixing spermatic heat and female reproductive fluids. In spontaneously generated kinds, it is materially caused by the heat of the sun, which in some environments can cook up biochemical compounds to the point of animating them (*GA* II.11.762a18–21; Falcon 2005). This renders the connection between causal antecedents and the life functions of spontaneously generated organisms external (*exo*, *Ph.* II.6.197b20, 8.199b15–17), incidental (*kata symbêbêkos*, *Ph.* II.5.197a5–6), and chancy (*to automatou*, *Ph.* II.6.197b18–20). The effects are good for their possessors. But there is no internal connection between the goodness of these effects and their cause. There is no final cause (Lennox 2000, 234; Depew 2010).

In Aristotle's judgment, Empedocles constructed his general theory of life by choosing bad examples as paradigm cases and misdescribing even these. His theory is not empirically plausible even in its own terms. Even if they managed to get off the ground under long-gone cosmic conditions, his organisms would have disappeared in our age of strife, since they are contraptions too loosely structured to have stuck together (*Ph.* II.1.192b27–33; *De*

An. II.4.416b6–9). For this reason, Aristotle stresses that, when their development is closely observed, organisms, even spontaneously generated ones, are not aggregations of separate parts. They are processes in which an originally indeterminate matter forms itself into a progressively more differentiated and articulated whole (*GA* II.1.735a10–25, II.4.740a1–24). If differentiation occurs in the right order, the process will reliably culminate in the generation of "another like itself," its end-point (*telos*) (*De An.* II.4.415a26–30). This is what the latter-day Aristotelian Harvey called epigenesis. To be sure, departures from the norm do occur. But far from undermining Aristotle's analysis, birth defects, early death, and spontaneous abortion underscore the contingencies to which the end-directed process of epigenetic reproduction is subject. They also induce in Aristotle a sense of wonder that lineages of organisms, each with its parts well adapted to each other and to its environment, remain descriptively constant generation after generation without being unconditionally necessitated (*Parts of Animals* I.5.645a17–25).

It is not only the end-directed reliability and self-propelled prowess of the reproductive cycle, however, that inspires awe in Aristotle. The fact that the biosphere is stocked with a full complement of lineages that, taken as a whole, exhibit a finely graded, mutually supportive scale of psychic powers running from reproduction through sensation, desire, and locomotion to cognition corroborates Aristotle's conclusion on independent, metaphysical grounds that the cosmos itself must be a perpetually self-generating continuum (*History of Animals* [*HA*] VIII.1.588b4–22; *Metaphysics* [*Met.*] XII.1071b2–12). If kinds higher on the *scala naturae* had to emerge from lower, or, a fortiori, if living beings had to emerge from nonliving, the biosphere could not have come into existence at all. Organisms, properly described, are too well integrated and too dependent on what has been called "top down" causality for that. Nor, for the same reason, could the *kosmos* as a whole have come into being from purely elemental processes. Since it obviously does exist, however, and, moreover, shows a great deal of order—the word *kosmos* means ordered—Aristotle concludes that the universe always has been and always will be.[10] This cosmological background, as well as his rather blithe assumption that environments are fairly constant or, when they are not, can be stabilized by the soulful (*psychikos*) agency of organisms themselves, makes it a good deal easier for Aristotle to argue that species forms and species lineages are reliably continuous. Still, his epigenetic account of reproduction builds in enough contingency to require us to dismiss any account that frames Aristotle as a typological essentialist. It is not a set of necessary and sufficient conditions for membership in a class that makes organisms what they are or that confers transgenerational stability on them. It is absurd to say that Aristotle held biology back for two millennia. On the contrary, modern biology began when Aristotle's theory of epigenesis was recovered.

THE THEOLOGICAL ROOTS OF THE "ESSENTIALISM STORY"

We have seen that, in rejecting chance, Aristotle did not deny that contingency dogs the etiology of living beings. We have also seen that the contingencies of Aristotle's embryology were buffered by his conviction that the universe is eternal, self-sustaining, and at its upper edges divine. The scope of contingency was greatly widened by Jewish, Christian, and Islamic creationism (M. J. S. Hodge and Radick 2009, 252). The clash between the conviction that God created the world out of nothing and the impulse of theologians to use Aristotle as a philosophical handmaiden (*ancilla*)—Orthodox Christian theologians remained happy with Neoplatonism—is an episode of such importance that the subsequent history of philosophy and science, including biology, can scarcely be understood apart from it.

For our purposes, it is less important to trace the detailed biological controversies that arose to mediate this clash than to appreciate the fact that, as the Middle Ages gave way to modernity and especially to the new mechanistic physics, the heightened theological contingency of the created world—God did not have to create it—made preformationism a more attractive scientific explanation of biological teleology than it would otherwise have been. This also explains why the recovery of the epigenetic approach in the later eighteenth century was less than perfect, typically incorporating into itself traces of preformationism.

Writing at the high tide of the European Aristotelian revival of the thirteenth century, Thomas Aquinas cleaved to "The Philosopher" as closely as he could, but even he had to admit that "the being of the world depends on the will of God, as on its cause" and that "it is not therefore necessary for God to will that the world should always exist" (Aquinas 1947, ST Ia, Q. 46, art. 1, co.). Aristotle's wonder that, among other things, species lineages reliably persist even if links in the chain are contingent was upgraded to Aquinas's *argumentum ex contingentia mundi* for the existence of God as first mover (Aquinas 1947, ST Ia, Q. 2, art. 3). The perceived fact that nothing around us has to exist makes it imperative to infer that, whether creation occurred once or is an ongoing process, a necessary being must ultimately stand behind nonnecessary beings.[11] Even then, however, Aquinas was perceived in some clerical quarters as too naturalistic. For this reason, scholastic philosophers tended thereafter to pour increasingly un-Aristotelian meanings into a Latinized Aristotelian lingo.[12] Voluntarists, for example, held that God, since he is all powerful, could make any sort of world he pleased, or none at all, as long as its arbitrary rules are not self-contradictory. He may govern this world through generally reliable natural laws considered as "secondary causes," as Aquinas says. But, because natural laws are really only God's positive laws, he can intervene in their workings at will. This claim is the root of the modern doctrine of miracles. It goes without saying that when David Hume or Charles Darwin challenged it, their public reputations suffered, except among a relatively small circle of atheists, who did these seminal figures no favors by refashioning them in their own image.

Voluntarism brought with it an epistemological shift toward nominalism, a doctrine about the genesis and nature of knowledge with no close counterparts in ancient skepticism. The biblical conviction that God knows and cares for each person, indeed for every sparrow (Matthew 10:29), undercut Aristotle's belief that God knows only universals by postulating individual forms (*Met.* XII.7.1072b14–31). The consequent idea that species terms and the higher-order generalizations that form the bulk of our languages are mere summaries of recurrent observed patterns, and so testify to our inability to see into the constitutive nature of things, went into persistent doubts that we can know "real essences" or find a "natural system" of classifying animals and plants, as opposed to identification-oriented, and hence pragmatic or utilitarian, artificial systems. The creationist stress on contingency that went into these currents of thought even colored the modern reception of Epicurean, Skeptical, and Stoic texts. Skepticism became more skeptical, Epicureanism more chancy, and Stoicism, which in antiquity saw God and the world as a single substance, more dualistic in the way creation by a transcendent God logically requires.

Against this background, the impulse of the Enlightenment was to increase our chances of knowing nature, including human nature, by freeing philosophy from its role as the "handmaid" of theology. It did so by using new technologies of observation to see the world better, and by recognizing that mathematics is the language in which natural laws are written. Still, we are so accustomed to the antitheistic, and not just anticlerical, thrust of the second, nineteenth-century phase of the Enlightenment that we sometimes fail to realize that its first impulse was to tighten God's connection to the world by showing him to be the creator of a rational universe that we are fully competent to know. For Gottfried Wilhelm Leibniz (1646–1716), for example, it is very far from the case that God was an "unnecessary hypothesis," as the physicist Laplace is said to have told Napoleon in 1802.

The theistic phase of the Enlightenment greatly affected natural history. It was probably John Ray (1627–1705) who was the first to apply logical concepts of definition by necessary and sufficient conditions to biological kinds. He has a claim to being the first species essentialist and species fixist (Wilkins 2009, 67, 93).[13] Not coincidentally, Ray was an intelligent design creationist, a Platonist associated with a school of philosophers at Cambridge University who were aware that it was a lot easier to logicize and mathematize the world on Plato's terms than on Aristotle's, and not least a preformationist. The preformationist cause was helped along by the even more logicist Leibniz's fascination with the infinite and with infinitesimals. Did not microscopes show how small living things could be? Could God not make them so small that they could all be contained in prior instances of each kind going back to the first? In our own species, we are given to understand that a series of little men ("homunculi") are latent in our reproductive organs. Just as a creator God had freely set down the mechanistic laws of physics at the beginning of the universe, so, too, rather than dispersing seeds into the environment in the way Augustine imagined, he encased into the prototypes of every species the purposiveness that all its successors would exhibit and that the decidedly nonpurposive laws of physics could not accommodate. These successors thereafter mechanically roll out: evolve in the original sense of the term (Richards 1992). Preformationism combines intelligent design with determinism. As a result, contingency and chance were pushed as far as possible from the world of living things. More importantly, this conceptual realignment had the unfortunate effect of assimilating contingency, whose opposite is necessity, to chance, whose opposite is purposiveness. Aristotle would have considered this conflation incoherent. For him, the contingency of the epigenetic process was the key to refuting Empedocles's appeal to chance.

The preformationist era did not last. Microscopy failed to turn up evidence of homunculi-like entities in either egg or sperm. Under the pressure of this failure (which in the new experimental culture counted as a black mark more than it might have at another time), Harvey's Aristotelian epigenesis made a comeback. The empirical embryological work of Caspar Friedrich Wolff (1735–1794) is generally recognized as a turning point. The fall of preformationism catalyzed other rejections. By the end of the eighteenth century, epigenesis had become associated with the heterodox possibility that nature is in some sense self-creative and goal-oriented. This perception stimulated calls for a new integrative and naturalistic science of life to be organized around the self-formative and self-organizing power of organisms. The science was to be called "biology," a new term (Treviranus 1802). This movement formed the milieu in which evolution in the sense of species transformism came into view for the first time. For some of its advocates, Johann Gottfried von Herder (1744–1803) and Jean-Baptiste Lamarck (1744–1829), for example, there was a link between living matter, species transformism, evolutionary progress, and sympathy with the ideal of upward mobility sponsored by late-eighteenth-century revolutionism (Desmond 1989). In contrast to fixist interpretations of the *scala naturae*, in which there is a clear definitional divide between taxonomically ranked kinds, the new biologists restored Aristotle's view that the scale of life is a continuum whose junction points are blurry enough to be classified differently depending on how you look at them (*HA* VIII.1.588b4–22).

The new paradigm of biology treated species as lineages, as Aristotle had, not as types. Still, its turn to epigenetic self-formation was in no sense a return to the eternity of the world. The imaginations of biologists, like those of almost everybody else, were still accustomed to the notion that only preformed design of some sort can keep the world from falling into the nothingness that, were it not for God's establishment of natural laws and occasionally his direct intervention, is its natural condition. Accordingly, even the most ardent advocates of epigenesis could not imagine how species continuity could be pulled off without help from something like what the natural historian Count George-Louis Buffon (1707–1788) called an "interior mold," which, by assembling dispersed "living molecules" (now released from preformationist encapsulation) in the way crystals serve as molds for new crystals, could ensure the functional integrity of organisms and the transgenerational continuity of species lineages (Sloan 1987, 1992; Buffon 1749).[14] Immanuel Kant, who took his cues in matters biological from Johann Friedrich Blumenbach (1752–1840), an early advocate of the self-formative or vital powers of living beings, gave his qualified philosophical blessing to the new paradigm, but he called it "generic preformationism" rather than epigenesis, because to him the latter term suggested a more radically self-creative, pantheistic romantic vision of nature that he associated with his renegade student Herder (Kant 2000, 291, *CJ* 5:423). For Kant, there is no preformed individual. But there are heritable species-specific (and racial) dispositions that are awakened in environments to which they are (pre)adapted (Look 2006; Zammito 2007).

Even after biology become evolutionary in the transformist sense, traces of preformationist styles of thinking never wholly disappeared. Medieval creationism had left a permanent mark. Living had somehow to emerge from nonliving. In fact, with the discovery of the Second Law of Thermodynamics in the middle of the nineteenth century, which showed that, in nature, disorder is inherently more probable than order, preformationist tendencies increased. From that time forward, the hunt was on for an "aperiodic crystal" that would work counterentropically to bring into existence life forms and sustain them (Schrödinger 1944). After Crick and Watson found the very thing they were looking for in the structure of DNA, a quasi-preformationist slant was imparted to twentieth-century molecular Darwinism.[15] Organisms, it was and still is said, are the output of evolved, entropy-bucking "genetic programs," cybernetic offspring of Buffon's interior molds.[16] It is here that the assimilation of contingency to chance in the preformationist era combines with a recent tendency to assimilate chance to stochastic randomness in a way that exerts a powerful influence on contemporary biology. Mutations may not be individually predictable, even in principle, but they are necessitated en masse by the higher probability of disorder than order in an entropic universe. Necessity takes the form of computable genetic algorithms, and chance is reduced to random mutation in the running of these programs. From this picture has arisen an interpretation of Darwinism as "random variation plus natural selection." I will suggest in the next section that this is not a very good interpretation, since it does not capture the nuanced relation between chance, determinism, and purposiveness that makes Darwin and Darwinism important "in the long run."

CHANCE, SELECTION, AND ADAPTATION IN DARWIN

It is against the background of the later-eighteenth-century rise of biology as an ascendant paradigm faced with a growing challenge to explain the origins of species that we should situate Darwin.[17] A good place to observe him dealing with the questions about contingency, determinism, and teleology that were philosophically at issue in the evolutionary biology of his day is in his warm and frank correspondence with the American botanist Asa Gray, with whom he exchanged almost three hundred letters over more than three decades. Their correspondence reveals how Darwin's theory of natural selection kept evolving after the appearance of the *Origin*, to embrace more contingency, more chance, and, paradoxically, more biological purposiveness.[18]

Darwin appreciated Gray because from the start he understood natural selection better than most early readers of the *Origin*. Gray knew that "without the competing multitude no struggle for life; and without this, no natural selection and survival of the fittest, no continuous adaptation to changing circumstances, no diversification and improvement" (Gray 1884, 378). Accordingly, Darwin relied on his American pen pal to help him beat back a hostile and misinformed interpretation of natural selection that began circulating, especially in the United States, almost immediately after publication of the *Origin*. When the language of adaptation is diagnosed as nothing more than a metaphorical echo of intentional design, what is left of natural selection is just chance fixation of traits resulting from the greatly enlarged time frame that Darwin had stipulated. In contrast to the patient replies he gave to technical objections by his professional colleagues, some of which genuinely stumped him, this rather Empedoclean interpretation, which was designed to make creationism look good and natural selection to look silly, exasperated Darwin and alienated him from those who traded in it.[19] Complaining to his American correspondent about views reportedly expressed by Gray's Harvard colleague Francis Bowen, a natural theologian, Darwin wrote, "The chance that an improved . . . pouter-pigeon should be produced by accumulative variation without man's selection is almost infinity to nothing; so too with natural species without natural selection" (C. Darwin 1860b, DCP #2998).[20]

Gray agreed. But almost immediately he himself began toying with the idea that God, having made natural selection a law of nature at the dawn of creation, might have intervened in its working so that evolution could get from local adaptations, whose dependence on and restriction to particular and highly mutable environments Gray understood perfectly, to an ordered sequence of species culminating in human beings.[21] To maintain the *scala naturae*, Gray argued in a review of *Origin*, why couldn't God bias the variation on which selection works in a progressive direction (Gray 1884, 87–177)? "You lead me to believe that variation has been led along certain beneficial lines," Darwin objected. "I cannot believe this" (C. Darwin 1860b, DCP #2998; Lennox 2010, 2013).

Darwin had theological reasons for not believing it. Having just taken God off the hook for natural evils, he didn't feel like re-indicting him (Beatty 2013). But there were also scientific reasons. By the time he wrote the *Origin*, Darwin had abandoned his earlier notion that environmental stress triggers just enough undirected variation for selection to reequilibrate otherwise perfectly adapted species with their environments (Ospovat 1981). He wrote the *Origin* only after he had assured himself by closely studying barnacles and other classes that variation in natural populations is abundant and adaptedness is entirely relative to environmental circumstances.[22] When Gray made his proposal, Darwin was already beginning to recognize the implications of these discoveries for envisioning how chance variation is related to adaptive natural selection (Lennox 2010, 2013). Increasingly, he thought that although heritable variations are small, continuous, and ubiquitous, their opportune occurrence is not part of any equilibrating mechanism, making them more fortuitous than he had earlier believed. Even when they happen to be available, moreover, there exists no more than an enhanced chance that organisms fortunate enough to possess them will survive, reproduce, and pass on their advantages to descendants. His exchange with Gray pushed Darwin further in this direction. He recognized that, to be effective, natural selection must give chance a free hand. Thus liberated from the residual pull of natural theology, Darwin's most basic idea became agnostic in tendency, to use Thomas Henry Huxley's weasel word, and so moved in a direction opposite from the one toward which Gray had hoped to nudge him.[23]

Still, greater appreciation of chance, as opposed to intelligent design, whether direct or indirect by way of created laws, was not the only effect of Darwin's encounter with Gray. In the course of responding to the shockingly bad reading that mistook natural selection for nothing more than chance plus time, he came to realize that since the process of natural selection is necessarily gradual, working continuously over many generations on an abundance of minutely differing and fortuitously available variations, its heritable effects spread *because of* the adaptive functions they increasingly serve. Even relative adaptedness builds on earlier successes and so results in very beautiful "contrivances," as Darwin called them. The carefully observed work of his later years—especially on insect-devouring plants, orchids, and earthworms—is more adaptationist in spirit than the *Origin* (Lennox 1993, 1994, 2010; Beatty 2006a). The adapted traits of these species presuppose a world full of contingencies, shot through with chance, and unguided toward any destination in particular. But, gradually spreading through populations as they do because of their amplifying good effects, adapted traits and the organisms that functionally integrate them are not at all like the contraptions cobbled together by Empedocles's primitive theory of natural selection. The life-preserving effect of Empedocles's chance setups may be good for their possessors, but, as we observed above, it is not because of that goodness that they came to be or to endure. The relationship between chance causes and good effects remains, as Aristotle says, "external" (*exo*). Ironically, however, because their traits result from a continuous process of discriminating useful from useless variation and passing on the former, Darwin's adapted organisms are more analogous to the teleologist Aristotle's than they are to the selectionist Empedocles's.[24] While significantly different, they too "come to be and to exist for the sake of" the life activities they promote (Lennox 1993, 1994, 2013; Depew 2010, 2015).[25]

Darwin's growing appreciation of the role of chance in the process of adaptive natural selection had two conceptual consequences. The conjunction of chance and selection stimulated his enhanced appreciation of the exquisitely functional nature of the "contrivances" they combined to evolve. In this sense, Darwin really was a (biological) teleologist (Lennox 1993, 1994, 2013). The chancy relation between variation and selection also implies that chance means no less and no more than the contingent nature of reproductively beneficial meetings between heritable variation, environmental utility, and natural selection. There is no guarantee, or even presumption, that appropriate variation will be available. The fact is, however, that Darwin was unable to make himself very clear about either of these points because, like Gray, he was still thinking in a conceptual framework too impoverished by the conceptual shifts I schematized in the previous section—the reduction of natural law to divinely ordained positive law, the modern assimilation of contingency to chance, and the confusion of biological purposiveness with intentional design—to express them very well. "I cannot think that the world as we see it is the result of chance," he told Gray, "and yet I cannot look at each separate thing as the result of design. . . . I am in a hopeless muddle" (C. Darwin 1860b, DCP #2998). In another letter he told Gray, "I am inclined to look at everything as resulting from designed laws, with the details, whether good or bad, left to the working out of what we may call chance. Not that this notion *at all* satisfies me" (DCP #2814, Darwin's italics). It didn't satisfy him for good reason. He really was in a muddle. He was stuck with an impoverished conceptual binary between chance and design that, at least since John Ray, his culture bequeathed to him and to Gray.[26]

DARWINISM AND THE PROBABILITY REVOLUTION

Hodge and Radick (2009) are right to place Darwin in a "long-run" that made his work possible and persuasive. We have seen that one aspect of this history was the recovery of epigenesis after early modernity's bout of creationist preformationism. Another, more proximate resource was the incipient probability revolution, which even in its infancy allowed Darwin to think about estimating chance*s* (M. J. S. Hodge 1987). "*If* variations useful to any organic being ever do *occur*," he wrote in an internal summary in *Origin of Species*, "assuredly individuals thus characterized will have the *best chance* of being preserved in the struggle for life; and from the strong principle of inheritance, these will *tend to* produce offspring similarly characterized" (C. Darwin 1859,

126–27, my italics). What Darwin and his early allies did not realize is that these chances and tendencies could be quantified and measured. Their successors, by contrast, in particular Francis Galton, Karl Pearson, R. A. Fisher, J. B. S. Haldane, and Sewall Wright, were highly attuned to a sea change in the history of science that, in their hands, resulted in a version of the theory of natural selection largely free of the stark choice between chance and design that hampered and puzzled Darwin and Gray. It is unlikely that Darwinism, with natural selection at its core, would have become the basis of a science of evolution in the twentieth century unless it had found its way to the probability revolution in ways that many other nineteenth-century ideologies, including the superficially Darwinized ideology promoted by Herbert Spencer, failed to do.

What held Darwin and his contemporaries back was their assumption that chance refers to our ignorance of the lawful conditions that in Enlightenment science fully determine every event. The best Darwin could do was suggest that the interacting causes of variation and its retention are too complicated to make predictions possible in practice (Beatty 2013). This left open the possibility, embraced by his bulldog Thomas Henry Huxley, that a divine intelligence (or a supercomputer) could indeed predict all individual events, since for Huxley chance went no deeper than the practical difficulties we have computing the forces that, for example, cause this rather than that facet to land face up when I throw dice (T. H. Huxley 1887). By embracing the probability revolution in ways that write contingencies, probabilities, and chances more deeply into the nature of things, Darwinians after Huxley and Spencer did better. It became possible to "tame chance," as Ian Hacking phrased it, by elevating one's attention from individual trajectories to a higher level at which, whatever is happening below, it is certain and predictable that a fair die will have precisely a one-in-six chance of landing with this or that side showing (Hacking 1990). The resulting revision of the theory of natural laws has transformed science in ways that are not captured by its popular image, which is still fixated on Galileo and Newton. As soon as he read the *Origin*, Charles Sanders Peirce, who was one of the first people in the United States to understand statistical mechanics, immediately grasped what was eluding Darwin and his circle. He went on to sketch an evolutionary theory based on assigning more reality to the path dependent workings of contingency and chance in the fabric of the universe than Darwin did (Peirce 1893).

Soon, mathematically adept Darwinians who were attempting empirically to validate the concept of natural selection were straining to see what reproduction would look like if you expanded Mendel's laws of inheritance, which range over particulate germ-line determinates located on chromosomes, to the level of an entire interbreeding population. In the process, Fisher in particular helped statistical science mature by going beyond mere averages to measuring the spread of variation or variance (M. J. S. Hodge 1992a). At the population level, genotypes and the organisms that carry them are represented as points in statistical arrays. As such, they are subject to the laws of probability. When Mendelian populations are reproductively unconstrained, it is by definition antecedently as likely that any particular organism will mate with any other as that a particular molecule will collide with any other in a container full of a gas. Unless some "factor," "force," or "agency" affects it, accordingly, the overall distribution of genotypes in a population will remain the same generation after generation rather than regressing to a mean, as Galton had believed. Natural selection is one such equilibrium-changing agency. Over transgenerational time it can shift the proportional representation of genotypes in populations in a nonrandom way.

In fact, the ascent to population thinking confirmed Darwin's hunch that adaptive natural selection is a continuous and gradual process, since mutations that have small effects are mathematically more likely to get an adaptive toehold than risky, single-step macromutational saltations or jumps (Fisher 1930). In this way, the shift to probability and population thinking has led to better understanding of natural selection.[27] But the same shift has also shown how strong a hand sheer chance has in evolution (McShea and Brandon 2010). In looking at the evolutionary process from the perspective of probabilistic population thinking rather than through the lens of the Enlightenment's determinism or, alternatively, the design paradigm that tugged at Darwin's and Gray's sleeves, the fathers of population genetics, especially Wright, were also able to see that, in principle, genotypes will spread through small interbreeding populations in random as well as nonrandom ways, that is, by genetic drift.[28]

The power of chance in spreading genetic variation has become especially salient since the discovery in the late 1960s that genes mutate at a clocklike rate and that many of them diffuse through populations in adaptively neutral ways (Zuckerkandl and Pauling 1965; King and Jukes 1969; Kimura 1983). The fact that populations of sapient humans in Africa contain more genetic variation, both individually and collectively, than populations on other continents is more indicative of the longer time our species has been in Africa, where we originated, than of the spottier adaptive utility of this variation. There is an element of chance, too, in gene flow between populations as they spread to other continents by emigration and immigration. Migrants carry with them only the genotypes they happen to have, not the entire array that defines the Mendelian populations from which they come.

In genetic drift, neutral mutation, and gene flow, the term *chance* retains only distant echoes of its ancient sense of value-laden effects accidentally achieved. So does Stephen Jay Gould's idea that evolution on the scale that paleontologists study is so untethered from adaptive natural selection that it is a matter of sheer chance whether whole phyla make it through the evolutionary bottlenecks that punctuate the history of life on earth (Gould 1989; Blount, this volume, chap. 10; Erwin, this volume, chap. 12).

By going out of his way to construe the phyla that made it through the Permian extinction that followed the Cambrian explosion as no better than those that didn't, Gould downplays the accidental goodness or aptness that Empedocles's theory of spontaneous generation stressed.[29] What comes to the fore instead is chance in the sense of randomness, stochasticity, and in some cases unpredictability so deep that it seems to have no causes at all. In a conceptual milieu influenced, however indirectly and impressionistically, by the dominant Copenhagen interpretation of quantum mechanics, the meaning that "spontaneity" still retained in nineteenth-century debates about spontaneous generation—a meaning that suggested "spontaneous" as a good translation of Empedocles's and Aristotle's *automatou*—has shifted toward "causeless." Mutation in the genetic material has also picked up this semantic shift. By its very nature, DNA makes unpredictable copy errors. Accordingly, mutations no longer bear any relation to environmental pressures that might once have been thought (by Darwin, for example) to cause spasms of undirected variation or to the adaptive benefits that a small fraction of them confer over transgenerational time. To be sure, point mutations in DNA do occur at a roughly steady rate, in obedience to the "molecular clock." Still, the unpredictability of what comes up at any given time and place seems to have confirmed to the point of excess Darwin's anti-Lamarckian postulate that variation arises independently of its subsequent utility.

I say "to the point of excess" because the formula "random mutation plus natural selection" that is commonly employed to reflect what Darwinism looks like in the light of the molecular revolution is misleading and does scant justice to the greatness of Darwin's insights. Darwin's idea was that the occurrence of useful variation, whatever its etiology may be, is chance only in relation to the process of natural selection. In this respect, the chanciness of variation is part of the definition of natural selection, not something that occurs in total independence of it. The formula "random variation plus natural selection," however, suggests otherwise. In implicitly defining the chance element in Darwinism as randomness that occurs prior to and with no conceptual relation to the process of natural selection, it stresses selection's weeding out role, renders invisible its "creative" role in gradually adapting organisms to environments, greatly shortens the long chain of causes and levels between mutations and traits, and overlooks what I regard as Darwin's great innovation in the history of biology: his appeal to chance to explain, not explain away, the functional, goal-directed, and purposive characteristics of organisms.

It is not only molecular biologists and some fellow-traveling physicists who promote the idea that Darwinism is "random variation plus natural selection" (see Weinberg 1992). Creationists, not least advocates of intelligent design, like the formula because it encourages them to claim that "according to Darwinism our existence is a mere accident."[30] This is as offensive to contemporary Christians as it would have been to Augustine, Aquinas, or Kant. Accordingly, the Discovery Institute, which promotes intelligent design, takes the formula as a reductio ad absurdum of Darwinism. In this respect, it echoes the intended effect of Bowen's hostile misinterpretation of natural selection as the preservation of traits by pure chance.

Even Darwinians of the Strict Observance, who strive to see natural selection at work wherever they can, can admit that it is not the only process that results in evolution. In the absence of natural selection, evolution can arise from random variation in the process of mutation, from genetic drift, and from the purely fortuitous survival or extinction of clades. I make no pronouncements or even guesses about how much of evolution is chance and how much is adaptive natural selection. Still, in considering this and related questions in the other chapters of this volume, it will be good to bear in mind that chance plays the same role in contemporary versions of the theory of natural selection that it did in Darwin's own work. It refers to the initial disconnect between the causes, if there are any, of variation and the gradual shaping of that variation into adaptations by a process of differential retention that, while it is probabilistic, is neither random nor coincidental. So considered, the process of adaptation by natural selection invites us to appreciate anew, in a world that is constantly changing, inherently contingent, and full of random events, Aristotle's, Galen's, Harvey's, Kant's, Cuvier's, and Darwin's wonder at the pervasive purposiveness of organisms. To have rekindled a sense of wonder about biological purposiveness, untethered from cosmic teleology, is my best guess about the significance of Darwin's life and work in what Hodge and Radick call "the intellectual long run."

ACKNOWLEDGMENTS

My thanks to the editors and anonymous reviewers for help in preparing this chapter. Special thanks to Jon Hodge, Jim Lennox, Charles Pence, Grant Ramsey, and Phil Sloan. None of them should be (or probably would like to be) saddled with my opinions.

NOTES

1. By "absolutistic," Dewey meant "idealist." He had himself been schooled in idealism. His turn to Darwinism and his effort, beginning in the 1890s, to follow in the footsteps of William James in articulating a Darwinism of the mind was a revolt against his former self.

2. Winsor (2006) has argued that Mayr's opposition to saltationist systematists such as Otto Schindewolf gave rise to his opposition to typology in the late 1950s. His rhetoric escalated thereafter. He added the term "essentialism" to "typological," thereby pitting "typological essentialism" against the "population thinking" of his own Biological Species Concept (Mayr 1976). Turning historian, he then blamed the long reign of typological realism that he had conjured up on Plato (Mayr 1963, 5; 1982, 45). (Winsor finds no evidence that Mayr was aware of Dewey's 1909 speech.) Joeri Witteveen (2013) has shown that Theodosius Dobzhansky and George Gaylord Simpson, Mayr's colleagues in making the modern evolutionary synthesis, all started using the population-typology distinction about the same time, but from the start meant different things by it.

3. For Amundson (2005), the theme of development is important not only for evolutionary theory's past but its future as well. The "evo-devo" revolution is showing that the population-genetic Darwinism of the modern synthesis can solve some but not all evolutionary problems, including some of the most important.

4. Citations to Aristotle's works are provided using the standard abbreviations and Bekker numbers. Readers interested in an English translation can consult that of Barnes (Aristotle 1984).

5. This is not to say that Galen and Harvey were entirely faithful to Aristotle's thinking. Both took the analogy between organisms and artifacts more seriously than their master, in part because they thought that all, not just some, organic parts have proper final causes or, in evolutionary terms, are adaptations. Unlike Aristotle, too, they thought that female as well as male reproductive organs contained contributions to the formal cause of generation. They certainly got that right.

6. Immanuel Kant, for example, dismissed Democritus's and Epicurus's views as "obviously absurd" (Kant 2000, 263, *CJ* 5:391).

7. Richards (1992) stresses this aspect of biological order in ascribing to Darwin the view that lower beings exist for the sake of humans even if they are moved along

by external forces like competition and selection. Making use of ideas like evolutionary arms races, which ratchet up the skill sets of successive kinds, and convergent evolution, in which similar environments shape similar solutions to adaptive problems, contemporary adaptationist Darwinians tend to affirm ex post facto evolutionary progress (Dawkins 1996; Conway Morris 2003; see Ruse 1996). It can be argued that the design paradigm still informs this interpretive framework, albeit at a distance.

8. The reason has to do with the recent triumph of phylogenetic systematics, or "cladism," which marks off taxa by their branching points (clades) and not by descriptive grades. The latter show traces of the Great Chain of Being that are slowly disappearing from biology and more slowly from our cultural imagination as a whole. See Hull 1988.

9. It is true that bacteria tend to generate variation that will be useful to them in certain circumstances, a process perhaps too strongly called "directed variation." This phenomenon is a result of an evolved mechanism but nonetheless puts pressure on the universality of Darwinian theory, which is heavily weighted toward metazoan phyla, in which the sequestration of heritable variation in eukaryotic genomes precludes any such thing.

10. Even David Sedley, who argues as vigorously as he can that there is a creationist streak in Greek biology, admits that Aristotle is an exception (Sedley 2007, 170).

11. Aquinas took to heart Augustine's warning, born of his disillusioning experience with Manichean pseudoscience, that theology should not tie itself to scientific theories that, when they are replaced, tend to discredit any religion that leaves itself hostage to them (Augustine 1992, sec. V.5). For his part, Aquinas's stress on the ontological dependence (contingency) of the created world on God led him to remain resolutely neutral on precisely how creation takes place, and even on whether it unfolds in time at all. Although it offended against faith, he did not think that Aristotle's notion of an eternal world could be ruled out on rational grounds. He suspected Augustine of violating his own scientific neutrality in proposing that *seminales rationales* are placed into the world at one time and expressed at another. There is a rich body of speculation about what an Aquinas *redivivus* would think about evolution. See W. E. Carroll (2000).

12. This lingo misled seventeenth-century philosopher-scientists, including Galileo, into thinking that they were contesting Aristotle when they were challenging "the schoolmen."

13. Amundson (2005, 95) awards this honor to Carolus Linnaeus (1707–1770) rather than to Ray. But Wilkins thinks there was more piety than biology in Linnaeus's fixist pronouncements (Wilkins 2009, 95). For more on Linnaeus's creationism, see Müller-Wille and Rheinberger 2012, 32–33. For studies of various figures in early modern biology, see Smith 2006.

14. "Breaking with preformationism was, I would claim, crucial to recovering a genealogical concept of species that was also open to some historical change over time. . . . Buffon's critique of preformationism made possible his own lineage and historical concept of species" (Sloan 2013, 239).

15. The physicist Max Delbrück, claiming to have found anticipations of the genetic code in Aristotle's account of the movements of sperm in *Generation of Animals*, wrote a slightly tongue-in-cheek paper proposing that the ancient Greek philosopher-biologist should be awarded the Nobel Prize along with Crick and Watson (Delbrück 1971). Why Aristotle's authority mattered to him is a mystery to me. But his claim passed as gospel to many people who had never read *Generation of Animals*, and even for some who had. It is, nonetheless, an interpretation in which the father of epigenesis becomes anachronistically a quasi-preformationist advocate of genetics *avant la lettre*—and by association transfers his own sexism to the fathers of molecular biology.

16. Ernst Mayr seems not to have appreciated that his own pet idea that what natural selection evolves are "genetic programs" carries traces of the very preformationism whose residues he criticized in molecular genetics (Mayr 1988, 16–17). Why, unless it is fighting against the entropic tide, should evolution have to be mediated and controlled by something like computer code? A recent suggestion that genes are only one among many "developmental resources" and that information processing metaphors are not very helpful in exploring developmental dynamics points in the direction of a more thoroughly epigenetic theory (Griffiths and R. D. Gray 1994; Moss 2003). The resulting sense of "epigenetic" contrasts with the restricted meaning the term acquired in the wake of August Weismann's (quasi-preformationist) sequestration of the germ line, according to which it refers to everything that happens to a cell *after* meiosis. Molecular geneticists such as Crick, Watson, and Monod wholeheartedly embraced this restricted meaning, thereby conferring on molecular genetics an autonomy from a wide range of biological processes whose evolutionary importance is only now becoming recognized again.

17. *The Origin of Species* somewhat obscures how deeply Darwin was affected by the epigenetic strain in the evolutionary theorizing that preceded his own. He framed his greatest book as a refutation of the static argument from design advanced by the Anglican cleric William Paley. He did so to get a hearing in a Britain that was raised on Paley's brand of natural theology (Depew 2009). By close study of his notebooks and drafts, scholars have shown, however, that in slowly arriving at his theory of natural selection, Darwin was working through a wide range of Continental evolutionary thinkers and their British disciples, all of whom focused on the developmental process (see M. J. S. Hodge 1985; Desmond 1989; R. J. Richards 1992).

18. For a detailed account of the Darwin-Gray correspondence, see Lennox 2010. For its politically charged context, see J. Moore 2010.

19. For a representative sample of professional reviews of *Origin*, see Hull (1973). Those who trafficked in exaggerating the role of chance in natural selection included, in addition to Bowen, William Whewell, Master of Trinity College, Cambridge, and, much to Darwin's chagrin, the philosopher-astronomer John Herschel. Lyell eventually came around to Darwin's view.

20. References to Darwin's correspondence include reference numbers for the letters in the Darwin Correspondence Project, www.darwinproject.ac.uk.

21. There was an element of strategy, both scientific and political, as well as of piety in Gray's proposal. He told Darwin that he had to" baptize" his theory in order to allow its main message to penetrate in northern states that, having embarked on a crusade against slavery, needed a strong dose of progressive theodicy in order to see the project through (Gray 1862, DCP #3489; J. Moore 2010, 571). Gray was a radical Republican; Darwin was no less antislavery but was a nervous nelly about the war.

22. "Natural selection tends only to make each organic being as perfect as, or slightly more perfect than, the other inhabitants of the same country with which it has to struggle for existence. . . . It will not produce absolute perfection" (C. Darwin 1859, 201–2).

23. This leaves unaddressed the question of evolutionary progress (teleology writ large) within the bounds of Darwinian principles. For a variety of views, see Richards 1992; Ruse 1996; Ruse, this volume, chap. 5; J. Moore 2010. It is important to bear in mind that even though human beings are not the preset goal of the evolutionary process, they are, at least from a biological perspective, an ex post facto result of natural selection's tendency to favor cognitively powerful species as less complex niches, occupied by less complex organisms, become unavailable. The distinguished midcentury population geneticist Theodosius Dobzhansky argued that cognition flourishes in culturally mediated niches (Dobzhansky and Boesiger 1987).

24. Darwin was willing to endure the ill repute attendant on "mere hypothesizing" in order to recommend his neopangenetic model of how parents pass on to offspring newly acquired variant traits in a way that can keep lineages tuned to changing environments. Like Aristotle and Harvey, Darwin thought of reproduction as "a form of growth" (C. Darwin 1875). But he also attempted to imagine how it is possible for successful variations to be incorporated by flowing through the bloodstream into the developmental process and so into reproduction. After August Weismann convinced the world that the germ line is sequestered from somatic cells, Darwin's model met with difficulties insuperable enough to provoke neo-Darwinians to suggest that Darwin was merely toying with generation theory. He wasn't (M. J. S. Hodge 1985).

25. It would seem that among these differences is that for Aristotle mere life exists for the sake of realizing the highest faculties of each kind. The perceptual capacities that give a higher rank in the cosmic order to animals are for Aristotle incipient forms of cognition, which gives humans a rank only a little below the gods. For Darwinism, by contrast, higher psychological capacities might be little more than instruments of life itself. There might be something to this, but I really don't think the contrast is quite so severe.

26. At one point Darwin and Gray almost broke out of the straitjacket of chance versus design. In the 1870s, Gray wrote, "Darwin's great service to natural science [is] bringing back to it Teleology; so that instead of Morphology versus Teleology, we shall have Morphology wedded to Teleology" (Gray 1884, 288). Darwin replied, "What you say about teleology pleases me especially" (C. Darwin 1874, DCP #9483). It is clear from the context that Gray and Darwin were both referring to the adaptedness of traits, not to large-scale monad-to-man teleology. They were, in fact, edging up

on the possibility that selection, as an entirely natural ex post facto process in which traits spread through interbreeding populations because of their positive effect on the life chances of the organisms that have them, might count as a limited kind of natural teleology. No call is thereby made on the future to bring into existence something that isn't the result of standard causes operating from the past. But this possibility—a possibility well explored by philosophers of biology in our own time (L. Wright 1973; Brandon 1981; Neander 1991; Lennox 1993, 2013)—soon regressed, especially on Darwin's part, to the old dispute about directed variation.

27. It is hard not to see the structure of the diploid chromosome as a beautifully evolved platform by which Mendel's laws, which are themselves contingent products of evolution (Beatty 1993), facilitate new genetic possibilities by breaking up, reforming, and multiplying gene combinations. Mendel's law of independent assortment ensures that genotypes at each locus on each chromosome will vary independently, making the chromosome a terrific randomizer and source of evolutionary novelty—and a reliable mechanism for passing on helpful novelties and disarming harmful ones. The law of segregation, for its part, makes room at each locus for dominant and recessive genotypes, thereby affording a way in which adaptive genetic combinations will be phenotypically expressed and successfully inherited, while harmful combinations will be eliminated or, as recessives, retained in gene pools as reserve fuel for natural selection to use in changed circumstances (Dobzhansky 1937). The chromosome, accordingly, is a persuasive witness to the deep interplay between chance and purposiveness in the evolutionary drama. It is one of the cases to which Dobzhansky was referring when he titled his essay, "Nothing in biology makes sense except in the light of evolution" (Dobzhansky 1973). Precursors of the diploid or two-locus chromosome are also witnesses. The diploid chromosome has the reliability required to coordinate complex metazoan life forms. Analogous mechanisms in microbial life have clever ways of inserting new genetic material directly into DNA segments, but they are so unreliable in creating stable lineages that it is doubtful whether the term *species* even applies at the level of *archaea* (Goldenfeld and Woese 2007). By the same token, diploidy, while reliable, is slow to adapt to very rapidly changing environments. So, in the few circumstances in which it has been possible, it has been supplemented by learning and cultural transmission.

28. I prescind from an important debate among philosophers of biology about whether genetic drift is a purely statistical phenomenon or, even if it results from a sampling process that is indiscriminate, has in every case determinate physical causes. See Matthen and Ariew 2002; Millstein, Skipper, and Dietrich 2009.

29. Gould's interpretation of the Burgess Shale fossils has been challenged by their most respected interpreter, Simon Conway Morris, whose adaptationism is so strong that it embraces a progressivist interpretation of the history of life on earth (Conway Morris 2003).

30. A post on the Discovery Institute website dated June 1, 1997, accessed on October 1, 2013.

CHAPTER 2

Chance and Chances in Darwin's Early Theorizing and in Darwinian Theory Today

Jonathan Hodge

This chapter takes on two topics: first, what Charles Darwin thought about chance and chances before and after he arrived at his theory of natural selection; second, what this narrow Darwin narrative suggests for the broader history and philosophy of natural selection theory since Darwin's time. Two continuity theses are defended: that there was no radical shift in Darwin's thinking about chance and chances when, late in 1838, he first formulated his theory of natural selection; and that the theory, construed by Darwin as a probabilistic and causal theory, has not changed since into a noncausal, statistical theory.

Continuity theses can be unexciting; but there is no suppression here of big issues. On chance and chances Darwin held the most common view of his day: that chanciness in causal theories is due not to any gappy indeterminacy in the causal order of nature but rather to gappy incompleteness in our knowledge of that order.[1] Since Darwin's day, various radical shifts in several sciences—most obviously in statistical and quantum mechanics—have made this ignorance view of chance rare, but no such developments have made the theory of natural selection no longer probabilistic-causal.

DECISIVE THEMES AND DISTINCTIONS: THE ACCIDENTAL, THE PROBABLE, THE ANCESTRAL, THE DESCENDANT, THE GEOGRAPHICAL, AND THE DEVELOPMENTAL

Recall next two invocations of chance and chances in the *Origin of Species* (C. Darwin 1859). There Darwin says that hereditary variations among individual organisms may arise by chance and that some of these variant organisms have better chances of survival and reproduction in the struggle for existence and so are naturally selected. These two invocations of chanciness suggest a contrast. In ascribing variations to chance, Darwin asserts, in conformity with an ancient threefold Greek distinction, that they are caused accidentally rather than necessarily or purposefully (for more on these distinctions, see chapter 1 in this volume, by David Depew). By contrast, in talking of individuals having differing chances of survival and reproduction, Darwin refers to different probabilities, with values implicitly falling between none and one. So the generation of hereditary variation in the length and strength of legs in wolves is due to unknown causal accidents and is a matter of chance; however, that individuals with longer and stronger legs have higher probabilities of success in living, breeding, and raising offspring is due not to mere chance but to the causal consequences of this variation.

This last invocation of chances as raised probabilities may seem to associate Darwin with the so-called probabilistic revolution, a modern development, mostly beginning in the seventeenth century, largely consummated in the nineteenth, and often dubbed the rise of statistical thinking (Krüger, Daston, and Heidelberger 1987; Krüger, Gigerenzer, and Morgan 1987; Gigerenzer et al. 1989). But this association misleads by suggesting that the young Darwin was responding to the latest new statistical thinking in the 1830s from Adolphe Quetelet and others, whereas his notebook theorizing at that time shows that he depended on no doctrines about chances not current in the previous century. So Darwin's theorizing was not applying to his explanatory tasks resources freshly minted by contemporary pioneers in statistical theory and practice (M. J. S. Hodge 1987; Gigerenzer et al. 1989, 135–44).

To understand why not, let the *Origin* prompt some general reflections about the pervasive chanciness in his theorizing; for, as often rightly insisted, all three of the main causal tendencies taken by Darwin to produce the process of natural selection—heredity, variation, and the struggle for existence—seem distinctly chancy rather than strictly determined; so that their interactive long-run outcomes, the adaptive divergent branching descents wrought by natural selection, must surely be chancy too. After all, the theory implies that what makes those divergent descents go in some directions rather than others are immediate, local influences external to the organisms themselves: soil, weather, migrations, invasions, predators, prey, and so on. Such are the influences, the circumstances, that make some variations advantageous in some species and other variations advantageous in other species, while the hereditary transmission of those variations is likewise far from completely reliable, only probable and reliable enough to allow this variation to be accumulated by selection over many generations. The theory invokes, then, no completely reliable directing influences within or between or around organisms; for it is the selective influences that mainly direct those changes, selective influences arising from the interactions of the organisms with environmental circumstances that change from one time and place to another in accidental, contingent ways entailing that the entire succession of life on earth could have been very different had these contingencies, these circumstantial cookies, crumbled differently.

Such general reflections on the *Origin*'s teachings suggest another contrast. In the *Origin*, Darwin's theory was not, it has rightly been said, a developmentalist theory (Bowler 1988). In developmentalist theories of evolution, the long-run phylogenetic changes over eons, from ancestral fishes to descendant mammals, say, are compared with the short-run ontogenetic changes whereby the fetus formed when two adult mammal parents mate develops into a very fishlike early embryo before developing into a much less fishlike newborn mammal. This comparison sanctions the assumption that the same laws and causes of development direct life's changes over both the long run—between ancestral fish and descendant mammal species—and the short run within the maturation of an individual mammal.

The absence of this developmentalist assumption in the *Origin* can aid our appreciation of how chance and chances entered into Darwin's thinking before he had his theory of natural selection. For consider an illustrative schematic contrast between the theorizing Darwin was doing from mid-1837 on—in the year and a half before he arrived at that theory—and the developmentalist theorizing elaborated in *Vestiges of the Natural History of Creation*, the book published anonymously by Robert Chambers in 1844. Imagine two old continental masses of lands, Westland and Eastland, separated by a wide ocean where a new island has recently arisen; and imagine the theorizing Darwin would be doing in 1837 in relating the new animal and plant species eventually inhabiting that island to the older species already inhabiting the two continents. Familiarly enough, Darwin would suppose that those island species are very probably descendants of earlier continental species that have arrived there as colonists. But elaborating that supposition would require him to distinguish various causes making it more likely that the colonists have come from one continent rather than the other. It could be that Westland is closer to the island, or that the prevalent wind travels from west to east, or that the ocean currents usually do. Perhaps Westland has more species than Eastland, even if Westland has fewer species that are good at getting to remote places and good at thriving in the conditions on the island. See how the accidental and the probable feature here. Which individuals of which species from which continent get carried by wind or by water in just the right direction to land upon the island is chancy in the sense of accidental; while if, obviously counterfactually, all else is equal, more will beat fewer, so the more probable supplier of colonists will be the continent with more species on it. For discussion of chance transport in the early Darwin, see Hodge (1987, 241) and, much more extensively, Johnson (2015, 27–47).

Now switch to what *Vestiges* says about Africa, South America, and the Galapagos Islands. The two continents both have many monkey species, but this resemblance is not due to common ancestry, because, it is insisted, there is no way monkeys could have migrated between the old and new worlds. Rather: life has originated with the simplest organisms on both continents; and then over eons it has evolved in each arena in accord with common laws of development to the same high levels of complexity, and so arrived independently in both regions at the structures and functions peculiar to the monkeys' particular high level of complexity. As for the Galapagos Islands, they are like many ocean islands in having had no mammal species originate on them. Why not? Because they have risen too recently above the ocean waters, so that life on those islands has not yet had time to develop from the levels of the aquatic fish and reptile species there to the highest terrestrial vertebrate levels peculiar to mammals (M. J. S. Hodge 1972).

So Darwin's causation is often chancy and horizontal, but *Vestiges*'s causation is mostly unchancy and vertical. This correlation of the chancy with the horizontal and the unchancy with the vertical is not surprising if one harks back to Aristotle (see again Depew's invaluable chapter). For Aristotle, a horizontal, coincidental conjunction—two friends happen, unplanned, to meet downtown, having gone there for different reasons—is an exemplary result from chance or accidental causation. By contrast, the progressive, maturational, quasi-vertical ascent of an egg turning itself into a chicken is, despite any occasional accidents, normally the exemplary work of reliable, necessitating, essential causation.

The processes that geological-biogeographical explanations invoked were much more contingent and chancy than those invoked by embryologists, and this contrast bears on Darwin's original favoring of common descent. Lyell and others had upheld a single (providentially chosen) place of origin for each species, and so a common descent within each species for all its subsequent members; and it was implied that success in migrating to any other suitable places was often a matter of lucky accidents. In Darwin's first favoring, in mid-

1836 or at the latest early 1837, of common descent and a single place of origin for any genus or family of species, such accidental migrational causation has influenced not just where any species is living now but where it originated, because such causation influenced where the species immediately ancestral to it chanced to land up.

CHANCE AND CHANCES IN DARWIN'S EARLIEST ZOONOMICAL THEORIZING IN SUMMER 1837

Turn now to the theorizing that opens Darwin's Notebook B in July 1837, a few months after he first accepted such supraspecific common descents and so became a convinced transmutationist (or evolutionist, in later lingo).[2] In two dozen pages, under the heading "Zoonomia," meaning the laws of life (as in the title of his grandfather Erasmus Darwin's biggest book), he sketches a system of theory modeled in its structure on Charles Lyell's synoptic version of Jean-Baptiste Lamarck's system (C. Darwin 1837, B 1–24). Like that synopsis, Darwin's system comprehends the course and causes of life's changes, first in the short and then in the long run. Darwin's systematic account deals initially with mostly horizontal and then later with mostly vertical changes. Within a week or two, however, he constructs a revised account of the vertical changes so that those changes are explicable as concomitants of longer-run extrapolations of shorter-run horizontal causes and effects.

Darwin gives an integrated treatment of a wide raft of issues that will preoccupy him for the rest of his life. Four clusters of these issues concern chance and chances: namely, first, the role of individual maturations in making possible intervarietal, intraspecific branching, adaptive divergences accompanying the horizontal migrations of any species facing new conditions on arrival in new areas; second, those continuations of intraspecific changes that result in longer-run, interspecific, congeneric divergences; third, the comparisons Darwin draws between the births, lives, and deaths of individuals and the origins, durations, and extinctions of species; and, fourth, the longest-run changes of all, the changes resulting in the divergences between the wider groupings: families, orders, classes, and so on.

Mating, with fertilization of females by males, and maturation in the offspring produced are, Darwin holds, the two features of all sexual reproduction that distinguish it from any asexual generation by grafting or budding; and any ontogenetic maturation recapitulates phylogenetic descent. It is this maturation with recapitulation that makes possible all adaptation in individuals, and so all hereditary adaptive variation, and so all adaptive diversifications

within and between species; and it does so because only immature organization is adaptively impressionable by environmental influences so that adaptive hereditary variations can be acquired. (This sounds like Lamarck because
here, as elsewhere, Darwin is following Lyell's synopsis of Lamarck's theorizing.) These adaptive hereditary variations are then chancy insofar as their
occurrence and character result from chance encounters with changed conditions due to migrations and other contingencies. So intraspecific, intervarietal
diversification is due to and directed by hereditary adaptations to local circumstances made by phylogeny-recapitulating maturations of individuals, but
it is not due to any lawful developmental tendencies directing and manifested
in those ontogenetic recapitulations of those phylogenies. And interspecific
adaptive diversification is wrought by reiterations over many more generations of this chancy causation for intravarietal divergences.

Being branching, this adaptive diversification can result in many descen-
dent species descending from a single common ancestral species. But, as in
Lyell, the total number of species is constant in the long run. So any additions
must be balanced by subtractions due to extinctions, the coming and going of
species, as in Lyell, being likened to the births, lives, and deaths of human individuals as disclosed by successive censuses. In Darwin, these additions and
subtractions contribute to a complex correlation: high levels of organization
correlate with more character gaps due to branching changes with many extinctions. Here, Darwin's first explanation of this correlation invokes his view
that the spontaneously generated infusorian monads, the simplest microorganisms of all, starting each line of organic progress, have vastly extended
but limited lifetimes lived out in each line; so, because these vast lifetimes are
limited, those lines that have gone highest, and so to mammals, must have
changed in species most quickly, owing, as Darwin says tellingly, to accidents
of position. Very soon, however, he rejects this view of vast but limited monad
lifetimes and revises the correlation, so that many branchings and extinctions
and gaps correlate not with level of organization but with group width, a class
being a much wider group than an order (C. Darwin 1837, B 24–44). With
this shift, from now on Darwin's theorizing about species births, lives, and
deaths includes no explanatory invocations of life's monadic startings. Nothing coming after those startings is explained as a consequence of any special
life-initiating causation.

Darwin's new explanation of this new correlation is retained over the
coming months while being integrated with three new conclusions about
maturational recapitulation: first, that the phylogenetically oldest characters
appearing early in ontogeny are the most strongly inherited and the least li-

able to change, and that the later stages in ontogeny are more impressionable and more easily modified in adaptations to new local circumstances; second, that new characters must be developmentally harmonious with ontogenetically earlier and phylogenetically older characters still persisting; third, that in phylogeny and so ontogeny, hermaphroditism precedes separate sexes. None of these three new views about maturational recapitulation makes Darwin's theory a developmentalist theory, because they entail no revision of his earliest views about maturations, adaptive variations, and changes in local circumstances. Nor is this a theory of ontogenetic innovations by terminal additions determined by lawful developmental tendencies; for innovations can arise, as he had always held, as preterminal modifications determined by the external circumstantial influences acting on impressionable maturing organization recapitulating recent phylogenetic changes late in ontogenies.

So Darwin's theorizing about change in the short and long runs is still grounded in his thinking about the matings and maturations peculiar to sexual generation. Consider next two distinctions preoccupying Darwin at this time. A puppy born with thick fur in a warm country is a monstrosity, even if that variation was caused as an adaptation to abnormal conditions in the womb. Such variations are both rare and disadvantageous enough to be soon blended out by crossing and eliminated by failures to survive and reproduce. A puppy growing thick fur on moving to a colder country is no monster but an adaptation, because this variation is elicited by the cold and advantageous in it (C. Darwin 1838a, C 62–66, 82–85). Darwin calls such variations necessary adaptations, in wording drawing ultimately on ancient Greek distinctions. The thicker fur is chancy in resulting from a chance encounter with colder climes; but the response made to those chance circumstances is not itself chancy but causally necessitated. So Darwin does not admit that there are chance adaptations. He talks as others did of chance or accidental peculiarities and variations, meaning, in conformity with the ignorance view of chance, that they are due to small, hidden causes effective prenatally. He does briefly consider whether such rare, infrequent chance peculiarities could sometimes be accompanied by more vigor in those individuals and so benefit some male deer, say, in fighting rivals and winning mates, and perhaps causing males in many groups of animals to be similarly armed with horns or tusks or spurs (C. Darwin 1838a, C 61). But at this time, Darwin does not see this explanation for resemblance in arms for fighting as suggesting how differences in adaptations to diverse circumstances are caused. To understand why not, consider his most exemplary imaginary adaptive diversification scenario at this time, still several months before he begins to move toward his theory of natural selection. In line with Lyell's account of Lamarck's views, Darwin thinks adaptive structural variations are often initiated by changes in habits and so in the uses of organs. Suppose a country is flooded for the first time and that all the jaguars there take up swimming for fish prey; then a new variety with webbed feet could arise through the inheritance of this acquired character: all the jaguars, the whole race as Darwin says, because otherwise the character will be blended out in crossings and so fail to persist. Here frequency and mating considerations are prominent in Darwin's thinking because variations arising as exceptional rarities cannot contribute to permanent adaptive change (C. Darwin 1838a, C 62–63). For more detailed documentation and exegesis, see Hodge and Kohn (1985).

NATURAL AND ARTIFICIAL VARIETIES, WINNING AND LOSING SPECIES, LUCK AND NUMBERS, PROVIDENCE AND PROGRESS, FREE WILL AND NECESSITY

This reasoning was reinforced for Darwin by the contrasts he was making from early 1838 on between two kinds of varieties in domesticated animals and plants: natural varieties and artificial varieties. The natural varieties of domesticated species are due to natural causes rather than to human artifice and are local varieties, isolated so as not to be interbreeding with others and diverging as they adapt slowly over many generations to local conditions of soil, climate, food, and so on. By contrast, artificial varieties are often monstrous, distinguished by variations arising as rare, maturational accidents only persisting thanks to the human art of picking, selective breeding, that has made varieties, often in few generations, that could never be formed and flourish without benefit of that human art. As Darwin read about the art of selective breeding, he became convinced at this time that species formation in the wild was to be compared with natural variety formation in domestic species and contrasted—yes, contrasted—with the making of artificial varieties by selective breeding (M. J. S. Hodge and Kohn 1985).

This comparing and contrasting cohered well, in Darwin's view, with other comparisons and contrasts concerning winning and losing among species. Winning for a species is not merely surviving but having, through splittings, branchings, and rebranchings many offspring species, many descendant species, whereas losing is extinction, ending without splitting. In understanding why winning species win and losers don't, Darwin considers two prominent human families and the causes that might lead one to have many descendants in the future while the other does not. The causes are various and are effective because their consequences for individuals also have consequences for the family's descent into future generations: the presence or absence of hereditary diseases, or of aversion to marriage, and so on (C. Darwin 1837, B 41, 146–49; Ariew 2007, 2008; M. J. S. Hodge 1987).

Darwin's emphasis on accidents and probabilities in the lives of species is often opposed by him to providentialist views that he is rejecting and replacing. A new island does not get the terrestrial species that are to live there from some previsionary, provisionary causation. They arise, rather, as descendants of migrants from older lands that happen to be close enough to supply ancestors that happen, moreover, to be numerous enough and lucky enough to travel and colonize successfully and abundantly. Again, an aberrant species—a ground-dwelling woodpecker species, say—is more likely to arise in a group with many rather than few species. Generally, then, causal explanations invoking greater or lesser luck and higher or lower numbers often make providentialist explanations unnecessary (M. J. S. Hodge 1987; C. Johnson 2015, 27–47).

There are, however, other contexts where Darwin—still very much a theist and a deist, a believer in God but not in the Bible as his word—is explicitly providentialist in his theorizing: most conspicuously in his teleology of sexual generation as the natural lawful means instituted by God to ensure that adaptive change in the long run eventually includes the peculiar progress from lower to higher levels of organization that has led quite naturally to mammals and so to man and his social and moral life. This special life, he emphasizes, is not perhaps the sole purpose but is certainly one purpose of the prehuman progress; this special life is, then, due not to a special providence but to special results from the general providence of life with sexual generation (Ospovat 1981).

On the laws of sexual generation and its lawful long-run consequences, Darwin has his Newton emulation moments: likening what he is doing with these laws for natural, unmiraculous adaptive descents to what Newton has done for planetary orbits with his laws (C. Darwin 1837, B 101). The laws of sexual generation are not, however, developmental laws. The higher levels of life are not invariably developed from lower levels in phylogeny in conforming to the same laws manifested in ontogeny, where fish form is lawfully destined to develop into mammal form. In Darwin's branching descents, only one line of fish species has had mammal descendants, and it did so presumably because of exceptional circumstance, as none of the other many fish lines have had such descendants.

Notice here a contrast with Lyell's version of Lamarck's system. In that version, branching adaptive descents are ascribed to the inheritance of characters acquired in responding to diverse local circumstances, while the rise of organization from monads to mammals in every line of escalation is credited not to that external, circumstantial causation but to the workings over eons of internal fluid motions distinguishing all living from all lifeless bodies. Darwin, by contrast, has no such division of causal-explanatory duties, for he invokes the same external, circumstantial causation working in successive sexual generations to explain both adaptive diversification and organizational progress. In doing so, he makes life difficult for himself and for his theorizing, because, although he does not want invariable progress in all lines of descent, he does want reliable progress in some. What is more, he is wary of countering Lyell's view that there has been no overall progressive change in the planet's condition since the oldest known fossil-bearing rocks were laid down; so he is wary of explaining any progress in life as an adaptation to a progressive improvement of the planet's habitability in passing from a fiery nebular beginning to its current calm and fertility. In the notebook years, he holds consistently to the view that the tendency of any characters to become more strongly and powerfully embedded in hereditary conditions over long generations makes progress possible, because it ensures that newer variations do not erase earlier ones, so allowing the cumulative changes required by progressive advances. Beyond this, he does not settle on any one explanation for progress being a reliable concomitant of adaptation over the very long run; but he does remain committed to thinking that a general theory of progressive change should be a corollary of a general theory of adaptive change (Ospovat 1981).

His theorizing about progress does not therefore lead Darwin into any special views about chance, chances, and necessity, or to any views not in play in his adaptive diversification theorizing. By mid-1838, he is explicitly committed to what was then called necessitarianism and later determinism, and no less explicitly to the ignorance view of chance. Free will, he says, is to mind as chance is to matter; in both cases there is the illusion of indeterminacy, but it is only an illusion owing to our ignorance of the causal necessitation (C. Darwin 1838d, M 27, 31; C. Johnson 2015, 189–204; Noguera-Solano 2013). Darwin needed no special support for this analogy, because he was now explicit, too, in his materialism about the mind as the workings of the brain, and so his ignorance view of chance included mind by comprehending brain. This determinism and ignorance view of chance allowed, indeed encouraged, Darwin to take chanciness more, not less, seriously in his theorizing, because to do so was to take seriously any lawful causation and any gaps in his knowledge of it. His own theorizing about animal and plant individuals engaged such theses most directly when considering the roles of chance conditions, conditions encountered accidentally, even unobservably, whether prenatally or later in individual maturations; and when considering their causal roles in the births and lives and deaths of species as quasi-individuals. These themes in Darwin's mid-1838 thinking can enhance any understanding of what happens next, in the fall and the winter following and so indeed in the rest of his life, and even in the long decades of Darwinian theorizing since.

CHANCE AND CHANCES AND THE CONSTRUCTION OF NATURAL SELECTION THEORY (SEPTEMBER 1838–MARCH 1839)

Darwin's theory of natural selection was arrived at in complex conceptual shifts, starting with the late September 1838 reading of Malthus through to the first formulation of the artificial selection analogy around late November, and on to the acceptance early in the new year that chance or accidental variations can be accumulated by selection in adaptive changes (M. J. S. Hodge and Kohn 1985). These shifts obviously included new thoughts about the contributions made by accidental and probabilistic causation to the effecting of adaptive change; but they did not require Darwin to think in new ways about the underlying causal sources of chanciness in individual and species births, lives, and deaths (M. J. S. Hodge and Kohn 1985; M. J. S. Hodge 1987).

The reflections prompted late in September 1838 by Darwin's reading in Malthus are cryptic (C. Darwin 1838b, D 134–35; Ariew 2007, 2008). Three points are, however, hardly contestable. First, various innovations in Darwin's theorizing did not take place then, but later: only later does he arrive at the analogy between natural and artificial selection; only later does he take chance variations to contribute to adaptive changes; only later, indeed, does he integrate any new conclusions about the consequences of Malthusian superfecundity (the tendency for reproduction to outrun food supply) with his long-standing views on how hereditary variations are due to sexual generation with its individual maturations. Second, most of the notebook sentences prompted by that reading deal with the causes of species extinctions; and, on extinctions, Darwin is not here coming to new conclusions but is going back to views of Lyell's that Darwin, way back in 1835, had dropped in favor of an inherent mortality for each species. Third, when Darwin, in one closing sentence, does discuss the consequences of superfecundity for the winning species, those not going extinct, he says that there is a sorting out of fitting structure and so an adapting of structure to changing conditions. However, it is not clear whether this sorting out, this retaining (wool sorters talked of sorting out the good wool that they kept), is thought of as going on among conspecific individuals or varieties or interspecifically among species. Darwin was seemingly unconcerned with those distinctions at that moment, as is suggested by a comparison he makes: Malthus had argued for a providential teleology and imperialist theodicy of superfecundity, for he urged that the consequences of superfecundity for intertribal conflict in ancient times had made invading and invaded peoples less lazy and more energetic. Likewise, says Darwin, with animals: the final cause, the benefit intended by God, of population pressure is the adaptive improvement of animal structures; but he does not say whether this is due to competition within or between species, or to both (M. J. S. Hodge and Kohn 1985).

Darwin's own metaphor for these pressures cites wedges as used in splitting logs. Consider the role of this metaphor in the return to Lyell's views. Often, the various species in an area are all maintaining their numbers and ranges. But, Lyell had argued, changes in conditions, such as climate changes, could cause some species to gain advantages over others, and to expand their numbers and range, so causing terminal losses of population and ground in those losers. Darwin saw Malthus as vindicating this claim, because superfecundity makes even slight changes in conditions effective in initiating such imbalances. Lyell had explicitly supposed that the species in any area are exploiting resources to the full. No species can expand unless one or more others contract. Darwin's wedging metaphor is to emphasize this assumption. The width of a wedge at the log's surface stands for the numbers of individuals of a species now living. If this wedge is to go further in, driven in by the power of reproduction making for population increase, it must force one or more wedges, other species, partly or entirely out of the log. Inward and outward forcings are variously caused: a predator species gains or loses numbers, so affecting prey numbers, or a winter has been exceptionally mild or severe. Throughout, Darwin sees Malthus vindicating Lyell about extinctions, as in the second, 1845 edition of his *Journal of Researches*, where Darwin rehearses the reasoning more explicitly and less metaphorically (C. Darwin 1845, 173–76).

In September 1838, the added good news concerns the winners, the successful survivors of the changing conditions, the beneficiaries of the sorting out of fitting structure. But Darwin is dropping nothing in his long-standing theory about species formations, adaptations, and diversifications. This sorting theme is added to what he has been calling "my theory," without any subtractions. Notice too that a further notebook reflection on Malthus, a few days later, is easily overinterpreted (C. Darwin 1838c, E 3). Malthus had talked up ascribing risings and fallings in human populations to natural laws rather than to divine interventions. And Darwin cheers him on in this and reflects that he, Darwin, is doing likewise with species origins and extinctions: ascribing them not to divine interventions but to natural laws. So Darwin's assimilating of his theorizing to Malthus's only concerns here this general issue of natural laws and divine interventions. There is no reason to read Darwin as thinking that Malthus has taught him his first ever insight into some special populational or statistical explanations and that he is resolving to extend this new insight about demographic risings and fallings to species formations and species extinctions (Ariew 2007, 2008).

The Malthus reading prompted no shift in Darwin's views as to how sexual generation, with its matings and maturations, ensures adaptive change in altered conditions. However, Darwin does soon reflect that only a structural variation that is adaptive for the whole lifetime of an individual will be retained in the Malthusian crush of population over many generations. Variations adaptive only to fetal conditions will not be, while retained variations, eventually becoming strongly hereditary, can be accumulated in prolonged progressive changes. Note that Darwin's long-standing preoccupation with sexual reproduction leads him here to individualism about the causes and consequences of advantageous and disadvantageous variation in this integration of his Malthusian insights with his earlier beliefs about adaptation and progress (C. Darwin 1838c, E 9).

In late November, he throws light on long-run changes through his first explicit contrast between two principles explanatory of change in the short run (C. Darwin 1838e, N 41-45). One principle is familiar: an adult father blacksmith, thanks to the inherited effects of his habits, has sons with strong arms. The other has no exact precedent: any children whom chance has produced with strong arms outlive others. The contrast is direct. Chance production means here, as it has all along for Darwin, production by small hidden and rare causes effective prenatally, so that the opposite of chance is postnatal habits. What is new arises from Darwin's recent Malthusian concerns: namely, his conviction that those products of chance that have the same benefits as the effects of habits can contribute to adaptive change, because, although rare, individuals with such beneficial variant structures will survive over future generations at others' expense. However, Darwin acknowledges a difficulty in deciding which adaptive structures—and instincts, because these principles apply, he notes, to brain changes—have been due to which of the two principles. Although variations due to habit changes are directly contrasted with chance variations, variations due to habit changes have always been understood by Darwin as more chancy than the kind of developmental variations invoked by *Vestiges*. Recall how the jaguars became web-footed in Darwin's earlier conjecture. Their change in habits and so in structures was not just occasioned but determined in its character by a local, circumstantial contingency—flooding happens—and not as any lawful, developmental necessity. And there is still this contingent chanciness in Darwin's extrapolation of the blacksmith-style inheritance to any diversifying adaptive changes that wild animals make in response to local alterations in conditions, alterations having impact on the lives of individuals and species in so many accidental, horizontal conjunctions, chancy environmental encounters with the external causes of those adaptive changes in instincts and structures.

A few days after the two-principles moment, Darwin makes the notebook entry that comes closest of all to marking the moment when the theory of natural selection is finally formulated for the first time. For he is again considering principles; this time there are three, and they can, he declares, explain everything. Strikingly, none of them is new: that grandchildren resemble grandfathers; that there is variation in organisms in changing conditions; and that fertility exceeds what food can support; in sum, these are heredity, variation, and the struggle for existence (C. Darwin 1838c, E 58). Darwin may well have wanted these three principles to subsume the earlier pair, so circumventing the difficulty of deciding which adaptive changes to ascribe to which of that pair. Notice too that this three-principles moment draws no analogy between the struggle in the woods and a farmer selectively breeding livestock. So this is not the moment when Darwin's construction of his theory of natural selection is completed under that designation.

CHANCE AND CHANCES IN NATURAL AND ARTIFICIAL SELECTION

Very soon after the three-principles moment, however, with prompting perhaps from some comparisons between wild predatory canine species and sporting breeds among domestic dogs, Darwin is embracing for the first time just such an analogy (C. Darwin 1838c, E 63). In doing so he reverses himself on his long-standing conviction that species formations in the wild were to be compared with natural variety formation in domestic species and contrasted with varieties formed by the art of selection. So in this shift he rejects his explicit former belief that there was nothing like artificial selection going on in the woods, deciding now that yes, indeed, there is. For two months, since September, he had held Malthusian sorting to be going on out there; what he now sees is that this sorting, this natural selective breeding, is—not in its causes but in its effects—strikingly like artificial selection but unlimitedly more effective.

Observe what Darwin was and was not doing in his new reasoning back and forth between artificial and natural selection. He did not reason thus: domestic races are made by artificial selective breeding; wild species are like domestic races; similar effects have similar causes; therefore species formations in the wild are made by a cause like artificial selective breeding. No, he argues that wild species are made by the causation entailed by the concatenation of the three principles of heredity, variation, and superfecundity. This causation is selectional in its effects but much more powerfully so than is artificial selection; therefore, since like causes produce like effects, this causation can produce results alike in kind but much greater in degree and can therefore cause not just the formation and adaptive diversification of varieties within a species, but the formation and unlimited adaptive diversification of distinct species.

With this argument in place from December 1838 on, Darwin can move by mid-March 1839 to an explicit statement that domestic varieties are made in the two ways he had long distinguished between; but he argues now that wild species are not made as natural, unselected varieties of domestic species are, but as artificial, selected varieties are made in domestic species (C. Darwin 1838c, E 118). By this mid-March moment, Darwin had made three changes to his thinking about variations. First, that initial rarity may not prevent variations from contributing to species formation. Whether arising from the inherited effects of use or arising by chance, they can contribute because their survival and reproductive advantage in the wild, like selective preservations on the farm, will raise their frequency over successive generations before blending can counteract their differences. Second, very small advantages can eventually have large cumulative effects. Malthusian superfecundity entails that slight new competitive advantages due to small recent changes in conditions can result in some species losing out totally to winning species with those advantages. Likewise, the struggle for existence ensures that very slight advantages for some slightly variant individuals will eventually make such small variations no longer rare but common. Third, artificial selection suggests how species can become adapted to circumstances not credibly supposed effective in eliciting the requisite variations. Any variations improving seed dispersal devices on plants are not elicited by external circumstances; but if they arise by chance, competition due to superfecundity will ensure that they are retained and accumulated and the devices improved (M. J. S. Hodge and Kohn 1985).

In all these ways, then, the selection analogy prompted Darwin to give initially rare chance variations decisive causal roles in the formation and adaptive diversification of species in the wild, roles he never gave them as long as he was contrasting and not comparing wild species with artificial varieties. However, concerning how chance variations are caused, and why it is appropriate to call them that, his views have not changed. Equally, there has been no change in his views on how artificial selective breeding can work effectively with chance variations. What has changed are not his views about chance and chances, the accidental and the probable, in the lives and deaths of individuals and species, but his views about how the accidental and the probable cooperate causally in the same way on the farm and in the wild. The variations making some adaptive change possible may arise by chance, and in so doing they may seem unpromising initiations of such a change; but if they give those variant individuals a better chance of surviving and so of reproducing, then this change will proceed over successive generations.

As for Darwin's thinking about chance and chances over the rest of his life, there is no momentous shift (Beatty 2006a, 2013, 2014; Lennox 2010; Noguera-Solano 2013; C. Johnson 2015). In the 1840s and 1850s, the main developments in his theorizing concern pangenesis, sexual selection, divergence, and the recapitulation of phylogenies in ontogenies. Pangenesis was Darwin's general theory about all generation, from healing tree bark to mammalian procreation. The theory, as constructed in the early 1840s but only published in 1868, had no new implications for the theory of natural selection, which is not mentioned in expounding pangenesis; in allowing for the production and inheritance of chance and of acquired variations, the theory added and subtracted nothing bearing on selection, whether artificial or natural. Sexual selection theory invoked the same cooperation between the accidental and the probable that natural and artificial selection theory invoked. The divergence theorizing developed new insights about the effects of natural and artificial selection, but it did so without requiring revisions in Darwin's views on the role of accidental and probable causation in the workings of these two processes (Ospovat 1981; Kohn 2009; Nyhart 2009). On the relations between ontogenies and branching phylogenies, Darwin did hold that the development of individuals in each species recapitulates the distinctive changes leading to that species from the common ancestral species; so different changes are recapitulated in different lines of descent. These recapitulations do then allow inferences back from present ontogenies to past phylogenies, but there is no lawful predetermining of future phylogenies as an extrapolation of present ontogenies, no such predetermining, then, as is central to *Vestiges*. What determines for Darwin the future changes leading on from any species extant today is not its past phylogeny as recapitulated in ontogeny, but whatever future adaptive changes will be wrought by natural selection favoring different specializations in different circumstances and conditions and so in different lines of descent. Once again, then, we see Darwin developing a theory of progress not from causal-explanatory resources or developmentalist resources additional to those invoked in his theory of adaptive divergence, but rather as corollaries and amplifications of his theory of natural selection (Ospovat 1981; Nyhart 2009).

Darwin's life after the *Origin* included no very consequential revisions to his views in the '40s and '50s. In his book on orchids, he got into chance and design issues. These discussions deepened and widened Darwin's insights and enhanced his critics' understanding of what he was inviting readers to agree and disagree with; but the discussions did not take his thinking about the accidental and the probable in directions he had not taken earlier (Beatty 2006a, 2013, 2014; Lennox 2010; Noguera-Solano 2013; C. Johnson 2015). What the discussions did bring out was that Darwin's thinking about the accidental and the probable was then and remains now hard to place in relation to the heritages descending from the diverse Greek traditions associated with Plato, Aristotle, the Stoics, and Epicurus (M. J. S. Hodge and Radick 2009; Depew, this volume, chap. 1). The emphasis on chance variation manifestly aligned him with Democritean and so Epicurean, atomist, and Lucretian antecedents, and some critics simplified matters for themselves by treating his theorizing as the latest horse from that stable. However, anyone focusing on the comparisons of artificial and natural selection could see that Darwin was sometimes echoing Abrahamic monotheists who, in drawing on Plato, had viewed nature as the art of God, a seventeenth-century aphorism inscribed approvingly by the young theist Darwin in an early notebook (C. Darwin 1838b, D 54) but to be found too in the medieval Dante. The difficulty with sustaining this take on natural selection is that by contrast with the art of the Craftsman in Plato's *Timaeus*—and, following that Platonic precedent, in centuries of monotheistic theorizing—natural selection was not readily construed as ensuring the physical, bodily realization of an intelligible, immaterial, timelessly preexisting plan, since it is just working at any time and place with the variations arising then and there and is conditioned in its working by the local, temporary circumstances to which it is adapting species. But is Darwin's account of progress not an account of a prior plan being fulfilled?

Yes, he does sometimes write as if it might be, but this fulfillment has been achieved by wasteful and cruel and unreliably chancy means that are hardly rational options for a good, wise, and omnipotent God. These issues about progress and its theological and ideological themes could and should take us into socioeconomic contexts. Traditionally, these contexts would be sought in the so-called industrial revolution; but there are compelling reasons for not always going there, and for going instead to other forms of capitalist life: agrarian, financial, and imperial capitalisms especially (M. J. S. Hodge 2009).

Attending once more to Darwin's selection analogy can give us an agrarian transition to the last sections of this chapter. For any examination of the philosophically suggestive continuities in the history of natural selection theorizing since Darwin can usefully begin with a few commonplace reflections on that analogy. What is common to a stud farm with race horses and wild grassy plains with horses preyed on by big cats is obviously selective breeding. And what is meant in saying this? Well, speediness is favored in both arenas. But there is this one common element and two crucial differences between farm and prairie: while in both arenas speediness is probabilistically causally relevant to survival and reproductive success (hereafter often shortened to reproductive success), the causation itself differs. On the prairie, speediness lessens chances of capture by predators; on the farm, it makes being culled or gelded or kept away from mates less likely. And the outcome differs, too: in the wild, better adaptation of the species to its local circumstances; on the farm, better adaptation to the breeder's end, her owning race winners. So what is common is that the differential reproduction is not fortuitous, not accidental, not chance, in that it is probabilistically caused by the differences distinctive of the hereditary variants. And the same is true if Darwin compares selective breeding for strength and weight on a draft horse farm with the wild, prairie process, or indeed with selection of waxy leaf coatings reducing moisture loss to the advantage of plants in a desert. The common nonfortuitousness of the causations that such comparisons all invoke was always definitive of Darwin's theorizing and has been no less so for all causal-explanatory construals and deployments of natural selection ever since.

PHILOSOPHICAL SUGGESTIONS FROM THE DECADES SINCE DARWIN

On many historical and philosophical issues about natural selection theory, sophisticated technical details are decisive. However, those details are decisive because one concept is fundamental to all causal-explanatory interpretations of natural selection from Darwin to today: nonaccidental, nonfortuitous differential reproduction. Consider some bugs feeding on green foliage; suppose half the bugs are red, the rest green, and this color variation is strongly hereditary. And suppose predation by color-sighted birds is the main cause of bug deaths. Now consider why for biologists today, as for Darwin, what is going on in this population is an exemplary case of natural selection suitable for introductory, conceptual-explicatory purposes. Obviously, it is because what occurs is causally nonfortuitous differential reproduction of hereditary variants. For the hereditary organism differences in color among individual organisms are probabilistically causally relevant to numbers of offspring contributed to the next generation. Obviously again, it is not sheer luck or accident that there is a correlation between being green rather than red and leaving more rather than fewer than average numbers of offspring, for being green rather than red contributes to the causation responsible for the greater offspring contribution. The proportion of green over red is going up, obviously yet again, because of the probabilistic causal consequences of having this property in this environment. And if all this seems far too obvious, note that two very different, definitional explications of natural selection, given by Richard Lewontin and by David Hull and often invoked as canonical, both fail to demarcate selection from drift; for neither includes the requirement that the property difference contributes to the causation responsible for the reproductive difference (M. J. S. Hodge 1987; Godfrey-Smith 2009). Note too that such selection is what is now commonly called, following Sober, selection *for* as distinct from selection *of* (Sober 1984). Suppose the thicker fur on some wolves is positively causally relevant to survival; then it is selected *for*; but suppose that the hereditary causation making for thicker fur also makes for darker fur; then there will be selection *of* darker fur even if it is not probabilistically causally relevant to survival and is not selected *for*. Selection *of* is causally dependent on selection *for*; and so analyzing selection *for* has priority and analysis of selection *of* can be left for another time.

To confirm that such cases of nonfortuitous differential reproduction of hereditary variants, cases of selection *for*, were for Darwin exemplary of the concept of natural selection and are still so today, consider, by way of contrast, causally fortuitous differential reproduction. Darwin was explicit: not all hereditary variation is probabilistically causally relevant to survival and so to reproduction (C. Darwin 1859, 81). Some differences do not make any difference; and these variants will fluctuate in their frequency over generations with, he implied, fluctuations due to unlucky deaths and lucky survivals that are causally fortuitous and so not counting as cases of natural selection.

Today, organism differences that do not make for survival and reproductive differences feature in elementary exegeses of drift; but the decisive comparisons and contrasts remain what they were in Darwin's day. If the bird predators are not color-sighted but color-blind, then any increase over successive generations in the relative frequency of red or of green bugs will be due to fortuitous differential reproduction and so due to drift rather than selection, as it will also if the main cause of deaths is not predation but encounters with falling hailstones; the same holds for slight differences in protein molecules that are causally irrelevant to survival and reproduction. Note, too, that there can be artificial drift, or at least artificial increases in drift, most obviously from interventions reducing population numbers, but more instructively by arranging for more deaths to be due to fortuitous causation, from fake hailstones and the like.

All differential survival and reproduction, fortuitous or nonfortuitous, has causes and effects, and so the differential survival and reproduction in cases of drift is no less causal than that in selection. The contrast between selection and drift is not between a causal and a noncausal process, but between a causally nonfortuitous and a causally fortuitous process. That changes due to drift, unlike changes due to selection, can only occur in a finite population, and that changes due to drift are more likely in small populations is irrelevant to this causal contrast: for the contrast is between the nonfortuitous differential reproduction—whether in a finite or an infinite population—and any fortuitous differential reproduction in a finite or an infinite population. Fortuitous differential reproduction can occur among some finite numbers of individuals in an infinite population, thanks to color-blind predation, say, or hailstone encounters, even though it will not cause any change of the infinite population. (Here I draw on correspondence with Roberta Millstein.) Note that there is no need in introductory conceptual explications of the nonfortuitousness of selectional causation, and of the fortuitousness of drift causation, to introduce any terms—such as *random* or *sampling* or *forces* or *vectors* or *factors*—plucked from statistics or from any other branch on the tree of science. And note too that in any further conceptual analysis of what it means for differential reproduction of hereditary variants to be nonfortuitous, it is unnecessary, indeed unhelpful, to invoke any concepts of fitness; for what is edifying will be integrations of biologists' insights about hereditary variation with general philosophical clarifications of probabilistic causal relevance. Obviously, fitness concepts will have a place in such integrations; but they are best brought in only when the concept and theory of natural selection has been explicated without invoking them.[3]

Notice also that two ways of comparing and contrasting selection and drift are inadequate because they do not invoke this distinction, coming down from Darwin, between nonfortuitous and fortuitous differential reproduction. First, it might seem that selection can be distinguished as differential reproduction with an outcome matching the prior expectation, whereas with drift there is a mismatch. On this view, with color-sighted predation, if the frequency of green bugs goes up, that is selection, while if it stays the same or goes down, that is drift. But this demarcation is faulty in not recognizing that in a finite population (and obviously there are no others out there), even an expected outcome that is far more likely and more often due to selection just may have arisen this time by very improbable but not impossible luck, by drift. Over some one run of a few generations, the birds may, by luck, have picked off red and green bugs in equal numbers, perhaps because more greens than reds happened to move when the birds happened to be looking in their direction, while hailstones happened to hit more red than green bugs. Plainly, it would take close enquiry into the causal details to confirm that this had happened. And this need for this enquiry is fatal to another, second, take on selection and drift: namely, that they are not ultimately to be contrasted at all, because drift is due to differential reproduction owing to causes we have so far failed to discern, so that with complete knowledge (echoes of the ignorance view of chance) there would be cases of selection but no cases of drift (A. Rosenberg 1994, 67–73). What is wrong with this view is what the very improbable but not impossible drift case shows: namely, that establishing that drift, not selection, was going on required extra, decisive, detailed knowledge, not uninformed ignorance, of the causes and effects of the fortuitous causation.

Obviously, more advanced explications of selection and drift as causal processes will take in all kinds of complications and will require qualifications of these introductory comparisons and contrasts: complications and qualifications associated with frequency dependent selection, selection for offspring number variance, and so on and so forth (Godfrey-Smith 2009). However, none of these more advanced topics require the elementary explications to be repudiated as erroneous, and so there is no need here to anticipate such repudiations.

Consider next a recent thesis that asserts in so many words that, yes, camouflage differences in bugs can cause evolution but, no, natural selection cannot, because natural selection is a statistical trend that is an effect and not a cause of evolution (Matthen and Ariew 2009). It may be possible that this statisticalist assertion can be saved from being an analytic non sequitur by some heavy revisionist conceptual lifting, but only at the price of incurring a pretty massive burden of historical revisionism.[4] In distinguishing various components in that burden, it will be as well to start with a component already in plain sight. For Darwin, some statisticalists seem to agree, natural selection was indeed a cause of evolution; it is, they say, only long after Darwin's day that it has become a statistical trend and effect, and not, of course, because animals and plants are no longer what they were in Darwin's day, but because the concept of natural selection has been transformed by shifts in the theory of evolution by natural selection (Ariew 2007, 2008; Ariew, Rice, and Rohwer 2014; Lewens 2010). But this historical thesis surely runs counter to standard textbook practices. As the bug predation example showed, the same kind of real or imagined instances of nonfortuitous differential reproduction of hereditary variants, instances that were prime exemplars of natural selection causing evolutionary changes in Darwin's time, are prime exemplars of exactly this today. Given this continuity, to say that camouflage differences can be a cause of evolution but that natural selection cannot, is to say that some particular process can cause evolution but the kind of process of which it is an exemplary instance cannot be held to do so. But to say this is equivalent to saying that a change from rain to sunshine can cause warming and drying of the ground but a change in the weather cannot.

This reflection on current conceptual-explicationary practices indicates that there is no quick way to counter the thesis defended here of conceptual continuity concerning natural selection in Darwin's day and in ours. Conversely, defending this continuity requires philosophical engagement with at least a half dozen clusters of issues. The task of elaborating philosophical resolutions of these issue clusters may be clarified historically.

Forces Issues

It is sometimes said by statisticalists that natural selection was a force for Darwin but is no longer a force for biologists today, so there is a discontinuity in the ontology of natural selection then and now (Ariew 2008). The trouble with this view is that there is nowhere in Darwin a sustained, coherent comparing of natural selection with any force in any branch of physics. Darwin and his followers knew that among the best examples of physical forces was gravitational attraction and that this force was understood as lawful and that the law—the inverse-square law with proportionality to mass product—invoked masses. But Darwin never attempted to identify a law that is to natural selection as the gravitational law is to the gravitational force, nor did he (nor, it would seem, did anyone since) identify for his theory any analogue of mass. The analogy

Darwin took most seriously was the analogy, the relational comparison, the proportionality, that asserted that the struggle for existence is to wild animals and plants as a human selective breeder is to animals and plants under domestication. This analogy presupposed and implied no analogies with forces in physics. Darwin himself made no sustained explicit comparisons and contrasts between selection, natural or artificial, and any force such as the force of gravity. Darwin said that he understood the term *natural selection* much as geologists understood their term *denudation*: as naming an agent and signifying the result of several combined actions, which does not sound much like Newtonian inertia or gravitational or electrostatic attraction (M. J. S. Hodge 1992b).

If history is any guide, statisticalist opponents and causalist defenders of comparisons between natural selection and forces in classical, statistical, or quantum mechanics should not expect to achieve clarifications of the main issues concerning the theory of natural selection as a probabilistic, causal, and empirical theory. For, as so often in conceptual comparisons, the very possibilities for clarification are limited because there does not exist the requisite prior consensus: in this case a consensus about the concept of force itself, as anyone can confirm who has looked at the relevant unconsensual literature since some causal-philic Australian philosophers wrote about forces so instructively in the 1980s (Bigelow, Ellis, and Pargetter 1988; Wilson 2007).

A sense of history can alert us to some enduring tendencies that may throw indirect light on what may be at stake here. The physicist Ernst Mach and his philosophical protégé Karl Pearson (an ardent noncausalist-selectionist evolutionary theorist), as positivist opponents of all causes (Pence 2011, 2015) including forces, and so opponents of all attempts at causal explanation in all science, have descendants today in neopositivist, neo-Duhemian, causal-skeptic empiricists such as Bas van Fraassen. Suppose it could be shown that statistical, noncausal construals of selection have—or should have—displaced causal construals in the twentieth century; then the question would arise whether that shift is due to new developments in biology or in statistics quite independent of positivist and empiricist views about all science, or whether this shift has been indebted, if only indirectly, to such views. However, as is argued next, no such shift has, or should have, taken place, so Pearson's Machian take on natural selection has not prevailed.

Struggle for Existence Issues

No one did more to bring statistics to bear on selection than R. A. Fisher and Sewall Wright. But despite disagreements over many biological questions, they were in agreement not only about their mathematical results but also in seeing their mathematical work as statistical analyses of causation, not as statistical replacements for causal analysis (Okasha 2009; Millstein, Skipper, and Dietrich 2009; M. J. S. Hodge 1992a, 2011).

Fisher set aside as misleading and redundant some of Darwin's causation. Darwin dwelled on the Malthusian tendency of populations to outbreed their food resources and invoked this tendency as causing the struggle for existence out there in the woods. In Fisher's writings, and, following him, many expositions since, this invocation of Malthusian struggle and any notions of nature red in tooth and claw are explicitly dropped as irrelevant to the concept of natural selection. However, Fisher was always a devout causalist in all his natural science (and, in statistical mechanics, an enthusiastic Boltzmannian realist and atomist, and therefore opposed to any Duhemian, positivist, energeticist rejection of atomism), so this shift away from Darwin's Malthusian views did not lead him to a statisticalist rather than causalist take on natural selection. Indeed, everything Fisher says about natural selection, including his theological take on it, is consistent with what has been said here about the concept of nonfortuitous differential reproduction of hereditary variants as the thread linking Fisher back to Darwin and on to evolutionary biology today. As for Wright's breeding strategy studies; path analyses; and natural philosophical debts to Spencer's homogenizing and heterogenizing causal factors and moving equilibria, and to Clifford's panpsychism: these influences and inspirations all gave him his own reasons for being an ardent causalist about selection and about natural and social scientific theories generally. His categorization of selection as a causal factor in evolution may have been much more abstract and comprehensive than anything Darwin offered, but it marks no shift to a construal of this factor as a statistical trend rather than a causal process: selection, he said, is a wastebasket category, including all causes of directed change not involving mutation or the introduction of hereditary material from outside, and comprising components as diverse as differential viability, dispersal beyond the breeding range, fertility differences, and so on (M. J. S. Hodge 1992a).

Mathematical Representation Issues

Clarifying what Fisher and Wright were and were not doing with their mathematics naturally requires reflecting on mathematics itself. On most interpretations, the relations between the premises of a mathematical proof and any theorem derived from them are understood to be logical relations and so, like the relations between premises and conclusions in any argument form, not causal. So, in that sense, mathematics as such and in and of itself is not causal. But it does not follow that if there are now mathematical representations of natural selection and its consequences, then selection and its consequences must be not causally related. Familiarly enough, mathematical representations of causal processes are found throughout science. Causal modeling is mathematical modeling of causal natural science concepts and theories and of their deployment in causal explanation (Millstein, Skipper, and Dietrich 2009).

To understand mathematical representations of selection, it is best to study those representations themselves, rather than studying mathematical representations of billiard balls being plucked out of bags, apple prices going up and down in the shops, coin tossings, or any other domains where there are no instructive analogs to animal and plant populations with differential reproduction, nonfortuitous or otherwise. When one studies mathematical representations of selection, and what abstractions and idealizations they make, it becomes clear that the causal-explanatory content and import of the concept of selection does not have to be eliminated either as a precondition of the representing or as a consequence of it; this is why textbooks do not teach two versions of the concept of selection: one causal-explanatory and informally explicated in natural language and another that is explicated formally and mathematically and devoid of causal and explanatory content and import. And so there is not a further step that takes these two concepts to be mutually exclusive, and so no step leading to the conclusion that the second concept displaces and replaces the first.

Once again, the agreements and disagreements between Fisher and Wright are pertinent here, provided we insist on an appropriate account of their views. Often, their disagreement is epitomized as Fisher's fundamental theorem of natural selection versus Wright's adaptive landscapes; and that looks like a choice between some algebra relating the rate of increase in average fitness due to selection to the amount of genetic variance in fitness, and some geometry depicting how various factors such as selection and drift can collaborate in raising and lowering average population fitness so as to enhance prospects for sustained evolution. But the real divide was not originally about these two options. Both Fisher's theorem and Wright's landscape were arrived at only shortly before publication, and long after the two men had first diverged in their views about evolution, that divergence being between Fisher's mass selection and Wright's shifting balances as answers to the question of how evolution usually and optimally goes. This divergence was over the causation of evolution, a divergence over very general, abstract, idealized theories, but over causal-explanatory theories, not over any derivational-equational results or diagrammatical-depictional techniques. And it is this causal-explanatory theory controversy that is still carried on today, often with analytic and evidential resources not available to the two friendly (yes, for ten years from 1924 on) disputants who started it all (M. J. S. Hodge 1992a, 2011).

The Fitness Issues: Historical

Darwin had no concept of fitness, and yet he was able to talk quite cogently about natural selection. Dawkins does so in our day with scarcely any use of this word. Dawkins's reason for avoiding it is not that it has no clear sense but that it has far too many senses given it for different reasons by diverse tribes of biologists: mathematical population geneticists, behavioral ecologists, and so on (Dawkins 1982, 179–82). Any history of the concept of natural selection should, then, not let the tail wag the dog. Without the concept of natural selection, there would never have been certain concepts of fitness. But without any concept of fitness, the definitional explication and explanatory deployment of the concept of natural selection can get along just fine, as it did for several decades after 1859. The notion that unless we have an unproblematic concept of fitness, there can be no cogent concept or coherent theory of selection is mistaken. Also mistaken, then, is the assumption often made by causalist analysts of natural selection: namely, that unless fitnesses can be understood causally, natural selection cannot be. This assumption is as mistaken as the assumption some statisticalists seem to make: namely, that if fitnesses can be shown to be noncausal, then natural selection cannot be causal.

To prepare for avoiding these mistakes, it helps to take a historical reflection into account. Suppose fitnesses are taken to be causal: to have causes and effects in the world. Then how come Charles Darwin and Alfred Wallace did not know about them? Conversely, by what laboratory or field investigations made since Darwin and Wallace did these causes, and their causes and effects, hitherto undetected, become discovered and convincingly evidenced? The question is crass, but useful, because it brings out how misleading are its own presuppositions. And, as the best scholarship confirms, the history discredits those presuppositions. In the case of the fitness concept serving mathematical population genetics, fitness was construed not as a categorization for empirical findings about causation in the woods, but as an element in mathematical representations of differential reproduction in selection but not in drift, and regardless of whether that differential reproduction was a causal consequence of viability or of fertility differences or of both (Gayon 1998). These representations were constructed to meet ideals of generality, quantification, and abstraction, and so, too, were the various versions of the relevant fitness concept. However, it is fallacious to think that because mathematical relations are in themselves not causal, then any issues about the causes and effects of selection are irrelevant to clarifying possible roles for various fitness concepts in the causal-explanatory theory of natural selection (Millstein, Skipper, and Dietrich 2009).

The Fitness Issues: Conceptual

It helps, in beginning to clarify these roles, to ask where fitness differences fit in, that is to say, where they fit in a series of causal successions. A simplified version of Darwin's causal succession would read: differences in individual heredities cause hereditary differences in organism characters; some of these organism differences (ODs) cause reproductive success differences (RDs); and these RDs over generations cause diachronic changes in a species. A simplified version of what has descended to our era from this Darwinian causal succession might go as follows: genotypic differences cause phenotypic differences, which cause RDs, and so on. Now bring fitness differences in. But where? Well, going back to our red and green bugs, their fitness differences (FDs) belong surely between the ODs, the organism color differences, and the RDs that those color differences cause. So far so good. But notice that the FDs can not be nonvacuously construed as causal intermediaries, for it is otiose to say that these ODs, these color differences, cause the FDs, which then in turn cause the RDs. For this adds nothing coherent to the causal content of what we had before: ODs causing RDs. (Some interpreters of fitnesses as propensities—as probabilistic causal dispositions—seem aware of this point and are duly wary of making these dispositions causal intermediaries and so pseudoenhancements of the causal ontology, but others seem unaware and unwary.) But could the FDs be noncausal intermediaries? Yes, indeed. Construe them as expectational differences, as differences in reproductive expectancies analogous to life expectancies. Then the ODs would be determining, noncausally but evidentially, the FDs (which could be said to be supervening on those ODs if one likes that kind of thing), and then those FDs would be expected RDs and so predictive of actual, realized RDs, and these realized RDs would be caused, as before, by ODs. Two more cogent options should be recognized: identify FDs with ODs or identify FDs with RDs. Either way the FDs are not otiose and are causal because ODs and RDs have causes and effects. So there are three genuine options, but not a fourth, because construing

FDs as causal intermediaries that are caused by ODs and are causing RDs is philosophically mistaken, historically misleading, and biologically unhelpful.

As for FDs construed as reproductive expectancies analogous to life expectancies, such FDs would have no causal-explanatory import. Smith, a young smoker, and Jones, a youthful jogger, have different life expectancies, but if the jogger outlives the smoker, we do not explain this longevity difference as resulting causally from the life expectancy difference when they were younger, for this would be plainly pseudoetiological and pseudoexplanatory, no causal explanation at all. We ascribe the longevity difference to the difference between smoking and jogging as causes probabilistically relevant to longevity differences. Likewise, then, if green bugs are outreproducing red ones, we can have only a pseudoexplanation of this if we credit it to FDs construed as reproductive expectancy differences. But this is no reason for a causalist to oppose such a construal of FDs. For the causal explanatory content and import of natural selection have always, since Darwin, depended on what is known of the causal relations between ODs and RDs. So, opting to construe FDs as noncausal expectational intermediaries is entirely consistent with a causalist take on selection; and it entails no shift away from causalism toward statisticalism about selection itself. One lesson from history that some causalists and some statisticalists may do well to ponder is that there can be various ways of fitting fitnesses into an ontology of natural selection, but that none of them entail either enhancing or eliminating the causal character, content, and import of the concept of natural selection itself. A second lesson is that in any philosophical analysis of natural selection, whether causalist or statisticalist, an unhelpful way to start is a way commonly chosen today: namely to start by thinking about heritable fitness differences. To start this way is to start with two concepts, heritability and fitness, which are complex and contested and so likely to contribute obscurity and controversy rather than clarification and consensus to any analysis of natural selection. To start this way is also to deny oneself the clarificatory guidance that history can provide. History suggests that it is best to start by understanding how the older concept of natural selection could be articulated and applied without invoking those two recent concepts, of heritability and fitness, and that it is best to begin clarifying those two by analyzing what they did or did not add to that older concept, natural selection.

The Principle of Natural Selection (PNS) Issues

Some philosophers of biology have attempted to articulate a quite general, nontrivial, nontautological, nondefinitional truth about natural selection and fitness, a general truth with real causal-explanatory content and import. Here is the kind of formulation: If a is fitter than b in environment E, then (probably) a will out-reproduce b in E. Obviously, whether these attempts at general, nontautological truth succeed depends partly on what explication is given of fitness differences (Brandon and Rosenberg 2000). But this dependence makes the whole exercise dispensable in any analysis of natural selection that starts from the continuity over a century and a half of thinking about the causes and effects of differential reproduction. For this continuity suggests another quite different way of understanding how the theory of evolution by natural selection can be empirical and explanatory and not tautological and not explanatorily vacuous. On this approach, which derives from Darwin's one long argument in the *Origin*, one starts from a definitional characterization of natural selection as a causal process: namely, from a definitional characterization of natural selection as the causal process that is occurring when and only when there is nonfortuitous differential reproduction of hereditary variants, a definition that, if accepted as true, is accepted by definitional decision, not from empirical confirmation. Then one goes on, as Darwin did and many expositors have since, to ask empirical questions about the existence of natural selection in the world, in fields, forests, or wherever, further empirical questions about what evolutionary change it is competent to cause, and finally empirical questions about what changes it has been responsible for in the past, and so what products of past evolution are to be explained as resulting from selection. Four clusters of questions, then, about natural selection—definition, existence, competence, and responsibility—with the first, definitional cluster conceptual (and so unempirical), the other three empirical. In sum: What is it? Is it? Can it? Did it? If selection is defined as it is here, then those last three clusters of questions are empirical questions requiring answers grounded in factual inquiries and so in testable generalizations about heredity, variation, and reproduction in diverse circumstances, times, and places. For there is no inferring of existence from definitional essence in the style of some mythical from-essence-to-existence ontological proof of God's existence, nor any inferring of competence from existence, nor any inferring of responsibility from competence. Study the argumentational structure of the *Origin*, and see how instructively it maps onto the definitional decisions taken much more recently by selectionists and neutralists, and onto the clusters of empirical issues that have divided them, issues about what selection and what drift exist in the wild, what those processes can cause and how quickly at any time, and what each has wrought over past eons (M. J. S. Hodge 1987). If one does this, then the quest for a PNS saving the theory of natural selection from tautology charges will seem less illuminating, even unnecessary.

CONCLUDING REFLECTIONS

It may be helpful to recall those themes that connect the short-run history of Darwin's thinking about chance and chances with the long-run history of natural selection theory from his day to ours.

Before and after he came up with his theory of natural selection, Darwin worked with concepts of the accidental and the probable that were descending partly from ancient traditions and partly from eighteenth-century legacies. So no nineteenth-century innovations in statistical thinking played significant roles in his natural selection theorizing. Likewise, his view that chance signified our ignorance, lack of causal knowledge in us, rather than lack of causal determination in nature, was no new view, rather a long-standing commonplace. It would seem plausible therefore to expect to find that the concept of natural selection has been fundamentally changed since Darwin. For, since Darwin, there has been both a comprehensive rise of statistical thinking about natural selection and a widespread move away from the ignorance view of chance. So history would seem to be strongly on the side of a statisticalist rather than a causalist view of natural selection today; on the side, that is, of those who hold that current understandings of natural selection are seriously unlike Darwin's understanding, and so on the side of those statisticalists who hold that today natural selection should not be understood as a cause of evolution but as a statistical trend in the effects of evolution. Accordingly, if history really is on the side of today's statisticalists, then this history can offer scant comfort to their opponents, the causalists, who argue that, despite everything that has happened since Darwin, the theory of natural selection is still what he took it to be: a theory of causation, probabilistic causation, a theory of the main cause of evolutionary change.

Statisticalism and causalism about natural selection are currently opposed philosophical doctrines; and philosophers—at least those who are not Continental—may claim not to consider history. However, both statisticalists and causalists often want history on their side. The present chapter, by a historian, does not engage the philosophical issues dividing these two camps. However, it does take sides in arguing that—despite what the previous paragraph suggested—causalists have history more on their side than do statisticalists. In taking sides the chapter invokes four clusters of issues, all integrating conceptual analyses and narrative testimony.

First: Darwin constructed his theory as a theory of probabilistic causation, not as a theory of forces, but by combining a concept of chance as accident with a concept of chances as probabilities. Chance hereditary variations are due to causal accidents, and some of these chance variations cause higher chances of survival and reproduction in the individuals varying that way. Notice, then, that the ancient notion of the accidental comes in twice: once in a positive way in understanding variations and once in a negative way in understanding chances as probabilities. Because the accidental advantageous chance variations have the causal effects they do, it is no accident that their chances, their probabilities, of survival and reproduction are higher. Any such probable differential survival and reproduction may then be called nonfortuitous, because in ancient and modern authors alike, concepts of the accidental and of the fortuitous often coincided.

Second: Darwin clarified this naturally nonfortuitous causation by comparing and contrasting it with the nonfortuitous causation at work in artificial selection. Biologists today continue to do so. Biologists today also clarify all selectional causation by comparing and contrasting it with the fortuitous causation constituting genetic drift, often doing so in order to clarify the issues dividing selectionists and neutralists (who ascribe more of molecular-level evolution to drift than selectionists do) over the causes of evolutionary change at the molecular level. Statisticalists are strikingly unconcerned with comparing and contrasting natural and artificial selection, and strikingly unconcerned too with the selectionist-neutralist controversies. In these unconcerns they do not have history on their side, seeing that these two clusters of issues are direct descendants today of causal-explanatory issues that go back over a century and that developed from Darwin's own causal-explanatory theorizing.

Third: statisticalist philosophers like to find Fisher, Haldane, and Wright—the founders around 1930 of modern Mendelian, mathematical population and evolutionary genetics—on their side. But this alliance is harder to establish than statisticalists might think. All three founders were strong causalists about science in general and about natural selection in particular. Before them, the Machian positivist Pearson had construed natural selection theory as a statistical theory that he took to be, like all physical theory, not causal and so not causal-explanatory, indeed not explanatory at all, but only predictive. Fisher, Haldane, and Wright did not sign up to Pearson's positivism; they saw their statistical analyses of selection as contributing to a causal-explanatory, not to a noncausal-predictive, agenda for evolutionary theory.

Fourth: so it is, after all, the causalists who have history more on their side than statisticalists do. But in letting history inform their countering of statisticalism, causalists may do well to be wary of making four moves: (1) They should be wary of force talk, as it has few if any favorable precedents in selection theorizing since Darwin. (2) They should be wary of making their explications of selection dependent on prior explications of fitness concepts. If, instead, they give priority to a causal explication of selection, then they can come to see that a causalist take on selection can fit very coherently with a noncausal, expectational, statistical take on fitness. Note that this coherence gives no comfort to statisticalists about selection, for it prevents them from arguing that since fitness is a statistical concept, so must selection be too. (3) Causalists who do go for causal takes on fitness should be wary of construing fitness differences as unacceptably otiose causal intermediaries supposedly mediating causally between phenotypical trait differences and reproductive output differences. (4) Causalists should be wary of thinking that they can strengthen their position by trying to articulate some nontautological, explanatory general lawlike principle of natural selection. Positivist views may have encouraged such nomological quests, but those biologists who have been causalist selection theorists have rarely joined these quests. History suggests that the most useful generalization about all natural selections is not a statement of law or a mathematical equation, but rather a definition of selection as a causal process, specifying what are causally necessary and sufficient conditions for its proceeding.

As for historians of science: they should be wary of presuming to tell philosophers and scientists what to think.

ACKNOWLEDGMENTS AND
BIBLIOGRAPHICAL COMMENTS

My profound thanks to Charles Pence and Grant Ramsey for their many generous and patient contributions to the planning and writing of this chapter, and for advice on its relations to earlier papers of mine often going over the same ground, particularly the paper of 1987, which gives more extensive analysis, argumentation, and documentation for many of the historical and philosophical topics taken up here. The present chapter has benefited decisively from the suggestions of two anonymous reviewers. It is much indebted, too, to discussions in conversation or correspondence with Marshall Abrams, André Ariew, Abhijeet Bardapurkar, John Beatty, Frédéric Bouchard, Robert Brandon, David Depew, Jean Gayon, Peter Gildenhuys, Stuart Glennan, Bruce Glymour, Chris Hitchcock, Philippe Huneman, Tim Lewens, Françoise Longy, Mohan Matthen, Roberta Millstein, Ricardo Noguera-Solano, Samir Okasha, Will Provine, Alirio Rosales, Elliott Sober, Michael Strevens, John Turner, Joel Velasco, and Denis Walsh. Curtis Johnson's excellent, comprehensive book on Darwin on chance reached me only after this chapter was close to completion, but I benefited from reading some of its chapters in earlier draft versions. For a discussion making points very like those made here about the accidental, the nonaccidental, and the probable in Darwin's and in current conceptions of natural selection as a causal process, see Peter McLaughlin's paper (2007), which is instructive also on selection for and selection of. Roberta Millstein, Robert Skipper, and Michael Dietrich are preparing a major book on drift, and I have benefited from seeing draft versions of some chapters. Will Provine's important book on drift came to hand only recently, but again I learned from earlier partial drafts. For a recent, well-received book on the philosophical analysis of natural selection, one goes to Peter Godfrey-Smith's (2009) masterly monograph. On selection (if not on drift), I'd like to think that what I have said here lines up well with that book's views. Godfrey-Smith's book does meticulous justice to the complexity of many issues that I have ignored or, worse, treated simple-mindedly. On the history side, I have tried to ensure that what I have argued for here coheres with the significant overlap between the two most authoritative studies of the history of Darwinian theory from Darwin to today, studies philosophers can often benefit from as much as historians (Depew and Weber 1995; Gayon 1998). I think the present chapter coheres well also with the sustained and novel analysis of Darwin's and later selection theorizing in Michael Strevens's book *Tychomancy*, although I still lack a sufficient grasp of that analysis to make decisive comparisons. Many of the historical and philosophical issues treated here are treated much more extensively in our editor Charles Pence's doctoral dissertation, available online, which also came to hand too recently for decisive comparisons to be made (Pence 2014). A forthcoming volume, *The Oxford Handbook of Probability and Philosophy*, edited by Alan Hajék and Christopher Hitchcock, will include chapters relating concepts of chance and chances in evolutionary biology to general analyses of probability. I have benefited from seeing a draft of one such chapter, on concepts of fitness, by Roberta Millstein. André Ariew has given as the title for a book he is preparing: "How Statistics Changed Natural Selection." I have not seen any of this text, but from valued, friendly discussions with the author, I am confident that it will refute almost everything I have argued for here.

NOTES

1. Two instances of this view of generation and variation in animals and men: "I ask . . . is the sex of the embryo produced by accident? Certainly whatever is produced has a cause; but when this cause is too minute for our comprehension, the effect is said in common language to happen by chance, as in throwing a certain number on dice" (E. Darwin 1794, 2:265). "Varieties of form or colour, as they spring up in any race, are commonly called accidental, a term only expressive of our ignorance as to the causes that give rise to them" (Prichard 1826, 2:548).

2. In accord with standard practice, references are to Darwin's mostly alphabetic notebook designations and to his notebook page numbers.

3. With apologies for a self-serving reminiscence: If I recall correctly, in conversation in 1982, Elliott Sober and I found that we both thought natural selection best interpreted as a causal process. I made the case for strengthening the arguments for this causal process interpretation of selection by combining it not with a causal-explanatory take on fitness differences but with a noncausal, expectational take that made fitness differences like life expectancy differences. Elliott was initially quite unpersuaded, but he came around in time to endorse such a combination—of causalism for selection and noncausalism for fitness—in his 1984 book *The Nature of Selection* (Sober 1984, esp. 95–102). Years later, in 2002, Dennis Walsh and three allies, duly dubbed "statisticalists," opposed that book's views in insisting on a noncausal, statistical view of both selection and fitness (Walsh, Lewens, and Ariew 2002; Matthen and Ariew 2002). Elliott Sober has not been convinced that they are right. Nor have I. He and I—one more continuity thesis here—continue to accept the 1982 combination, which, as he notes in his book, may well have a precedent in Fisher's 1930 book *The Genetical Theory of Natural Selection*. Recently, however, he may appear at least to have qualified his 1984 acceptance of this combination (Sober 2013).

4. There appears to be no one article or book devoted to surveying the statisticalist-causalist controversies. Sober's causalist views in his 1984 book were targeted from the start by the first pioneers of statisticalism (Walsh, Lewens, and Ariew 2002; Matthen and Ariew 2002), who especially opposed his view of evolutionary theory as a theory of causes understood as forces—or as seriously like forces, at least. Several papers by Millstein defend causalism (Millstein 2002, 2013) without defending force analogies; Brandon and Ramsey (2007) do likewise. However, Millstein and Brandon disagree about how to distinguish selection and drift (Millstein 2005); Millstein's views derive in part from Beatty (1984). My own views align me with Millstein on the distinction. Some but not all defenses of causalism have invoked the so-called propensity concept of fitness, with propensities construed as probabilistic-dispositional and probabilistic-causal. Robert Brandon has explained to me that in his invocations of propensities, they are not construed as additions to the causal ontology, and so are not construed as additions that mediate causally between phenotypic trait differences and reproductive output differences. I am not sure that all those embracing propen-

sity takes on fitness are as ontologically wary as Brandon is. For recent papers that have extensive citations to other relevant contributions to the continuing statisticalist-causalist controversies, see Otsuka et al. 2011; Otsuka 2014; Ariew, Rice, and Rohwer 2014; Ramsey 2013; Hitchcock and Velasco 2014. For samplings of current views of how concepts of chance in evolutionary biology may relate to various philosophical analyses of chances and of probabilities, see Sober 2009; Brandon and Fleming 2014; Millstein 2011.

CHAPTER 3

Chance in the Modern Synthesis

Anya Plutynski, Kenneth Blake Vernon, Lucas John Matthews, and Daniel Molter

The modern synthesis in evolutionary biology is taken to be that period in which a consensus developed among biologists about the major causes of evolution, a consensus that informed research in evolutionary biology for at least a half century. As such, it is a particularly fruitful period to consider when reflecting on the meaning and role of chance in evolutionary explanation. Biologists of this period make reference to "chance" and loose cognates of "chance," such as: "random," "contingent," "accidental," "haphazard," or "stochastic." Of course, what an author might mean by "chance" in any specific context varies.

In the following, we first offer a historiographical note on the synthesis. Second, we introduce five ways in which synthesis authors spoke about chance. We do not take these to be an exhaustive taxonomy of all possible ways in which chance meaningfully figures in explanations in evolutionary biology. These are simply five common uses of the term by biologists at this period. They will serve to organize our summary of the collected references to chance and the analysis and discussion of the following questions:

- What did synthesis authors understand by chance?
- How did these authors see chance operating in evolution?
- Did their appeals to chance increase or decrease over time during the synthesis? That is, was there a "hardening" of the synthesis, as Gould claimed (1983)?

What was the synthesis? When did it begin, and when did it end? Who were the major participants? And what did they synthesize? There is a good deal of debate among historians of biology about the meaning of the synthesis: its aims and scope, its participants, and when it began and ended (see, e.g., Burian 1988; Gayon 1998; Cain 1994; Smocovitis 1994b, 1994a, 1996; Depew and Weber 1995; Largent 2009; Cain and Ruse 2009; M. J. S. Hodge 2011; Delisle 2009, 2011; Provine 1971). We grant that there is no single way to tell the story of the synthesis; so we choose to be ecumenical and include as many potentially relevant figures as possible and as permissive a periodization as possible (though we must be selective in our discussion of representative figures, due to limitations of space). Arguably, the earliest influential document of the twentieth century that qualifies as synthetic is R. A. Fisher's "The Correlation between Relatives on the Supposition of Mendelian Inheritance" (1918), which established the compatibility of a multifactorial (or Mendelian) theory of inheritance with a "biometrical" view of quantitative traits, or traits with continuous distribution, which selection might gradually shift over time. This laid the groundwork for what is sometimes called the orthodox view of the synthesis, as explained and defended in Mayr and Provine's (1980) volume, according to which the major figures of the synthesis came to broad agreement on the view that gradual selection on minor genetic variants is largely responsible for the diversification and adaptation we see today. Macroevolution—the divergence of species and lineages—does not require appeal to causal factors over and above those deployed in microevolutionary theory, according to this orthodox view. The major texts of the synthesis provided arguments and evidence in support of this general theoretical agreement.

The "orthodox" view, of course, foregrounds agreement upon empirical and theoretical principles of evolution, but it is also surely the case that the synthesis was constituted by self-conscious actors, who aimed at the formation and institutionalization of the discipline of evolutionary biology. In the 1940s, Mayr launched the Society for the Study of Evolution and its associated journal, *Evolution*; he also was instrumental in organizing important meetings, drawing together biologists with different areas of specialization from around the globe to establish a more interdisciplinary biology, and establishing centers of research and common curricula. That is, the synthesis was not only a scientific change, but also a sociological and institutional one (Cain 1994; Smocovitis 1994b, 1994a). The name "modern synthesis" was coined by Julian Huxley (1942), whose book *Evolution: The Modern Synthesis* was both a comprehensive overview of then current biology and an articulation of a research program. The theoretical work and organizational events of the synthesis spanned roughly three decades, from about 1920 to 1950.

This period exhibited a fluctuation of views about the role(s) of chance in evolution, eventuating in a "hardening," or emphasis on selection (Gould 1983) that continued well past the synthesis period. The stabilization of the modern synthesis view on chance (insomuch as there was a stable view) was established firmly in the 1940s, and the major texts of this period all significantly drew upon the population genetic documents published during the "early phase" of the modern synthesis, primarily accomplished by Haldane, Fisher, and Wright in the 1920s and 1930s. Significant further work was done in the 1940s and '50s, making clear the relationships of ecology to evolution and of paleontology to systematics.

In contrast to the "orthodox" view just described, Cain (2009) proposes that we abandon the "unit concept" of the evolutionary synthesis altogether and suggests that we shift focus to a wider range of transformations in the biological sciences during the 1930s. Historians' prior assumptions that the synthesis was a single event, bounded in time, that there was a single theory endorsed by synthesis participants, and the very idea that "conjunction is meaningful," he claims, are all mistaken. Instead, Cain argues that we should focus on "organizing threads" of research into the nature of species and speciation, experimental taxonomy, and the shift from object- to process-based biology. We agree that historians ought to be wary of identifying the synthesis exclusively with agreement on what we have called the orthodox view above. Cain usefully suggests that we should think of the period as one involving "problem complexes." He identifies four: selection, variation, heredity, and divergence. To be sure, all four problems were central matters of interest for biologists during this period. But there are different and perhaps equally informative ways to decompose the synthesis. For instance, Gayon's comprehensive *Darwinism's Struggle for Survival* (1998) focuses on different yet nonetheless core problem complexes: defending "Darwinism" from its (perceived) detractors and integrating genetics with evolution. To be clear, we don't see these approaches as in opposition, but simply as focusing on distinct aspects of the same historical period, akin to Wimsatt's view that the same system can be decomposed in multiple ways (Wimsatt 1972). Not unlike the fruit fly, the modern synthesis is a complex historical process that can be decomposed in many different ways.

We contend that a variety of problem complexes, goals, questions, and methods—theoretical as well as pragmatic or institutional—is not incompatible with general agreement on certain core views about evolutionary change. While the retrospective emphasis on classical population genetics may have been oversold (as historians of population genetics themselves have argued; see, e.g., the new preface to Provine 1986 in the 1989 edition), there was certainly agreement upon a core set of commitments, what Burian has called a "meta" theory (Burian 1988) or *research program*, in service of addressing a suite of questions about the evolution and diversity of life. What was important to authors of the late synthesis was not what the models of classical population genetics required, but what they permitted. The core elements of the classical models describe how evolution would progress, under certain assumptions, leaving open to empirical investigation how frequently those assumptions are met in the world. Thus, different synthesis authors were free to disagree on the relative significance of selection and drift. Nonetheless, we contend that there was a shared family of commitments about *the operation of chance in evolution throughout the synthesis*, at the level of segregation and assortment of genes, patterns of mating, probabilistic processes like selection, and "random" events (such as floods or earthquakes) that might lead to ecological or geographical isolation. We turn now to a characterization of these distinct forms of chance.

FIVE SENSES OF CHANCE

There were (at least) five different senses of chance at play in the synthesis.[1] First, one might assume appeals to chance in science to be making metaphysical claims about the world as *fundamentally indeterministic*. Only rarely, however, is the question of determinism or indeterminism addressed overtly during the synthesis. Dobzhansky (1956, 1962, 1967) and Haldane (1942) wrote works toward the end of their careers that touched on philosophical issues such as free will and indeterminism, and interest in such questions was a lifelong passion of Sewall Wright (Provine 1986). However, through 1950, when appealing to chance, most synthesis authors were silent about "quantum indeterminacy." Indeterminism in physics did play an important role for Fisher (1934), and we will discuss this in further detail below. But for the most part, when discussing chance, such authors are referring to events and processes at a relatively macro scale: segregation of genes, isolation of small subpopulations, and so on.

Second, the term *chance* is sometimes used interchangeably with *random*. There are more and less precise senses of *random*; the most precise sense is the notion of a *random variable*. Random sampling from a uniform distribution results in outcomes that are equiprobable; sampling from nonuniform distributions results in outcomes that are not equiprobable. When speaking of random mating or random sampling of alleles in the process of meiosis, most synthesis authors appear to be referring to a sampling process whose outcomes are assumed to be equiprobable. Or, in such cases, the outcome (of a mating with type X or Y, or allocation of allele A or a) is spoken of as random, simply in the sense that two (or more) outcomes are equiprobable. Other times when authors referred to random mutation, they meant "random with respect to fitness" (i.e., mutation was not "directed" or a response to environmental challenges, as some Lamarckians claimed). Sometimes, however, when an event like mutation was spoken of as random, the author may have meant that its chance of occurring is unpredictable (in the epistemic sense) or, in contrast, is due to indeterministic processes (ontic). The ambiguity may have been deliberate, as so little was known about the mechanisms underpinning mutation at this point (Sahotra Sarkar, pers. comm., April 13, 2013). That is, given the limited understanding of the structure of the gene and the causes of mutation at the molecular and submolecular levels at the time, the authors indeed meant that they were unsure about the relevance of quantum indeterminacy (though see, e.g., Sloan and Fogel 2011 for a discussion of the "Three-Man Paper" and the role of radiation in mutation at the time).

Third, and most often, *chance* is frequently used as a *proxy for probability*. For instance, possession of a trait might raise (or lower) the chances of some outcome. All synthesis authors speak of natural selection as a matter of probabilities, or as probabilistic in this sense: even exceptionally high fitness does not guarantee survival or reproductive success, but only increases in organisms' "chances." Likewise, the chance (probability) that a random gene combination is adaptive was thought to be very low (S. Wright 1932, 358). The chances of various outcomes are thus spoken of as "high" or "low," when outcomes are *unequally probable*.

Fourth, events such as floods, storms, meteorites crashing into the earth, and volcanoes are sometimes spoken of by synthesis authors as chance, random, or, interchangeably, contingent events. Such events are chancy in the sense that they are rare and they *result in (usually nonfortuitous) outcomes for organisms, lineages, and species, which are unusual, "unlucky" or not predictable*, given existing biogeography, survivorship, or ongoing ecological circumstances. Volcanic explosions, mutations, lightning strikes, and other random events in this sense are uncorrelated (either as events or in their effects) with other causes that shape evolution (e.g., selection). In other words, such events *disrupt current trends*—by dividing landscapes, wiping out resources, or eliminating or isolating groups that would otherwise be physically continuous and interbreeding.

Fifth, and finally, *chance* is often used to refer to *outcomes in contradistinction to, or "opposing" selection*. Wright speaks of both drift and mutation as "chance" factors that "oppose" selection (S. Wright 1932, 359). Admittedly, this is in part due to the fact that the outcomes of drift and mutation are random, in the second sense defined above. For a particular trait and environment, selection will have a predictable "direction," whereas drift and mutation result in outcomes that are (relatively) unpredictable. Such chance outcomes are not "directed," either toward some desired (adaptive) outcome or toward any outcome in particular. Most synthesis authors assumed that mutations were most often deleterious. So the direction of mutation *was* predictable in the sense that it was generally assumed that adaptive mutations are rare. Second, the direction of drift at the population level is a reduction in heterozygosity, as small sample sizes of finite populations are likely to have a smaller representation of variation than the original population sampled. In this sense, at least, drift results in a "predictable" outcome. Nonetheless, many synthesis authors often assimilated "non-directional" changes in gene frequency with whatever "opposes" the direction of selection. This, though, should not be interpreted as the claim that drift is "whatever we cannot explain"; synthesis authors were at pains to identify which empirical facts are of relevance to testing claims that one of the two factors was at work, and how.

In sum, what synthesis authors meant by chance in any particular instance was context dependent. A chance event was often defined in terms of a contrast, for example, with a "directional" cause, process, or tendency, or a predicted outcome. Appeals to chance events or outcomes were often in *contradistinction to, or "opposing," a particular predicted outcome.* Given the rarity of adaptive mutations, one often cannot predict, for example, when and where a fortuitous mutation will come about; nor can one predict whether isolation of a small subpopulation will yield a fortuitous gene combination. Such events might then be spoken of as "due to chance," which is just to say that the author did not know precisely when particular such events would occur, though he or she could predict that, as a general category, such events are rare. And an author could model chance processes as akin to processes of random sampling, one of the insights that theoretical population geneticists brought to the table.

AIMS AND METHODS

Our aim here is to survey the uses and meanings of these terms in the context of the early and later synthesis, as well set their use(s) in a larger context in order to address the questions we laid out in the introduction.

We document below uses of the term *chance* and its (loose) cognates *random, probable, by accident*, and so forth in the major texts of the synthesis authors. We realize that the choice of major texts is somewhat arbitrary. However, in part due to limitations of space, we opted to focus on those texts that we see as representative of key views on the role of chance by major figures in the synthesis period: Fisher's *The Genetical Theory of Natural Selection* (1930), Wright's 1931 and 1932 papers; Dobzhansky's first (1937) and third (1951) editions of *Genetics and the Origin of Species*; Mayr's *Systematics and the Origin of Species* (1942); Simpson's *Tempo and Mode in Evolution* (1944); and Stebbins's *Variation and Evolution in Plants* (1950).[2] We have chosen to compare early and late editions of Dobzhansky's *Genetics and the Origin of Species* because they illustrate in a striking way a shift of perspective on the relative significance of chance in evolution over the course of what Gould (1983) has called a "hardening" of the synthesis.

Understanding how and in what sense chance was understood by synthesis authors requires more than a summary of their particular views; also required are a careful comparison and contrast of their distinct uses of the term and its cognates and an understanding of its relevance to the process of evolution. The synthesis authors referred to chance not infrequently, but, as we've said, with meanings that depended importantly on context. By and large, the most common use of the term *chance* is simply as proxy for *probability* or *probable*, as in, "the chances of fixation of a novel gene are low in a small population." The second most common is as a proxy for *random*. "Chance," or "random," mating occurs when distinct types of organism in a population may be equally likely to mate with one another.

Many authors treat senses four and five above interchangeably. That is, authors are often unclear in any case whether "chance" might refer to unusual events that disrupt current trends—such as floods, earthquakes, and so forth, or "forces" that act "contrary" to selection. For instance, Stebbins argued that certain correlations between various morphological characteristics are so strong as to rule out "their origin by chance," which is to say that they are unlikely to be due to chance fixation of alleles and more likely to be due to selection. In other words, when a correlation between environment and trait is strong, synthesis authors were more likely than not to expect an explanation that might appeal to natural selection.

Finally, one general trend we noticed in the literature surveyed below is that the frequency of appeals to chance in the sense of drift as an explanatory factor in evolution appears to shift back and forth from the early to the later synthesis. While early synthesis authors like Fisher and Haldane thought drift to be a relatively unimportant factor in adaptive evolution, Wright's emphasis on drift influenced Dobzhansky's first edition of *Genetics and the Origin of Species* (1937), which exhibits an emphasis on chance factors, particularly in speciation. Dobzhansky's appeal to chance factors (e.g., isolation and drift), in explanations of speciation and nonadaptive differences between species, was far more common in the 1937 edition of *Genetics and the Origin of Species* than in the 1951 edition. A shift in emphasis toward selection over drift as a major factor in evolution in later editions supports Gould's (1983) thesis that there was a "hardening" of the synthesis, or progressively greater emphasis on adaptation and selection. With this overview in mind, we now offer a historical perspective on how each synthesis author viewed chance in evolution.

THE EARLY SYNTHESIS

This period—roughly 1918–35—is when Haldane, Fisher, and Wright developed a general mathematical representation of Darwinian evolution in populations, on a Mendelian theory of inheritance. Evolution was represented as changes in the relative frequency of genes, due to selection, mutation, migration, and drift, or random sampling of alleles, from one generation to the next. All saw themselves as resisting views that they understood as in tension with orthodox Darwinism: Lamarckian, "orthogenetic," and other "directed" views of evolution. Such theories invoked factors outside of mere selection, mutation, migration, and drift as shaping the direction or character of change in populations over time. All argued that selection on slightly varying characters was sufficient (over evolutionary time) to generate the diversity and adaptive variation we find currently. However, Wright believed that isolation of subpopulations and "random fluctuation of gene frequencies" enabled populations to move to novel "adaptive peaks," escaping suboptimal gene combinations (S. Wright 1931, 1932).

Fisher

Fisher was a "synthesis" thinker in the broadest possible sense; his aim was not only to synthesize Darwinism and Mendelism, but also to discover the fundamental "unifying" laws of biology, on analogy with physical laws, such as the second law of thermodynamics. Indeed, he often compares his enterprise with physics and compares the statistical properties of genes in populations to the aggregate behavior of molecules in a gas (Fisher 1922). As M. J. S. Hodge (1992a) has argued, Fisher was strongly influenced by two nineteenth-century figures, Boltzmann and Darwin, whom he saw as ushering in a new, "indeterministic" scientific worldview—one that admitted of probabilistic explanations, or explanations that took the action of aggregative effects in the context of populations of organisms or physical systems to be explanatory. Fisher took Boltzmann's second law of thermodynamics to "transmute probability from a subjective concept derivable from human ignorance to one of the central concepts of physical reality" (Fisher 1932, 9). Fisher compares and contrasts his fundamental theorem of natural selection with the second law of thermodynamics:

> The fundamental theorem proved above bears some remarkable resemblance to the second law of thermodynamics. Both are properties of populations, or aggregates, true irrespective of the nature of the units which compose them; both are statistical laws; each requires the constant increase of a measurable quantity, in the one case the entropy of a physical system and in the other the fitness, measured by m, of a biological population. As in the physical world we can conceive of theoretical systems in which dissipative forces are wholly absent, and in which the entropy consequently remains constant, so we can conceive, though we need not expect to find, biological populations in which the genetic variance is absolutely zero, and in which fitness does not increase. (Fisher 1930, 37)

In other words, in his fundamental theorem, Fisher thought he had discovered a fundamental principle of biology, akin to the principle of entropy, though he also believed there were "profound differences" between the two. Biological systems are impermanent, whereas energy is never destroyed; fitness is qualitatively different for every organism, but entropy is the same across physical systems; fitness may be increased or decreased by changes in the environment; entropy changes are irreversible, while evolution is not.[3] And entropy leads to progressive disorganization, while evolution leads to "progressively higher organization" in the organic world.

This last comparison, between evolution as a force for progressive organization, and entropy leading to disorganization, was the core of what Hodge calls Fisher's "two-tendency" view of the universe. One tendency admitted of creative change, another of progressive loss of order. Fisher saw the second law of thermodynamics and the fundamental theorem as similar in that both are statistical laws that capture a fundamental dynamic between population-level properties—either of physical system or biological populations. Whereas entropy captures the dynamics between energy and time, the fundamental theorem captures natural selection's dependence on the chance succession of favorable mutations.

It is no accident that Fisher saw this parallel between physics and biology. He had postgraduate training at Cambridge with James Jeans in statistical thermodynamics and was particularly influenced in his thinking about evolution by innovations in physics, such as Maxwell's work on the properties of gases. As M. J. S. Hodge (1992a), J. R. G. Turner (1987), Morgan and Morrison (1999) and Depew and Weber (1995) have argued, Fisher transposed this mode of thinking, for example, about the physics of gases, from physics to evolutionary biology. Probabilistic models were, for Fisher, not an expression of ignorance, but a way of representing stochastic change in aggregative systems, such as populations of organisms or molecules in a gas (Fisher 1922).

Fisher's major work, *The Genetical Theory of Natural Selection* (1930), was a sustained defense of his interpretation of Darwin's view: namely, that selection acts gradually on "Mendelian" factors, or mutations of small effect. Fisher mentions chance eighty-one times in the *Genetical Theory*. The most common sense of chance to which he appealed was as proxy for probability. In the *Genetical Theory*, chance plays an important role in evolution in at least three senses. First, on the "particulate" or Mendelian theory of inheritance, chance enters into evolutionary explanation insofar as there is indiscriminate sampling via the random combination of genes. Second, Fisher (1922, 1930) argued that drift, or what he would call the "Hagedoorn effect," that is, "fortuitous fluctuations in genetic composition," played a role in evolution, though he tended to assume that effective population sizes were large, and so the rate of loss of alleles due to drift was low. The notion of effective population size is derived from the Wright-Fisher model: a simple model for the representation of change in populations over time, which assumes constant population size, nonoverlapping generations, and no mutation, recombination, selection, or population structure. Essentially, the model is a "null" model, representing change over time as entirely due to a random sampling process (with replacement) from one generation to the next. This model was taken to capture the variety of ways in which populations are subject to "random" fluctuations in allele frequencies: for example, fluctuations in population size, extent of inbreeding, overlapping generations, and/or spatial dispersion. Wright (1931) and Fisher independently calculated the probability of fixation of one or another variant of a gene under such a simple sampling regimen.

Third, and finally, with respect to selection, Fisher emphasized that fitness is a probabilistic cause of change. Fisher remarks:

> We are now in a position to judge of the validity of the objection which has been made, that the principle of Natural Selection depends on a succession of favourable chances. The objection is more in the nature of an innuendo than of a criticism, for it depends for its force upon the ambiguity of the word chance, in its popular uses. The income derived from a Casino by its proprietor may, in one sense, be said to depend upon a succession of favourable chances, although the phrase contains a suggestion of improbability more appropriate to the hopes of the patrons of his establishment. . . . It is easy without any very profound logical analysis to perceive the difference between a succession of favourable deviations from the laws of chance, and on the other hand, the continuous and cumulative action of these laws. It is on the latter that the principle of Natural Selection relies. (Fisher 1930, 37)

Here Fisher is quite clear that natural selection is in many ways the exact opposite of "mere" chance, in the sense that it yields outcomes that are unequally probable. Just as casinos can profit from gambling (on average), so too, selection can act to change populations over time.

Perhaps the best way to understand Fisher's view on the role of chance in evolution is to consider his 1934 paper "Indeterminism and Natural Selection." Fisher there argues that current physics endorses an "indeterministic view of causation," by which he means both (1) that appeals to probabilistic causation have become far more common in science, for example, in statistical mechanics, and (2) that physicists endorse the "principle of indeterminism," or quantum indeterminacy. His central argument in this paper is that an indeterministic view of the natural world is a more "unified" and comprehensive basis for modern science: "Besides unifying the concepts of natural law held in diverse spheres of human experience, the view that prediction of future from past observation must always involve uncertainty, and, when stated correctly, must always be a statement of probabilities, has the scientific advantage of being a more general theory of natural causation, of which complete determinism is a special case, possibly still correct for special types of prediction" (Fisher 1934, 104). That is, Fisher saw in physics a model for the lawful treatment of aggregative behavior in biology; probabilistic causes are, it seems, no less "real" than deterministic ones, provided we consider causal behavior of aggregative systems. Fisher holds that indeterministic behavior at the micro level could give rise to probabilistic behavior at the macro (see also M. J. S. Hodge 1992a). Fisher argues that "only in an indeterministic system has the notion of causation restored to it that creative element, that sense of bringing something to pass which otherwise would not have been, which is essential to its commonsense meaning" (Fisher 1934, 106–7). By "that creative element," Fisher appears to be referring to both to human agency and the capacity of natural selection to create genuinely novel adaptations; he is also replying to the work of authors like Bergson and Smuts, whose views of "creative" evolution required a special agency. Fisher contested that the creative elements of evolution did not require a jettisoning of mechanistic explanation, but of determinism. In a deterministic world, the human capacity for choice (and purposive behavior in animals, generally) is not real, but rather must be *illusory*. Yet, Fisher takes the purposive behavior of animals to be evident "not as an epiphenomenon, but as having a real part to play in the survival or death of the organisms that evince them" (1930, 108). In other words, since purposive behavior is a genuine cause of evolutionary change (organisms act and interact with one another and their environments, thus shaping their own genetic fate) and genuinely purposive behavior is inconsistent with determinism (on his view), indeterminism appears to be the only view consistent with evolutionary explanation. The argument is a sort of transcendental deduction of the necessity of indeterminism. M. J. S. Hodge (1992a) and J. R. G. Turner (1987) link Fisher's vision for a "creative" biology to his political (eugenic) and religious (Christian) commitments. Although these political and religious views do not directly influence his views about the role of chance in evolution, they certainly influenced his views about the significance of selection as a power for improving not only society, but also complexity and adaptation in the biological world.

Haldane

Haldane, like Fisher, was a "synthesis" thinker, though Haldane's biological interests ranged further. At one time or another in his life, he made major contributions to biochemistry, biochemical genetics, human genetics, statistics, theories of the origin of life, and evolutionary biology. Undoubtedly, his greatest contribution was his quantitative or statistical interpretation of evolution by natural selection in light of Mendelian genetics as expressed in his series of papers entitled "A Mathematical Theory of Natural and Artificial Selection" (beginning with Haldane 1924a, 1924b, and continuing until 1934), in his book *The Causes of Evolution* (1932), and in his synthetic overview of genetics, *New Paths in Genetics* (1942). In those works, he sought to defend Darwin's theory of natural selection as the primary cause of adaptive evolution. Along the way, he gives some telling commentary on his views on fundamental physics, indeterminism, and implications for eugenics and human freedom.

In all these works, Haldane mentions chance very infrequently, only sixteen times in *The Causes of Evolution*, and the vast majority of these times, he is using the term *chance* as proxy for *probability*. The infrequency with which Haldane discusses chance suggests that he regards it as having a relatively less significant role in adaptive evolution than selection. In fact, Haldane is yet more explicit: "in a numerous species," the reduction of genetic variation as a result of drift would take "a long period even on an astronomical, let alone a geological time scale" (Haldane 1932, 117). For example, Haldane considers Elton's (1924) appeal to chance in explaining the evolution of the arctic fox in Kamchatka. Elton argued that on a fairly regular basis, modern arctic foxes suffer catastrophic population loss, or regular incidences of random extinction. As a consequence, effective population size for the arctic fox is very low. So the effect of drift should be large. In effect, Elton argues that random extinction is a significant cause of evolutionary changes in the fox. Haldane objects to this line of reasoning on the grounds that even with such drastic reduction in population size, the chance of loss of an allele is small, and ecological conditions could change dramatically throughout. In his view, such highly variable ecological conditions likely played a more significant role in the arctic fox's current state than drift. In sum, Haldane claims that "random extinction has probably played a very subordinate part in evolution" (Haldane 1932, 117).

One last point is worth mentioning. On the role of chance in mutation, Haldane makes the following (somewhat obscure) comment: "Muller has discovered how to control the rate of mutation, and it is clear that mutation is accidental rather than providential" (Haldane 1942, 20). Haldane here is making reference to the role of radiation in inducing mutation. With the publication of the paper "On the Nature of Gene Mutation and Gene Structure" by Timoféeff-Ressovsky, Zimmer, and Delbrück in 1935 (known as the Three-Man Paper; translation in Sloan and Fogel 2011), most geneticists, Haldane included, at this time would have known that irradiation increases the mutation rate, proportional to the applied dose, though exactly how this happened was relatively unknown. What Haldane means by "accidental" here, in contrast to "providential," seems to trade on two senses of the term: first, that whether a given mutation occurs at one place or another on the chromosome is a matter of chance, and second, that mutations are by and large likely not to be advantageous (see Merlin, this volume, chap. 7, for a discussion of recent research on mutation). So, while this is far from explicit, Haldane seems to think the chance of an advantageous mutation's arising is more or less a product of indeterministic forces, or, at least not "providential" ones.

Wright

Sewall Wright is best known for his "shifting balance" theory of evolution, which he outlined in a series of papers between 1930 and 1932 (S. Wright 1930, 1931, 1932). Wright's arguments in these papers were hugely influential in both the early and the late synthesis. Particularly influential was his metaphor of the adaptive landscape, a multidimensional representation of fitness (in both individuals and populations) as a product of different combinations of genes. Wright's views about the relative importance of chance in evolution are best seen within the context of his shifting balance theory.

The shifting balance theory was Wright's answer to what he understood to be the "central question" of evolution. Wright noticed in his work on the evolution of coat color in mammals, and in his work at the USDA, that mass selection is effective but does not result in optimal traits. How then can genuinely novel gene combinations arise? Wright conducted experiments that suggested that inbreeding in general leads to a decline in fitness but can also lead to novel trait combinations. He became convinced of the advantages of combining selection with inbreeding within herds, followed by crossbreeding. The isolation of small subpopulations, followed by migration and thus shifting an entire population to new "adaptive peaks," was thus a way to answer the problem of how novel adaptive gene combinations might come about. Hodge nicely sums up Wright's central question (and answer) thus: "Under what statistical or populational conditions is this cumulative change most rapid, continual and irreversible, with or without environmental variation or change? Wright's answer is: When a large population is broken up into small local subpopulations with only a little interbreeding among those subpopulations, and when there is inbreeding, random drift and selection within those subpopulations, and when one or more subpopulations having individuals with selectively favored, superior gene combinations exports those individuals to other subpopulations and so contributes to transforming the whole population, the entire species" (M. J. S. Hodge 2011, 31).

As is clear from the above, chance plays multiple roles in Wright's shifting balance scenario, at various *levels* as well as *stages* of the evolutionary process. There is chance fixation of novel gene combinations in subpopulations, as well as the chance spread of these novel adaptive combinations through the entire population.

In the 1931 and 1932 papers, Wright mentions chance a total of thirty-two times; 63 percent of the time he refers to chance as probability and 37 percent of the time he refers to chance as randomness. He uses the word *chance* far and away more than anyone else, and he also uses it at higher rates. Although Wright speaks of chance playing a role in evolution in many different ways, he is most distinctive among all the early synthesis authors in seeing a role for drift in adaptive evolution.

Wright defines drift as the process by which "*merely by chance*, one or the other of the allelomorphs may be expected to increase its frequency in a given generation" (S. Wright 1931, 106, emphasis added). Similarly, in his 1932 paper, Wright defines drift as the process whereby "gene frequency in a given generation is in general a little different one way or the other from that in the preceding, *merely by chance*" (1932, 360, emphasis added). Thus, "merely by chance" refers to two distinct classes of causes: segregation/recombination, and reproductive stochasticity: "If the population is not indefinitely large, another factor must be taken into account: the effects of accidents of sampling among those that survive and become parents in each generation and among the germ cells of these" (ibid.). By "accidents of sampling," Wright is referring to the fact that in diploid organisms, one of two copies of an allele is randomly passed from each parent to an offspring as a result of recombination during meiosis. Alternatively, reproductive stochasticity concerns the possibility that, just by chance, some parents may have more offspring than others. As Wright puts it, "The conditions of random sampling of gametes will seldom be closely approached. The number of surviving offspring left by different parents may vary tremendously either through selection or *merely accidental causes*" (1931, 111, emphasis added). By "accidental causes," Wright appears to be referring to what we have called "contingent" events—for instance, living on the north rather than the south face of a mountain.

LATER SYNTHESIS

What we are calling the later synthesis was the period from roughly 1935 to 1950, during which there was an institutional reorganization of the field of biology, which involved the founding of the Society for the Study of Evolution and the associated journal *Evolution*, as well as the organization of a variety of interdisciplinary conferences and book series, to promote evolution as a subject of study and to link existing disciplines—systematics, paleontology, and genetics. Though participants disagreed about which questions are most central to evolutionary biology, as well as which answers are most likely, all generally agreed on the consistency of the new genetics with a "Darwinian" view of evolution. What this Darwinian view amounted to was subject to various nuances, but all agreed on what Gould (2002) has called the "fundamental principles of Darwinian logic": that selection acts by and large on individual organisms, that selection leads to both genetic changes in populations (microevolution) and speciation (macroevolution), and that the very same causes of evolution in populations were responsible for the divergence of species and lineages. Moreover, all saw themselves as responding to various "opponents" to this "Darwinian" view, including anti-evolutionists, as well as "orthogenicists" and neo-Lamarckians. The core commitments of many of the latter synthesis authors were as follows:

- First, they saw their work as providing a "Darwinian" alternative to "directed" or "orthogenetic" views of evolution, according to which evolution has a predetermined direction.
- Second, they held that the origin (mutation) and sorting (recombination) of genes are in some sense chance or random processes.
- Third, all viewed natural selection as a probabilistic cause of adaptive change in populations.
- Fourth, all took the current distribution of species and adaptations as, in large part, a matter of contingency, both in terms of when and where mutations arise and are sorted in meiosis, and in terms of which environmental challenges are presented, that is, whether "contingent" events like storms, floods, and natural disasters were more or less in operation in the ecology and evolution of any lineage.

Dobzhansky

Genetics and the Origin of Species (1937) is a survey of biological work from genetics, population genetics, ecology, and natural history, as it bears on the fundamental problem of species' origins. In the first edition, Dobzhansky sees a significant role for chance in evolution, a role significantly diminished in the third (1951) edition. Dobzhansky speaks of chance operating in the spontaneous modification of chromosome structure, the distribution of chromosomes during meiosis, and migration patterns and isolation of populations; and as featuring significantly in the fixation and loss of genes through random mating in small isolated populations.

One of Dobzhansky's central goals in 1937 was to defend the view that microevolution is sufficient for macroevolution. Dobzhansky argued that mutation is necessary, but no single mutation is *sufficient* for species level change. He argues:

> Species differ from each other usually by many genes; hence, a sudden origin of a species by mutation, in one thrust, would demand a simultaneous mutation of numerous genes. Assuming that two species differ in only 100 genes and taking the mutation rate of individual genes to be as high as 1:10,000, the probability of a sudden origin of a new species would be 1 to $10,000^{100}$. This is not unlike assuming that water in a kettle placed on a fire will freeze, an event which is, according to the new physics, not altogether impossible, but improbable indeed. (Dobzhansky 1937, 40)

Dobzhansky has a particular target in mind in this argument: the saltationist or "mutationist" view, defended by authors such as DeVries and Bateson. Interestingly enough, the argument itself appeals to improbability, and this was a rationale that many synthesis authors used to discredit defenders of "mutationist" or "orthogenic" theories of evolution. Though his book came out in 1940, Goldschmidt was roundly demonized by synthesis authors (particularly Mayr) for defending "saltational" evolution, or evolution by major "macromutations" (see, e.g., Goldschmidt 1940; Gould 1982).

While single mutations could not suffice to generate novel species, at the population level, Dobzhansky thought chance played a significant role. Dobzhansky notes that to some extent it is just a matter of luck whether a mutation becomes established in a population or whether it is lost: "A majority of mutations turning up in natural populations are lost within a few generations after their origin, and this irrespective of whether they are neutral, harmful, or useful to the organism. The numerous mutations which persist are the 'lucky' remainder which may be increased in frequency instead of lost" (Dobzhansky 1937, 131).

In other words, only a few offspring will inherit a newly mutated gene, and in species with a stable population size and a high rate of reproduction, it is a matter of luck whether those offspring will pass on the inherited mutation. Consequently, the vast majority of mutations, even those immediately beneficial, are weeded out after a few generations. Only "lucky" mutations persist. While Dobzhansky does not use the term *genetic drift* in the first (1937) edition, he does follow Wright's lead, arguing that when a small population becomes isolated, simply by chance, some genes will become fixed and some will be lost: "Each of the colonies with very small breeding populations will soon become genetically uniform owing to the depletion of the store of the hereditary variability they once possessed. It is important to realize that in different colonies different genes will be lost and fixed, the loss or fixation being due, as we have seen, simply to chance" (Dobzhansky 1937, 134). In later editions, Dobzhansky refers to this process as "Drift" or the "Sewall Wright effect."

While Dobzhansky grants that mutations arise "by chance" and that the cause of mutation is unknown ("the name 'spontaneous' constitutes an admission of ignorance of the phenomenon to which it is applied" [1951, 38–39]), he does speculate on the causes of mutation (39). Dobzhansky cites a number of experiments in which mutation rates were increased with the application of radiation, iodine, potassium iodide, copper sulfate, ammonia, potassium permanganate, lead salts, and mustard gas (43). However, he is uncertain about the mechanism of mutation and hesitates to speculate regarding whether the ultimate explanation involves indeterministic causal processes, stating only that the outcomes are unpredictable: "X rays, ultraviolet rays, and the chemical mutagens mentioned above seem to be unspecific, in the sense that they increase the frequency of change (or destruction) of apparently all the genes of an organism. There is no way to predict just what genes will be found changed" (45). Dobzhansky seems to downplay the role of chance in later editions of *Genetics and the Origin of Species*. Beatty (1987) has argued, however, that Dobzhansky always viewed drift and selection as complementary and that his later emphasis on selection merely amounts to a shift in position on the relative *significance* of selection and drift. Beatty suggests that while Dobzhansky's empirical work on various laboratory species played an important role here, his primary motivators were his personal views in opposition to the "classical" versus the "balance" view of natural variation.[4] According to the classical view, often associated with eugenic ideology, most highly adapted populations are genetically uniform, suggesting that genetic variation in evolving (human) populations is ultimately detrimental. Dobzhansky strongly disagreed with this classical view and was firm in his belief that variation is absolutely necessary for the long-term survival and success of the human species. Thus, his ideological commitments in this context may have played an important role in his assumptions about the natural distribution of variation in most populations.

Mayr

Mayr's unique viewpoint was that of a systematist and a biogeographer. Mayr believed systematics and biogeography could yield insights into evolution that population genetics alone could not. In this sense one primary aim of *Systematics and the Origin of Species from the Viewpoint of a Zoologist* (1942) was to demonstrate the import of systematics and biogeography—or, perhaps better, the tradition of natural history—to the ongoing evolutionary synthesis. To achieve that aim, Mayr hoped to show how this perspective was essential to (1) explaining speciation, (2) providing additional evidence for gradualism, and (3) strengthening the case of the biological species concept.

Mayr argued that the "zoologist viewpoint" was better equipped to identify a *necessary condition* on speciation: nonbiological isolating mechanisms. Mayr argued that sympatric speciation was next to impossible and supported by little evidence, while allopatric speciation was much more common and supported by a rich body of evidence. For Mayr, if there were no barriers to random dispersal and mating, populations were unlikely to diverge. By necessity, then, portions of such populations must be reproductively isolated (either geographically or biologically) if new species are to arise. For Mayr, most cases of speciation will occur *after* a climatic (or geographic) event splits a single population.

Mayr uses the term *chance* merely eight times in the entire book, and in the majority of those cases, he refers to chance as proxy for *probability*. However, for Mayr, contingency does play a role in *geographic isolation* (most often the result of a climatic event). Both geographic and biological isolation (the result of selection on reproductive isolating mechanisms) are *necessary* to speciation, according to Mayr: "There is a fundamental difference between the two classes of isolating mechanisms, and they are largely complementary. Geographic isolation alone cannot lead to the formation of new species, unless it is accompanied by the development of biological isolating mechanisms which are able to function when the geographic isolation breaks down. On the other hand, biological isolating mechanisms cannot be perfect, in general, unless panmixia is prevented by at least temporary establishment of geographic barriers" (Mayr 1942, 226). If we attribute the production of geographic barriers to chance entirely, then chance plays an equally important role to speciation events as does natural selection of biological isolating mechanisms. Again, while he does not comment on the *relative* role of such chance events, there is room to argue that they played a fundamental role to Mayr's view of evolutionary speciation.

Simpson

In his *Tempo and Mode in Evolution* (1944), George Gaylord Simpson sets himself the primary task of synthesizing paleontology and genetics. To effect such a synthesis, he focuses on explaining the various causes of the "tempo and mode," or rates and patterns, of macroevolutionary change. Among those causes, Simpson places special emphasis on chance, which he refers to nineteen times in that single work. Of those mentions, 45 percent involved chance as probability, and 51 percent involved chance as randomness. In this section, we explicate what causal role Simpson attributed to chance to explain differences in tempo and mode.

By *tempo* Simpson means "rate of evolution," which he defines as rate of change of gene frequencies in a population relative to some absolute unit of time such as years or centuries. He favors defining rate of evolution as the "amount of morphological change relative to a standard," which might suggest a similar but not identical rate of genetic change (Simpson 1944, 3). The *standard* Simpson proposed for measuring rate of morphological change is taxonomic in nature. Specifically, Simpson proposed measuring rate of morphological change by dividing the number of successive genera by their total duration (17). For example, the line of successive genera starting with *Hyracotherium* and ending with *Equus* consists of eight genera and has a duration of approximately 45 million years. So the rate of morphological evolution, according to Simpson's standard, is 5.6 million years per genus, or .18 genera per million years. Using this standard of measure, Simpson describes the rates of morphological change for several genera. In this way, he establishes that there are differences in the tempo of evolution for different fossil groups or lines of successive genera.

Simpson identifies various *modes* of evolution: micro-, macro-, and mega-evolution. The modes are individuated by taxonomic rank. So microevolution involves differentiation within a species but no "discontinuity"—that is, "branching" at the level of species or speciation. Macroevolution involves differentiation and discontinuity, including speciation as well as branching at the level of genera. Finally, megaevolution involves events at the micro- and macro- level as well as discontinuity at the level of higher taxonomic ranks such as families and orders (Simpson 1944, 97–98).

Simpson claims that population size plays "an essential role" as "a determinant both of rates and patterns of evolution" (Simpson 1944, 66). He shows that population size affects tempo, which in turn affects mode of evolution. Simpson reasons in the following way. First, he accepts Wright's argument that large populations will exhibit high *variability* (i.e., *potential* for variation) but low *variation* and, consequently, will exhibit a slow tempo of evolutionary change. Selection will be either weak or strong in large populations. If it is strong, then even though the population is *variable*, selection will eliminate any *variation* that crops up. On the other hand, weak selection, Simpson argues, "tends to end in a static condition of fixed gene ratios" (67). In such cases, the mode will likely be microevolution.

Second, for Simpson, chance plays a role by increasing the tempo of evolution in intermediate-sized populations and, thus, produces discontinuities or branching events at the level of species or genera. This occurs for just the reasons Wright (1931) outlined. Intermediate-sized populations will exhibit variability similar to that of large populations. However, they will also be more susceptible to drift. So, although selection acts to eliminate variation, drift counters selection and maintains variation. Hence, intermediate population size is more likely to produce adaptive and discontinuous evolution (i.e., macroevolution).

Finally, Simpson considers small populations with little variability. Small populations face the same sort of difficulty as large populations: little variation available for evolutionary change. However, an adaptive mutation in a small population has a greater chance of fixation, or the "utilization of mutations in small populations is more efficient, that is, a single mutation has a much greater chance to survive or to become universal in the population and can do so much more rapidly" (Simpson 1944, 68). Simpson also appeals to Elton's (1924) *Random Extinction Model* and Wright's (1931) *Shifting Balance Theory*. Random extinction events like floods or catastrophes reduce population size drastically. Shifting Balance Theory involves variants going to fixation faster in small populations, so the tempo of evolution is significantly higher. Simpson argues that the number of successive, discontinuous genera produced by such rapid change may lead to higher taxonomic discontinuities at the level of family and order (i.e., megaevolution).

Stebbins

Stebbins's *Variation and Evolution in Plants* (1950), published in the Columbia Biological Series, is the last in a series of texts often identified as the "core" texts of the synthesis. The volume is also in many respects more comprehensive or synoptic than the earlier volumes, frequently drawing upon evidence and argument from the authors of both the early synthesis (Wright, Haldane, Fisher), and the latter synthesis. Stebbins's goal in this book is to summarize advances in genetics, cytology, and the "statistical study of populations" and their import for the evolution of plants. The book is in some ways more comprehensive even than this, as Stebbins often pauses to compare and contrast evolution in plants and animals, discussing similarities in genetic "systems," modes of speciation, and the role(s) of population size and structure on evolutionary trends. Much like the *Origin*, the book opens with a discussion of variation and its causes and moves from a discussion of basic systematics and trends in variation among and between plants to their bases in environmental plasticity and genetic mutations. Second, Stebbins discusses the experimental and "historical" evidence for natural selection, pausing to consider causes of adaptive and apparently nonadaptive characteristics as well as correlation of characters. Third, he devotes a lengthy chapter to genetic systems as factors in evolution (where "genetic systems" are "internal" factors influencing the rate and nature of recombination and thus the rate and direction of evolution, which Stebbins argues is influenced largely by distinctive gene combinations) and moves on to the roles of isolation, hybridization, and various forms of polyploidy in speciation. The closing chapters in the book review long-term trends in the evolution of distinctive karyotypes and morphology, drawing upon both genetic and paleological data.

Stebbins takes a relatively equivocal view about the relative importance of selection and drift in evolution. He emphasizes that there is much that is as yet unknown about the intensity of selection, heritability, and the sizes of interbreeding populations in nature (Stebbins 1950, 39, 145). Even with this caveat, however, Stebbins indicates that "in cross-breeding plants natural populations are rarely maintained for a sufficient number of generations at a size small enough to enable many of their distinctive characteristics to be due to random fixation." In other words, Stebbins takes random fixation of genes in small isolated populations to be rare; for example, it may occur in cases where species are either confined to "highly specialized habitats" or reduced to small population sizes due to extreme environments or unusually drastic reductions in population size. Stebbins is clearly familiar with Wright's arguments, and he characterizes "random fixation of alleles" as instances of the "Sewall Wright effect," "undoubtedly the chief source of differences between populations, races, and species in nonadaptive characters" (144–45). However, he seems to think most populations of crossbreeding plants are relatively large, so that the chance of "random fixation" of genes is likely less significant than the effects of selection in such populations. Nonetheless, Stebbins does note that in the tropics, or island populations, there would be greater opportunity for nonadaptive differentiation due to random fixation.

Stebbins mentions chance twenty-nine times in the volume. The vast majority of instances are referring to chance as a proxy for "at random," as in "chance fixation" of alleles in isolated populations. Stebbins views such chance factors as potentially playing an early role in the differentiation of species and genera. Divergence, he argues, is most likely to result from special or unusual environmental agents that lead to isolation via "chance reassortment of different combinations of genetic factors" (1950, 508). Such appeals seem to echo Wright's shifting balance model, discussed above, transposed into the context of speciation and divergence. The second most common appeal is to probability, as in: "maximal chance for cross-pollination." The least common appeal is to "contingency." Most of his appeals of this sort involve a contrast with known or suspected causes (e.g., selection).

CONCLUSION

While it is clear from the above that the synthesis authors disagreed on many questions, they agreed upon the following points:

- Mutation is the ultimate source of variation, and mutations arise "by chance," where this is understood as "by and large assumed to be of deleterious effect" (or, not "directed" toward adaptation).
- Meiosis is a source of "random" variation (in sexual reproduction), the 50-50 chance of receiving alleles from either of two chromosomes.
- Isolation of small subpopulations is a source of "random" variation, in the sense that isolates may in many cases be treated as "random" samples from parent populations.
- Inbreeding is a source of chance gene combinations; that is, isolation of small (and thus genetically unique) subpopulations can be a source of evolutionary novelty.
- Drift can be represented as the sampling of alleles from a finite population, such that changes in population size are the main factors yielding an increase in "random fluctuations" in gene frequency.
- Drift and isolation may play a role in adaptive evolution and/or speciation, with new gene combinations arising in small isolated subpopulations.
- While it is unknown what the causes of mutation are at the submolecular level, the role of radiation in inducing mutation suggests, but does not prove, that indeterminism may play a fundamental role in evolutionary change.

- Contingent events play a significant role in macroevolutionary change—for example, due to catastrophic events such as geological and/or climatological changes, yielding extinctions, isolation of species/genera, and/or variable rates of macroevolutionary change.

In sum, synthesis authors shared a set of core commitments about the role(s) of chance in evolution. They agreed that chance plays an important explanatory role in evolution and that appeals to chance are not simply an acknowledgment of ignorance. Rather, appeals to chance (and its cognates) were to be interpreted as proxy for appeals to probability, random sampling, contingent events, or events in contrast to selection. Where they disagreed was about the *empirical* question of the relative *importance* of this or that chance factor in evolution. Some placed greater emphasis on drift than others, and, as we have argued, there was a pendulum shift from early to late synthesis; while Fisher and Haldane emphasized selection, Wright ushered in an emphasis on drift, which was later superseded by the "hardening" of views in favor of selection and downplaying the role of drift. To be clear, there was not a philosophical or conceptual transition in the synthesis, only a change in empirical views regarding whether and to what extent drift (as a matter of fact) was an important factor in evolutionary change. As we hope to have demonstrated, reading the works of authors of the modern synthesis is invaluable as a way of reflecting on core issues in the philosophy of biology. We have considered at least three interrelated questions that the texts of the modern synthesis can help illuminate:

The sense(s) in which evolution is a "probabilistic" theory.
The matter of whether (and in what sense) drift is a cause of evolutionary change.
The ways in which chance, contingency or accident is understood to play a causal role in evolutionary change, according to the synthesis authors.

Do these views provide us with insights into current debates among philosophers of biology, for example, about whether or not we ought to view natural selection (or drift) as causal? We begin with a caveat. It is our view that one should be extremely wary of reading philosophical claims back into historical texts, when these are not made explicit. Only rarely did these figures engage in explicit commentary on causation, metaphysics, indeterminism and determinism, or their relevance to biology. When they did, they were often quite circumspect. Given how little they knew yet about mechanisms of inheritance in 1930, for instance, Haldane was very careful *not* to speculate as to how and why mutations arise, or whether their "random" character had anything to do with fundamental indeterminism. He was very careful to clarify that by the expression "mutations arise by chance," he meant "by chance *with respect to fitness.*" In other words, mutations were not by and large fitness enhancing and did not arise in response to environmental conditions, as some Lamarckians claimed. When speaking of "random" events, synthesis authors were by and large very careful to specify what they meant, noting what empirical conditions would be of relevance to assessment of such claims. For example, Haldane explains: "By a population 'mating at random,' I do not mean one practicing sexual promiscuity, but one in which an individual is no more and no less likely to mate with a relative than an unrelated person, and no more and no less likely to mate with a person heterogeneous for the same recessive gene as himself than with a homozygous normal" (Haldane 1942, 149). That is, terms like *chance* or *random*, for these authors, by and large referred to very specific conditions, or facts about mating regimes, geographical isolation, or environmental contingencies, not "ignorance." So one should not read appeals to chance as asserting theories about the fundamental nature of the physical universe, such as commitments to fundamental indeterminism.

Nor should we attempt to read metaphysical presuppositions into mathematical models. It is certainly possible to agree upon how to represent evolutionary change in populations in mathematical terms and yet fundamentally disagree on questions ranging from whether natural selection or drift is a cause or not, to whether or not causation is acting at one or another temporal or spatial scale (as long-standing debates about levels and units of selection attest). That is, very little of great metaphysical import should be inferred from the endorsement of the empirical adequacy of a mathematical model. It is how the model is *interpreted* that tells us about the commitments of the authors of the model. That is, the use of statistical methods and models does not, in our view, discourage a reading of these authors as endorsing the thesis that heritable differences in organisms make a (probabilistic) causal difference to survival and reproductive success. Moreover, the very fact that these authors disagreed about the relative causal significance of selection versus drift for evolutionary changes in gene frequency suggests that they interpreted their theories as causal theories and that one can (and should) see drift and selection as distinct causes of evolutionary change. As Hodge (this volume, chap. 2) remarks: Fisher and Wright saw their "mathematical work as statistical analyses of causation, not as statistical replacements for causal analysis."

However, given their very different scientific and historical contexts, we ought not to expect that Haldane, Fisher, Wright, Dobzhansky, or Mayr would necessarily appreciate how current debates in philosophy are motivated or framed. For instance, one current debate in philosophy of biology concerns whether we ought to interpret evolutionary theory in "causal" or "statistical" terms. It seems to us that synthesis authors would see this as a false choice. Natural selection is meaningfully spoken of as a cause, in the sense that differential survival and reproductive success is (probabilistically) caused by differences in heritable traits; but, of course, the overall effects of this process can be observed only across generations and in populations. So it is represented at the population level in terms of population level variables, such as mutation rates, selection coefficients, and so on. Thus, in some sense, one represents the causes of changes over time as a "statistical" outcome, but Haldane's, Fisher's, and Wright's use of "statistical" models to represent population level variation, selection, and drift is hardly grounds to reject selection or drift as distinct causal processes. Clearly Fisher, Haldane, and Wright took them to be "causes" of evolution and did not (at least as far as we can discern) have metaphysical worries about whether causal processes are possible at the population level. Arguably, Fisher's creation of the analysis of variance (ANOVA) makes an implicit commitment to the very idea of partitioning causal variables at the population level. Wright, having worked for the USDA, would be very familiar with the use of population level variables and the role of artificial selection in intervening on everything from oil content in corn to milk yield. Haldane's aim in his popular book *The Causes of Evolution* is to demarcate and identify the major *causes* of evolution (and rule out other proposed causes). That is, we largely agree with Hodge's thesis (Hodge, this volume, chap. 2; among others, e.g., Okasha 2009; Millstein, Skipper, and Dietrich 2009; M. J. S. Hodge 1992a, 2011) that there was continuity between synthesis authors and Darwin regarding natural selection as a probabilistic cause of evolution.

We also agree that the contrast between drift and selection, for synthesis authors, was not between noncausal and causal processes but, as Hodge puts it, between "causally non-fortuitous and causally fortuitous" processes. Indeed, much of the work of the synthesis was to explain and describe the differences between such processes and how they made a difference in actual populations. They aimed to establish exactly which kinds of empirical information would be of relevance to deciding the causes of evolution in any case. In fact, arguably one of the central aims of synthesis authors was to distinguish merely hypothetical from actual causes of evolution, to provide empirical evidence for the causes that they took to be central, and to provide a general mathematical framework for describing such causal processes, such that predictions and retroactive inferences about the relative role(s) of distinct causes could be precisely and empirically tested. We thus take it that they would also agree with Strevens (this volume, chap. 6) that there is certainly an objective distinction between drift and selection. Drift was not simply differential reproduction owing to causes we have so far failed to discern (pace A. Rosenberg 1994). The synthesis authors put a great deal of work into showing how to empirically distinguish the respective roles of drift and selection in actual populations. Establishing the role of drift requires, as Hodge puts it, "extra, decisive, detailed knowledge, not uninformed ignorance." As Strevens writes, "The great majority of serious evolutionary explanations citing drift are not . . . mere attributions of arbitrary deaths and wonky statistics to chance. They rather use mathematical models of evolutionary processes to make predictions about differential reproduction" (Strevens, this volume, chap. 6). That claim could well have been written by any synthesis author.

NOTES

1. While we do not follow their classification of senses of chance, we were certainly influenced in our classification by Millstein 2011; Gayon 2005; and Beatty 1984.

2. An initial draft included a section on Julian Huxley's *Evolution: The Modern Synthesis* (1942). While we think that this was without doubt a central book of the synthesis, for reasons of space, we could not include that section.

3. The claim that entropic changes are irreversible while evolution is not is misleading: entropic changes are indeed irreversible (which is why the second law is referred to as "Time's Arrow"), but evolution is also irreversible in its own sense. We may, for example, evolve large lizards again, but they will not be *T. rex* or *Brontosaurus*. Evolution can proceed from simple to complex and "degenerate" back to simple again (think of the branch of the annelid worms that became parasitic), but organisms are not really reversing the pathway or route by which they evolved. Thanks to Gar Allen for this comment.

4. We refer here to the "classical" versus "balance" views on whether populations are highly uniform at the genetic level, or variable. These views have been associated with Mueller and Dobzhansky, respectively.

CHAPTER 4

Is it Providential, by Chance?
Christian Objections to the Role of
Chance in Darwinian Evolution

J. Matthew Ashley

In 1874, Princeton theologian Charles Hodge penned the concluding sentences of his book-length review of Charles Darwin's *On the Origin of Species*, answering the question from which the book took its title:

> We have thus arrived at the answer to our question, What is Darwinism? It is Atheism. This does not mean, as before said, that Mr. Darwin himself and all who adopt his views are atheists; but it means that his theory is atheistic, the exclusion of design from nature is, as Dr. Gray says, tantamount to atheism. (C. Hodge 1994, 156–57)

Writing some 130 years later, Cristoph Schönborn, the influential cardinal-archbishop of Vienna and chief editor of the Catholic Church's universal catechism, echoed Hodge in strikingly similar terms:

> Ever since 1996, when Pope John Paul II said that evolution (a term he did not define) was "more than just a hypothesis," defenders of neo-Darwinian dogma have often invoked the supposed acceptance—or at least acquiescence—of the Roman Catholic Church when they defend their theory as somehow compatible with Christian faith. But this is not true. The Catholic Church, while leaving to science many details about the history of life on earth, proclaims that by the light of reason the human intellect can readily and clearly discern purpose and design in the natural world, including the world of living things. Evolution in the sense of common ancestry might be true, but evolution in the neo-

Darwinian sense—an unguided, unplanned process of random variation and natural selection—is not. Any system of thought that denies or
seeks to explain away the overwhelming evidence for design in biology
is ideology, not science. (Schönborn 2005)

This coincidence of views held by a nineteenth-century Presbyterian theologian and a twenty-first-century Roman Catholic theologian underscores a
persistent objection raised by Christians to Darwin's theory. The role given
to chance in the Darwinian telling of life's history is taken to render impossible a complementary account of God's purposive involvement in the world
(including the world of human history). This objection surfaces or (one sus-
pects) lurks just below the surface in theologies current in a broad spectrum
of Christian denominations—extending far beyond the narrower subset of
Young Earth Creationists.[1] The issue is a complex one in Christian theology,
for a number of reasons. As other essays in this volume make clear, the role
of chance in evolution is itself an extremely difficult matter to sort out, as is
the question of the degree of "trendiness" or directionality this role allows
us to infer from or attribute to life's history (see Ruse, this volume, chap. 5).
As Plutynski et al. (this volume, chap. 3) point out, among the scientists who
forged the modern synthesis, there are at least five different senses of what
"chance" meant and entailed. This essay will hold itself largely agnostic on
these questions. It will, however, argue that certain concerns over chance articulated by theologians after Darwin transcend the variety of ways that they
understood scientists and philosophers to be asserting its presence, so that
some progress can be made theologically by probing the assumptions that
underlie these concerns, while bracketing the resolution of the questions in
science and philosophy.

With this proviso in force, I make two claims in what follows. First, by
mapping the topography of the theological landscape in which this objection arises, I argue that one source of confusion is a multiple impingement of
concern over chance in theology. These concerns arise in different subfields
of the discipline of theology and engage topoi in which the stakes are taken
to be different by different theological schools. The arguments in these different subfields are often framed and adjudicated on different grounds, with
different kinds of evidence and different standards for what counts as persuasive. Finally, these subfields often involve the theologian in broader and
quite controverted disputes in Christian theology's negotiation of its situation in a modern or postmodern intellectual milieu (viz., over the relationship between faith and reason). A "solution" of the problem of chance in one

subfield may run afoul of concerns that arise in another. Second, I argue that a further impediment to a more creative adjudication of the problems raised by "chance" for Christian accounts of God's providential oversight of history is the continuing (usually tacit) hegemony of certain metaphors and argumentative tropes for elaborating those accounts. I conclude (without pretense to having solved the problem of giving due theological weight to the presence of chance in evolution) by making some references to premodern ways of conceiving and portraying divine agency in history. I do so on the conviction that these roads once taken in the Bible and in early Christian theology are worth considering today in the light of the reminder that modern evolution gives us of the irreducible messiness of all history.

MAPPING CONCERNS OVER CHANCE

Issues raised by the presence of chance in physical processes, however this is construed, have been addressed by Christian theologians in at least two different subdisciplinary fields of Christian theology. One such field is named by Catholic theologians fundamental theology (from *theologia fundamentalis* in neoscholastic and neo-Thomist theological manuals of the late nineteenth and twentieth centuries) and named by Protestants apologetics. Generally taken to be prior to theology proper, because it does not start from the sources of theology (Scripture, tradition, and religious experience), this subdiscipline deals with justifying the rational character of a discipline that is (1) taken to deal with matters that ultimately transcend reason and (2), partly because of this, requires assent to revelation, which is found for Christians in Scripture and (for Catholics) tradition. One division of this field works out the possibility and limits of rational knowledge of God from our experience, unaided by revelation (this subfield is often named natural theology; for a good introduction, see Re Manning, Brooke, and Watts 2013). Here we find the "argument from design," or "teleological argument," which argues from some construal of the presence of design, purpose, or goal-oriented process in the world to a Creator responsible for it. Perhaps most familiar during the period covered in this essay from William Paley's version in his *Natural Theology*, its lineage of course stretches much farther back.[2] The argument's philosophical cogency was contested long before 1859, and different versions have been proposed both before and since. Whether the presence of chance in evolutionary processes sounds a death knell to the argument depends on the version chosen and (once again) on how one understands chance in evolution.[3] This being said, not all theologians are so invested in natural theology's success, and of those who are, not all take the argument from design to be the best candidate for redeeming natural theology's potential. Karl Barth is one of the most relentless critics of natural theology, arguing that, even if successful, it gives us only the image of human rationality and not the wholly other God of Scripture. The Catholic theologian John Henry Newman is more sanguine about the result of rational argument to God's existence, but he had little use for the argument from design. Much depends on how a theologian assesses at the most general level the relationship between human reason and faith, of which conflicts or convergences with science form a subcategory.

The other area of theological investigation in which chance is raised as a problem concerns divine providence, a part of the theology of God and the God-world relationship understood now on the basis of how it is presented in the sources of revelation. Few assertions are more central to the whole symphony of biblical testimonies to God than the claim that God cares for and is intimately involved with God's creation, as a whole and in its particulars. Does chance make this claim untenable? Again, much depends on the account of chance one takes up, and also on how one conceives this providential oversight, although here the crucial decisions are not made solely on philosophical grounds, but on the basis of conclusions one draws from interpreting what Scripture has to say about God's agency, conclusions, as we shall see, that may or may not cohere with views of divine agency that are put in play in natural theology. To demonstrate the complications that result from the interplay between these two loci for estimating the impact of chance in theology, I consider in greater depth the two figures with whom I began.

Charles Hodge

Charles Hodge (1797–1878) is one of the most prominent representatives of the so-called Princeton School, itself one of the most influential approaches to harmonizing science and religion in in nineteenth-century US Protestant thought (see Noll 1983). He graduated from Princeton Seminary in 1819, returned the following year to begin teaching, and remained there until his death, almost sixty years later, forming generations of students. In addition to this impact on leading church leaders and theologians in the nineteenth century and beyond, Hodge's publications and lectures, along with other labors such as founding and editing what became known as the *Princeton Review*, make him one of the most influential American theologians of the nineteenth century. Hodge also had a lifelong interest in science and was committed to the harmony of science and religion, a commitment put to a severe test with the publication and gradual acceptance of Darwin's *Origin of Species*.[4] He wrote his critical review of *Origin* in 1874, at about the same time that he was completing his three-volume *Systematic Theology* (which continued in use in Protestant seminary classrooms until the 1930s).

Hodge did not deny the possibility of natural theology or the cogency of proofs for God's existence (including the argument from design); but neither did he hold that knowledge of God depended on the proofs. On Hodge's view, knowledge of God falls in the category of "intuitive truth," which "the mind perceives to be true immediately, without proof or testimony" (C. Hodge 1968, 1:192).[5] Various proofs for the existence of God are not "designed so much to prove the existence of an unknown being, as to demonstrate that the Being who reveals himself to man in the very constitution of his nature must be all that Theism declares him to be" (1:203).[6] And what does theism declare God to be? Hodge answers: "an extra-mundane, personal God, the creator, preserver and governor of the world" (1:204). Not every proof will demonstrate all or the same subset of these predicates. As a consequence, the most secure and comprehensive set of "facts" from which to infer and elaborate these predicates in their entirety is found in the Bible, which is to the theologian as the natural world and its facts are to the natural scientist.

What is Hodge's account of the argument from design? He begins by defining "design" in such a way as to highlight the features of theism the proof is meant to demonstrate, but also in such a way as to counterpose it to chance and set up an insuperable opposition between the two. Design entails "(1) the selection of an end to be attained. (2.) The choice of suitable means for its attainment. (3.) The actual application of those means for the accomplishment of the proposed end" (1968, 1:216).[7] The principal instances or analogates of design are a book or a machine (he is particularly fascinated with "Mr. Babbage's calculating machine"). Given the presence of design, so defined, the predicates that, on Hodge's reading, can be inferred (or, perhaps better, confirmed) from the argument follow almost analytically. Confronted by design in the created world of this sort, we are warranted to conclude to a source for that design that is (1) different from the world (thus, God is extramundane), (2) exhibits intelligence and purpose (hence is personal), and (3) is the creative source of that order in all its manifestations (creator of all that is) (1:216, 1:227). That this kind of design is abundantly manifested in the world of our experience is amply attested, Hodge asserts, going to the trouble of listing all the volumes in the Bridgewater Treatises for the skeptical reader who needs persuading (1:217n).

What the argument cannot do on its own is demonstrate God's ongoing involvement with the world as the one who preserves it and governs it. This requires the testimony of Scripture and an internal conviction that, Hodge claims, is universal (that is, this is an "intuitive truth"). Despite his endorsement of the Bridgewater Treatises and of Paley's *Natural Theology*, Hodge consistently rejects a deism that would find everything we need to know about God defended by their arguments.[8] For one tendency of the argument was to argue that the wisdom and power of the designer is even more firmly attested by the self-sufficiency of the created universe to bring about certain results simply by laws and structures created at the beginning, without further intervention. For Hodge, such a view might be consistent with the argument from design taken on its own, but it (1) "is obviously opposed to the representations of Scripture," (2) "is inconsistent with the absolute dependence of all things on God," and (3) "does violence to the instinctive religious convictions of all men," who "recognize God as everywhere present and everywhere active. If they do not love and trust Him, they at least fear Him" (1968, 1:576; he gives the same objections in C. Hodge 1994, 87–88).

With this we have passed from the section in which Hodge treats confirmatory proofs of theism to his section on divine providence, moving from a part of theology that relies (or is obliged to rely) on reason alone, to reason reflecting on revealed truth in Scripture. Providence divides into preservation— the doctrine that all things "owe the continuation of their existence, with all their properties and powers, to the will of God"—and government, which "includes the ideas of design and control. It supposes an end to be attained, and the disposition and direction of means for its accomplishment" (1968, 1:581). The crossing back and forth between these two theological loci reinforces Hodge's aversion to "chance." If the scripturally grounded doctrine of preservation offers what the argument from design cannot—the insistence that God is not the *deus otiosus* of deism—the argument from design offers the central interpretive concept for approaching what Scripture has to say about divine government, the concept of design. In transferring the concept of design as he defined it in his theistic argument for use in scriptural interpretation, however, Hodge intensifies the conflict between "design" and "chance." In doing this he distances himself from Paley's greater willingness in *Natural Theology* to consider chance as a part of a designed universe.[9] Because the designing agent in this case is the omniscient, omnipotent God, once an end is selected, the choice of a suitable means to attain the end must be such "as to render certain the accomplishment." The consequence is that "the doctrine of providence excludes both necessity and chance from the universe, substituting for them the intelligent and universal control of an infinite, omnipresent God"

(C. Hodge 1968, 1:582).[10] Such an emphasis on the continual, immediate, and effective control of the universe by God threatens to rob inner-worldly causality of anything other than only apparent efficacy, an extreme that Hodge also rejects.[11] Yet he seems far more worried about the other extreme (deism): "This banishing God from the world is simply intolerable and, blessed be his name, impossible. A God who does nothing is, to us, no God" (1994, 88). The Scriptural doctrine for understanding the presence of "contrivance" in the world "does not ignore the efficiency of secondary causes; it simply asserts that God overrules and controls them" (1968, 1:86). At the end of the day, the precise character of an "overruling" and "controlling" of secondary causes that does not strip them of their efficacy is, Hodge contends, inscrutable to us. It would be better if we simply made do with the bare facts of the Scriptural doctrine. "But men have insisted upon answering the questions, How does God govern the world? What is the relation between his agency and the efficiency of second causes? And especially, How can God's absolute control be reconciled with the liberty of rational agents? These are questions that never can be solved. But as philosophers insist upon answering them, it becomes necessary for theologians to consider those answers and to show their fallacy when they conflict with the established facts of revelation and experience" (1:582).[12]

Hodge combines, then, a strong view on transparency and lack of mystery when it comes to the *presence* of design and its reference to a creator (which, I have argued, drew from his reliance, however selective, on the argument as formulated in the style of William Paley and the Bridgewater Treatises) with an insistence on opacity when it comes to the *mechanism* by which God effects design—that is, an insistence on mystery precisely there. At first blush, an assertion that nondirected, random causes ("chance") could explain the apparent design of living beings undermined the former rejection of mystery, but could Hodge not have solved the problem by appropriating some modified form of Paley's attempts to allow chance into a designed world? I conclude that his commitment to the concept of design for interpreting Scripture (in his theology of providence) and his suspicion of probing further into how God "governs and overrules" secondary causes in nature (that is, an insistent affirmation of mystery in the second sense) gave him little recourse other than to attack this assertion in a way that seems unnecessary for the ends of natural theology.[13] This was a bind of his own making, in part because of the way he transferred the concept of design defined in the realm of natural theology to the realm of scriptural interpretation, and back again, further intensifying in the process its opposition to any role for chance in life's history.

Cristoph Schönborn

The issue of chance in evolution did not, in the main, enter into early debates in Catholic theology over evolution. For some, such as Catholic priest and scientist John Zahm, this could be because the matter could be safely ignored by the 1890s (when he was writing), given the general eclipse of Darwin's theory and the primacy it gave to natural selection and the authority of figures such as Asa Gray (whose disagreements with Darwin on a possible place for design in his theory are well known).[14] The more neuralgic points in Catholic debates had to do with the uniqueness of human beings as created in the image of God, a uniqueness whose marker was taken to be the possession of souls directly created and infused by God or, even more strong, a direct creation of the first human couple both body and soul (Genesis 1:27).[15] In the next century, Pierre Teilhard de Chardin famously ran afoul of Catholic Church authorities for unmooring original sin from a historical action by a first couple. About thirty years later, in one of only a few authoritative statements by a pope, Pius XII insisted only on monogenism and the direct creation of the human soul by God as conditions under which Catholics could entertain the truth of evolution.[16] When John Paul II visited the topic of evolution in 1996, he quietly dropped the issue of monogenism and focused only on human uniqueness (John Paul II 1996a; translated as John Paul II 1996b). No mentions of chance and providence.

All the more surprising when Cardinal Schönborn wrote his indictment of "evolution in the neo-Darwinian sense—an unguided, unplanned process of random variations and natural selection." Responding to the controversy over the *New York Times* op-ed piece, including strong objections from Catholic scientists such as Francisco Ayala, George Coyne, and Kenneth Miller, along with theologians such as John Haught and Denis Edwards, Schönborn followed up with a set of lectures given in the following year and collected into a book: *Chance or Purpose? Creation, Evolution, and a Rational Faith* (2007). The title signals the same counterposing evident in Hodge's work: *either* chance *or* purpose.[17] Throughout the book, he relentlessly presses an opposition between asserting the presence of chance in the processes that drive history and asserting divine guidance. He goes even further than Hodge in associating the assertion of the role of chance in creation with Gnosticism, the early opponent to orthodox Christianity. "While pagan antiquity, for the most part, 'divinized' the world, idolized it, in the period of the rise of Christianity a philosophical reaction, the so-called 'Gnosis' or 'Gnosticism,' devalued it. The world and matter above all, are the result of an 'accident,' of a 'fall' or deterioration. The material world is not actually something good, not something that was intended, supposed to be there, but is purely negative" (18, cf. 58).

As Schönborn limns it, chance is a broad category indeed, extending even to the metaphysical category of contingent versus necessary being (with thereby a gesture toward the ontological argument for God's existence, another staple of natural theology). It is also (as in Hodge's approach), a category set up almost a priori to be antithetical to the working out of purpose. As with Hodge, there is significant slippage and tension in Schönborn's treatments of the different areas of theology in which, on his reading, chance shows itself as problematic.

In fundamental theology, Schönborn draws on a venerable Catholic approach to understanding God's agency ("primary causality") in relation to creaturely agency ("secondary causality"), according to which God's agency is not exercised alongside of or in competition with creaturely agency. To assert the contrary would violate the cardinal principle of God's transcendence, reducing God to one causal agent among others, however preeminent in knowledge, power, and so forth. Rather, God's causality is exercised "through" creaturely causality, by sustaining and enabling creatures to exercise the agency proper to them.[18] Utilizing the traditional distinction between "preservation" and "guidance" that we saw above in Hodge's theology of providence, Schönborn presents the Thomistic doctrine as the philosophical basis for "continuing creation" (2007, 78–81). But he shies away from embracing this position fully when he moves to speaking about God's providential governance of creation, almost as if he had read Hodge's objections to it—that is, that it removes God from the world (see C. Hodge 1968, 1:579). Disavowing a God who intervenes in natural processes (Schönborn 2007, 29, 43–44, 49), as well as a deist God who stands back from them as they independently produce results for which this God purposed them (18, 27, 43, 49), Schönborn seeks a third category (in a way not unlike B. B. Warfield's definition of "mediate creation") by postulating a distinction between "contributory causes" (or "preconditions") and "creative causes" (81–84)—the former meaning the physical causes that make up (in Thomistic parlance) secondary causes, and the latter . . . well, it is not entirely clear what the latter are. Do they make up a third, mediate level of causality through which, as well, divine primary causality operates?[19] A form of direct divine intervention? At first it appears that this kind of causality operates only for the "leap" from inanimate to animate and from nonhuman to human.[20] Yet once a mechanism has been set up for establishing distinctions that cannot be transgressed by evolutionary causality (and the chance processes imminent to it), this causality also is ready-to-hand to be invoked for other such distinctions—such as between plants, animals, and humans (62) or a whole "republic of natures" (98). Schönborn takes all of these distinctions to collapse for a "Neo-Darwinism" that asserts a robust role for chance in the unfolding of natural history, which is, as a consequence (for him, almost, a priori), unguided. Despite his disavowal of an engineer God (98–99) as the only image corollary to an understanding of purpose and design in nature, and despite his assertion of a noninterventionist position in which divine causality works through created causality rather than above or alongside it, Schönborn does appear to be committed to a view in which any discerning of purpose, directionality, or value in the natural world requires positing some sort of causality that is distinct from the natural causality studied by science, and thus it leads him to supplement the twofold distinction between secondary and primary causality.

In Schönborn's 2005 op-ed piece, he cites the International Theological Commission (ITC), a group of theologians appointed by the pope to assist the Vatican in grappling with difficult doctrinal issues. The ITC composed a document on a number of issues dealing with theological anthropology and the theology and theological ethics of human stewardship of creation (International Theological Commission 2004). He quotes from this document to the effect that "an unguided evolutionary process—one that falls outside the bounds of divine providence—simply cannot exist." But this is not quite the whole picture, since the commission is more than willing to let issues of the role of chance in evolution be decided by scientific debate. It deploys the same Thomistic ontology of divine agency that Schönborn invokes to assert that even processes that truly proceed in a chance way can be understood as within the ambit of divine providence, and without introducing some third form of causality:

A growing body of scientific critics of neo-Darwinism point to evidence of design (e.g., biological structures that exhibit specified complexity) that, in their view, cannot be explained in terms of a purely contingent process and that neo-Darwinians have ignored or misinterpreted. The nub of this currently lively disagreement involves scientific observation and generalization concerning whether the available data support inferences of design or chance, and cannot be settled by theology. But it is important to note that, according to the Catholic understanding of divine causality, true contingency in the created order is not incompatible with a purposeful divine providence. Divine causality and created causality radically differ in kind and not only in degree. Thus, even the outcome of a truly contingent natural process can nonetheless fall within God's providential plan for creation. According to St. Thomas Aquinas: "The effect of divine providence is not only that things should happen somehow, but that they should happen either by necessity or by contingency. Therefore, whatsoever divine providence ordains to happen infallibly and of necessity happens infallibly and of necessity; and that happens from contingency, which the divine providence conceives to happen from contingency" (*Summa theologiae*, I, 22, 4 ad 1). . . . Divine causality can be active in a process that is both contingent and guided. Any evolutionary mechanism that is contingent can only be contingent because God made it so. An unguided evolutionary process—one that falls outside the bounds of divine providence—simply cannot exist because "the causality of God, Who is the first agent, extends to all being, not only as to constituent principles of species, but also as to the individualizing principles. . . . It necessarily follows that all things, inasmuch as they participate in existence, must likewise be subject to divine providence" (*Summa theologiae* I, 22, 2). (International Theological Commission 2004, sec. 69)

Whatever reservations one might hold about the opening sentence, the ITC is clear that the issue of chance in evolution needs to be decided by argumentation internal to science. Equally important: a process in which contingency is present in whatever measure can be understood on theological grounds to be guided, a view with which Schönborn is clearly uncomfortable, because, I would argue, he (like Hodge) is committed to a clear and transparent reading of God's purposes and providential governance in natural and historical processes. How this providential guidance is accessible to rational discernment and elaboration on the Thomistic view consistently propounded by the ITC is another question, and a difficult one. Schönborn is reacting to those who assert that absent a scientifically verifiable method for discerning this guidance, or its results, there are no rational grounds at all for asserting it.[21] In his *New York Times* piece, Schönborn cites the First Vatican Council (1869–70) to claim that "by the use of reason alone mankind could come to know the reality of the Uncaused Cause, the First Mover, the God of the philosophers." True enough; but this brings us back to the realm of natural theology. This movement back and forth between different subfields in theology, I argue, muddies the waters as much for Schönborn as for Hodge. However, even in natural theology, the Council does not assert that the argument from design is the path by which knowledge transpires; nor does it state that divine providential oversight of the world is in the same way accessible to human reason. That is, it makes the distinction between fundamental theology and theology of providence, and asserts different forms of rational perspicacity appropriate to each. As with Hodge, Schönborn appears to require the same degree of transparency of the cosmos's history to divine purpose that the Christian tradition has historically asserted (in the area of natural theology) regarding the existence of God. Chance is taken to occlude both; some assertion of divine intervention that can supplement or trump scientific accounts is the solution.

DISCERNING PROVIDENCE IN CHANCE PROCESSES

I have argued that the common ground in protests against chance in evolution, from two theologians who otherwise are quite diverse, can be understood both formally and substantively in ways that shed light on why this has been such a difficult problem to resolve. Formally, the problem arises from multiple points of impingement on theology of the assertion of chance in evolutionary processes: in the area of natural theology and arguments over whether and the degree to which knowledge of God can be derived from rational reflection on nature unsupplemented by revelation; in theological reflection on God's agency and on divine providence, occurring in theology proper, where appeal to revelation is both appropriate and necessary; and in theological explication of the implications of human uniqueness as created in the image of God. Hodge and Schönborn do not adopt (or fully adopt) options for dealing with chance that might make sense in terms of one point of impingement because of problems raised at others. Thus, for example, contemporaries both of Hodge and Schönborn asserted that if one were to accept that God allows God's power to be limited by the "relative non-directionality" of chance processes, then one gains a way of thinking about the presence of imperfections in the natural world (difficult to account for with a strong watchmaker argument), and even the presence of evil. It might even strike one as a fit way for God to proceed if God desired diversity (Edwards 2010, 64–66). Neither Hodge nor Schönborn takes this path because, I argue, it puts too much pressure on the "argument from design" in natural theology and on the exigency to have a tangible marker for human distinctiveness (which ends up entailing some sort of divine intervention, embraced confidently by Hodge and more indirectly by Schönborn since it violates his own stated preference for a noninterventionist God and the Thomistic solution to coordinating divine agency with creaturely agency).

On a more substantive level, both theologians require a high degree of order in the world, an order that is in turn transparent to the designs of a providential creator and governor. While averse to many deist interpretations of or extrapolations from Paley's famous argument, Hodge does embrace the strong commitment to perspicuous, artifact-like design that launches Paley's book, incorporating it into his definition of design.[22] Schönborn distances himself from God-the-mechanic but places just as much emphasis on order in the natural world (in which chance processes have to be quite strictly contained).[23]

It is true that one or two swallows do not a summer make, and a more comprehensive review of theological objections to (or acceptances of) chance would be required to substantiate this typology and determine what portion of objections belong to it; but I suspect that such a review would show that the number is not small. As an initial warrant for that claim, I have offered and argued here the remarkable coincidence of views from theologians separated by more than a century and from two different (nonfundamentalist) Christian denominations with quite different commitments on any number of issues, including the relationship between science and religion.

Short of a more exhaustive survey, perhaps a complementary warrant for my suspicion could be provided by describing, however tentatively, the difference it would make *not* to conflate approaches to concerns raised by chance in different theological areas, or to cross-correlate them without great care. As a preliminary, it is worth reiterating that it is important to be more precise about the sense in which chance is being used, so that one can be clear as to whether and in what way its presence in natural processes impinges on the definition and elaboration of different theological topics. To be sure, this requires greater clarity on the part of scientists and philosophers, and one wonders whether a robust enough consensus on chance will emerge to be of use to theologians. Minimally, it would require patience on the part of theologians while scientists and philosophers sort through the issues; conversely, it would allow a certain equanimity vis-à-vis the sorts of attacks that (rightly) irritated Schönborn (by the likes of Monod and Provine, who are heavily referenced in Schönborn's books and who assert an account of chance that, it seems to me, is not supported by the large, admittedly diverse, mainstream of scientists and philosophers of science).

More directly to the point, what we might learn from the two case studies just considered is that the cogency on theological grounds of insisting on transparency of processes in the natural world to divine purpose needs to be interrogated in the light of theology's own sources in Scripture and subsequent theological reflection rather than just assuming it from the rational argument that one uses in fundamental theology or apologetics. As my analysis of Hodge and Schönborn shows, insisting on the kind of transparency assumed by a Paley-esque argument from design backs theology into a corner where it is difficult to accept any kind of chance in natural processes, however minimal and however constrained by lawlike processes (as it surely is in evolution).

In this regard, I believe another look at how God's providential governance of history is portrayed in Scripture is salutary. Is it really the case that God's governance is given there as absolutely and transparently working itself out in history, with no room for chance? Robert Alter, in his little masterpiece *The Art of Biblical Narrative*, suggests otherwise (Alter 1981; see also Sternberg 1985). Alter argues that the authors of the historical narratives in the Hebrew Bible virtually invented this genre precisely in order to avoid setting up a dichotomy between their surety of divine oversight of history and the no less apparent messiness and contingency of history. "The ancient Hebrew writers . . . seek through the process of narrative realization to reveal the enactment of God's purposes in historical events. This enactment, however, is continuously complicated by a perception of two, approximately parallel, dialectical tensions. One is a tension between the divine plan and *the disorderly character of actual historical events* . . . ; the other is a tension between God's will, His providential guidance, and human freedom, the refractory nature of man" (Alter 1981, 33, emphasis added). And again:

> The monotheistic revolution of biblical Israel was a continuing and disquieting one. It left little margin for neat and confident views about God, the created world, history, and man as political animal or moral agent, for it repeatedly had to make sense of the intersection of incompatibles—the relative and the absolute, human imperfection and divine perfection, the brawling chaos of historical experience and God's promise to fulfill a design in history. The biblical outlook is informed, I think by a sense of stubborn contradictions, of a profound and ineradicable untidiness in the nature of things, and it is toward the expression of such a sense of moral and historical reality that the composite artistry of the Bible is directed. (154)

The genius of the biblical narratives lies in the ways that they deploy (indeed, on Alter's view, discover for the first time) diverse narrative strategies and tactics to succeed in this task. Philosopher Paul Ricoeur makes a similar point regarding the kind of intelligibility proper to narrative in general.

"To make up a plot," he argues, "is already to make the intelligible spring from the accidental, the universal from the singular, the necessary or the probable from the episodic" (Ricoeur 1985, 1:41). If we understood this better, would we be so disturbed by the ways that evolution inserts chance into life's history? Perhaps we have read the biblical narratives as providing too monolithic and unambiguous a reading of how God's purposes are revealed in history.

Many important Christian theologians have, moreover, been cautious in asserting too strongly a similar kind of openness to easy, univocal readings of what God is up to when it comes to history outside of Scripture. Augustine's *City of God* can be read, in one way, as a long argument against such readings, an argument that does not surrender for one moment an insistence that God is working through history, but also insists on history's opacity when it comes to God's purposes, precisely because it is a messy history, rife with the contingency of human choices, many of them sinful. In writing of Augustine's view of human history, Robert Markus contends that

> he was certainly never without a deep sense of God's ever-present activity in each and every moment of time, as in every part of space. He often thought of the whole vast fabric of human history as a majestically ordered whole, an extended song or symphony, in which each moment has its unique, if impenetrably mysterious significance. In this sense all history displays the working of God's providence. But in another sense only "sacred history" [e.g., God's actions described in Scripture] tells us what God *really* has done, what meaning events have within the economy of salvation. . . . Only "sacred history" will furnish clues to what God has *really* done—apart from such insights as he may grant to individuals *privatim* into his dealings with them personally. (Markus 1970, 17)

Clues. But only clues. And these clues cannot be elaborated without further ado into an assertion of a divine plan that runs roughshod over the messiness of history—certainly human history, as Scripture and postscriptural theological reflection insists, as well as the broader tableau of the history of life and on earth and even the history of the cosmos.[24] This perspective on divine providence, and the degree to which men and women can understand its workings, even on the basis of faith and those clues that Scripture can provide, seems an appropriate vantage point from which to be open to those insights that natural science can provide, even if they accumulate into a strong case against asserting the transparency of natural history to a preordained plan working itself out predictably, perspicuously, and irreversibly. On the other hand, if theologians are to be at the service of the faith that also nourished biblical authors and theologians such as Augustine, they can and must insist, pace Monod, Provine, Dawkins, et al., on the possibility of a narrative imagination, with an intelligibility proper to it, that can find meaning and hope in the singular, the contingent, the adventitious. While disagreeing on inner-Christian grounds that the universe is ultimately indifferent to our suffering, such a view can converge, again on inner-Christian grounds, with Stephen Jay Gould's understanding of "such a view of life as exhilarating—a source of both freedom and consequent moral responsibility. We are the offspring of history, and must establish our own paths in this most diverse and interesting of conceivable universes—one indifferent to our suffering, and therefore offering us maximal freedom to thrive, or to fail, in our own chosen way" (Gould 1999, 207).

NOTES

1. Indeed, the figures treated in this essay either predate this movement or vigorously distinguish themselves from it.

2. I refer, of course, to the famous "watchmaker argument" (Paley 1802). As I observe in several notes below, Paley was to some extent a victim of the success of the power of his causal metaphor of mechanistic design (watchmaking), which he articulated so well that subsequent readers, including Hodge, took it over without also taking into account the greater complexity of Paley's argument. I am indebted to one of the reviewers of this chapter for the salutary warning to do justice to Paley on this.

3. For an overview of the argument's history and the impact of Darwin, see Ruse 2003. For a historical survey of ways that reconfigured versions of the argument cohabited quite happily with Darwin's theory, see Brooke and Cantor 1998.

4. His Baconian approach, which he shared with other American advocates of so-called Scottish Common Sense Realism, is evident in his daily practice (from about 1830 on) of recording the daily temperature, wind direction, and cloud cover in Princeton. See the introduction to C. Hodge 1994.

5. Hodge's version of "common sense realism" understood the trustworthiness of such truths to be of the same order as truths drawn from sense knowledge. The same grounds by which we trust the latter warrant trust in the former. This is the epistemological anchor of his approach to the relationship between science and religion.

6. Later, he maintains that such proofs are necessary because (1) even self-evident truths are denied, (2) in our current (corrupt) moral state we are particularly prone to deny the existence of a holy and just God, and (3) "efforts are constantly made to pervert or contradict the testimony of our nature to the existence and nature of God" (C. Hodge 1968, 1:234).

7. Later in the section he counterposes design to "chance or the action of mere physical law" (C. Hodge 1968, 1:230).

8. To be fair to Paley, he does not assert this. He argues toward the end of his book that natural theology proves the existence of a transcendent, intelligent mind that produces and supports the world. This then whets the appetite for the curious theist (confirmed in her belief) to turn to Revelation for those other particulars "which our researches cannot reach, respecting either the nature of this Being as the original cause of all things, or his character and designs as a moral governor; and not only so, but the more full confirmation of other particulars, of which, though they do not lie altogether beyond our reasonings and our probabilities, the certainty is by no means equal to the importance" (Paley 1802, 542).

9. Paley is not ready to accept that there are chance processes in nature in the most rigidly stochastic sense, arguing instead that chance processes can either be described in the Aristotelian sense as the intersection of two unrelated lines of causality, either of which is deterministic taken on its own, or be attributed to an epistemic lack on the part of the observer—ignorance of the laws involved or an inability to master the complexity of their interaction. That said, he argues that chance makes sense in a universe designed for our good. For example, it is a sign of a benevolent designer that by and large our time and circumstances of death are governed by chance, since knowing the fixed day and hour would lead to the same horror in those who approach it as a condemned prisoner the night before his execution, robbing them of a joie de vivre, much to be desired, that is born of ignorance of the day and the hour of one's demise (1802, 513–21). Paley's arguments are valid for any number of accounts of chance.

10. One could perhaps argue that this view allows incorporation of some of Paley's views on the role of chance in a designed universe as long as one also sees chance in Aristotelian terms or as a symptom of ignorance by the observer.

11. He names this position a doctrine of "continuous creation" and objects that it "effectually destroys all evidence of the existence of an external world" (C. Hodge 1968, 579; see also 592–95).

12. Recognizing that Hodge's solution, despite what Hodge himself averred, made any form of theistic evolution untenable, his student B. B. Warfield tried to probe a bit further to establish the independence of secondary causes, introducing a third category of "mediate creation" between creation ex nihilo (and miracles) and the (more or less unsupervised) operation of secondary causes (Warfield 2000).

13. I think this helps us understand why Hodge's most forceful arguments against chance derive not so much from a commitment to Paley, however he construed Paley's argument, but from appeal to universal internal convictions (the third source of "facts" besides scientific analysis of the world and theological interpretation of Scripture): "Every man can see that his life has been ordered by an intelligence and will not his own. His whole history has been determined by events over which he had no control, events often in themselves entirely fortuitous, so that he must either

assume that the most important events are determined by chance, or admit that providence extends to all events, even the most minute" (1968, 1:586).

He obviously took the former alternative to be unthinkable.

14. Zahm wrote, "As to natural selection, it labors under difficulties which are apparently even more serious [than those confronting inheritance by use or disuse], and to such an extent is this true, that it may well be questioned if there is a single pure Darwinian now living" (1896, 196). Elsewhere he writes of a "higher, a subtler, a more comprehensive teleology than the world has ever before known," due to the theory of evolution (376).

15. Zahm, following St. George Mivart, thought it enough to assert God's unmediated creation of each soul, leaving the creation of the body to evolutionary processes. The Holy Office did not conclude it was enough, and while his book did not suffer the fate of Copernicus's, he was required to withdraw it from publication.

16. Thus, a belated vindication of Zahm's view: Pius XII 1950, secs. 36–37.

17. E.g., "Darwin seemed to be standing before this choice; either a Creator or chance" (Schönborn 2007, 64).

18. Schönborn expresses his appreciation for this approach at several points (2007, 49, 65 in particular). Edwards (2010) offers an extended argument for and development of this approach vis-à-vis modern science.

19. This reading is suggested by following the parallel with Warfield's notion of mediate creation.

20. This announces an ancillary concern, also present in Hodge, for erecting a definite and impermeable threshold of human uniqueness, for the sake of the biblical assertion that humans are created "in the image of God."

21. Schönborn quotes an assertion to this effect from Will Provine (Schönborn 2007, 28) and elsewhere, on somewhat different issues, Julian Huxley (27) and Jacques Monod (112–13), providing names for the "Neo-Darwinian ideology" he is contesting, while also (although not always consistently) distinguishing it from neo-Darwinian *science*.

22. And, as noted above, he incorporates it in a way that belies Paley's openness to some role for chance in a designed universe, expressed toward the end of the book.

23. While Schönborn does not belong among the "common sense realists," as does Hodge, it is remarkable to see his evocation of its mode of argument: "It is entirely rational to assume that there is a significance in the development of nature, even though the methods of natural science quite require this to be set in brackets. *Yet my common sense cannot be excluded by scientific method.* Reason tells me that there is order and a plan, meaning and purpose, that a clock has not come into being by chance, and far less still the living organism of a plant, an animal, or indeed a human being" (2007, 30–31, emphasis added).

24. It is thus interesting to note that the fragile linkage between belief in evolutionary mechanisms and in social progress that Michael Ruse notes (see Ruse, this volume, chap. 5) can also be found, *mutatis mutandis*, when it comes to the relationship

between a belief in providential "mechanisms" operating in natural and human history and a belief in progress. Augustine denied any such linkages; optimistic liberal theologians of the nineteenth century such as Albrecht Ritschl argued that they were self-evident; Karl Barth sided with Augustine in fiercely rejecting it. The thought of a figure such as Teilhard de Chardin spans (and is perhaps fragmented by) both of these fault lines, as do, perhaps, the theological syntheses of those who follow him.

CHAPTER 5

Does Darwinian Evolution Mean We Are Here by Chance?

Michael Ruse

Never use the word higher & lower—use more complicated.
—Scribbled by Charles Darwin in the margin of
his copy of Robert Chambers's *Vestiges*

EVOLUTIONARY THINKING BEFORE CHARLES DARWIN

Those of us who were raised in Christian homes (whether or not believers now) and who work on the history of evolutionary thinking are often struck—a turn of the head, a familiar phrase, a distinctive emotion—by how the two worldviews are so different and yet so much the same. It is as if the latter is the child of the former, each very different people and yet with uncanny and unsettling resemblances. Perhaps a better metaphor is that of the German *Doppelgänger*, although which is the good twin and which the evil I will leave for others to decide. Actually, if you look at the history of evolutionary thinking, going back to the beginnings in the eighteenth century, the time of the Enlightenment, this eerie echoing or mirroring is no great surprise (Ruse 1996, 2005). In many respects, evolutionary thinking came into being as a forward-looking alternative to a religion that many thought was near exhaustion. Both Christianity and evolution (back then it was usually called "development") tell of origins. Both Christianity and evolution explain change through time. And most importantly, both Christianity and evolution privilege our own species, *Homo sapiens*. It may not be quite all about us, but we are the center of the story (Ruse 2012). Museums are particularly egregious sinners—if sin it be—in this regard. The wonderful exhibit at the Museé d'Histoire Naturelle in Paris starts with bacteria and ends with its visitors on television. Amusingly, the American Museum of Natural History in New York City has a display in the basement explicitly eschewing this vision of life's history and has a display upstairs of mammals ranged in order from the most primitive to humans!

Christianity is theistic; it believes in a God who is not only creator but who, if necessary, is prepared to intervene in His creation and who is above all the God of Providence. Without His help we humans are as nothing and salvation is impossible. That is why Jesus had to die on the cross. Evolution, at least back then, was deistic; it believed in a God who shaped the universe and set it all in motion and who now lets it unfurl without further intervention and who is above all the God of progress. We humans, unaided, can make for a better life, through education, through science, through politics and more. In the still-relevant words of the most distinguished historian of the concept of progress: "The idea of human Progress, then, is a theory which involves a synthesis of the past and a prophecy of the future. It is based on an interpretation of history, which regards men as slowly advancing—*pedetemtim progredietes*—in a definite and desirable direction, and infers that this progress will continue indefinitely" (Bury 1920, 5). For the Christian, humans are special because they are made in the image of God. We have intelligence and a moral sense, unlike other organisms, and this sets upon us certain obligations, which unfortunately too often prove to be too much. But we are loved by God and that is why He cares for us. For the evolutionist, humans are special because they are the end point to which evolution has always pointed and which finally, progressively, it has produced.[1] If it is obligations you seek, then it is to further social and moral progress and to keep the system moving forward.

Erasmus Darwin, the late-eighteenth-century physician, friend of industrialists, supporter of the American Revolution, and grandfather of Charles, makes these points in the verse that he churned out in volumes as he traveled from patient to patient.

> Organic Life beneath the shoreless waves
> Was born and nurs'd in Ocean's pearly caves;
> First forms minute, unseen by spheric glass,
> Move on the mud, or pierce the watery mass;
> These, as successive generations bloom,
> New powers acquire, and larger limbs assume;
> Whence countless groups of vegetation spring,
> And breathing realms of fin, and feet, and wing.

Thus the tall Oak, the giant of the wood,
Which bears Britannia's thunders on the flood;
The Whale, unmeasured monster of the main,
The lordly Lion, monarch of the plain,
The Eagle soaring in the realms of air,
Whose eye undazzled drinks the solar glare,
Imperious man, who rules the bestial crowd,
Of language, reason, and reflection proud,
With brow erect who scorns this earthy sod,
And styles himself the image of his God;
Arose from rudiments of form and sense,
An embryon point, or microscopic ens! (E. Darwin 1803, 1:26–28, lines 295–314)

Erasmus Darwin explicitly tied his biology to his philosophy. This idea of organic progressive evolution "is analogous to the improving excellence observable in every part of the creation; such as . . . the progressive increase of the wisdom and happiness of its inhabitants" (E. Darwin 1794, 1:509).

The first part of the nineteenth century tells a similar story. Evolutionary thinking and the social doctrine of progress were entwined. It is hardly too much to say that without the latter the former would never have come into being. People believed in or hoped for social progress. They happily read it into the organic world. In those days, there was not much solid empirical evidence for evolution. No less happily, then, and in a good circular fashion, they used the belief in organic progress as justification for their beliefs about social progress! Interestingly, when you dig down beneath the surface of the arguments used by those who really hated evolution, as often as not, you find the real worries are about progress and the way in which it pushes aside Christian beliefs about Providence. A nice example pro and con is given by the differences between the two great early-nineteenth-century French biologists Jean Baptiste Lamarck and Georges Cuvier. The former espoused a form of evolution, one in which humans came firmly at the top of the chain. He was also an ardent social progressionist, a friend of that group of forward-looking French thinkers called the "philosophes." This paid dividends during the Revolution, for he thrived and flourished—although he did find it politic to change his name from the Chevalier de la Marck, to plain *citoyen* Lamarck! Cuvier, on the other hand, argued strongly against evolution. Philosophically, as one deeply influenced by the final-cause thinking of Kant and Aristotle, Cuvier—through his doctrine of the "conditions of existence"—could not see how undirected law could produce the intricate organization of functioning organisms. To this, he brought empirical evidence against evolution, for instance that the mummified forms of animals brought from Egypt by conquering French forces were identical to today's forms and hence disproved the hypothesis of change. Absolutely crucial also in Cuvier's opposition was that he was a senior civil servant, both under Napoleon and at the time of the Restoration, and as a good servant of the state hated unrest and dangerous philosophies. Born in a border province between Germany and France, Cuvier was a Protestant, but the literal reading of the Bible was never part of his motivating philosophy. Much more significant was that, although in some sense he could accept the idea of scientific progress—after all, his own work pointed to that!—he hated doctrines of social progress and the unrest to which he thought they led, and hence he stood with religion against them (Ruse 1999).

In the middle of the nineteenth century, we find the same linking of evolution and progress and the elevation of the status of humans. Robert Chambers, publisher and well-known writer for the public, penned (anonymously) an evolutionary tract, *Vestiges of the Natural History of Creation*. He was explicit: "A progression resembling development may be traced in human nature, both in the individual and in large groups of men. . . . Now all of this is in conformity with what we have seen of the progress of organic creation. It seems but the minute hand of a watch, of which the hour hand is the transition from species to species. Knowing what we do of that latter transition, the possibility of a decided and general retrogression of the highest species towards a meaner type is scarce admissible, but a forward movement seems anything but unlikely" (Chambers 1846, 400–402).

Humans were special. Then came Charles Darwin, who spoiled everything.

CHARLES DARWIN

Let's start positively with what Charles Darwin accomplished. In his *On the Origin of Species*, published in 1859, he set out to do two things. First, to establish the fact of evolution—all organisms came by a developmental process from other forms, probably just one or a few original organisms a long time ago, by natural processes. (He did not talk about the origin of life as such, but the implication certainly was that it too would have been natural.) Second, to propose a mechanism of change: natural selection. Invoking an argument by the political economist Thomas Robert Malthus, Darwin argued that because of limited supplies of food and space, there will be a struggle for existence (and even more importantly a struggle for reproduction):

A struggle for existence inevitably follows from the high rate at which all organic beings tend to increase. Every being, which during its natural lifetime produces several eggs or seeds, must suffer destruction during some period of its life, and during some season or occasional year, otherwise, on the principle of geometrical increase, its numbers would quickly become so inordinately great that no country could support the product. Hence, as more individuals are produced than can possibly survive, there must in every case be a struggle for existence, either one individual with another of the same species, or with the individuals of distinct species, or with the physical conditions of life. It is the doctrine of Malthus applied with manifold force to the whole animal and vegetable kingdoms; for in this case there can be no artificial increase of food, and no prudential restraint from marriage. Although some species may be now increasing, more or less rapidly, in numbers, all cannot do so, for the world would not hold them. (C. Darwin 1859, 63–64)

Darwin then pointed to the fact that whenever you have a population of organisms all in the same species, there are nevertheless differences between the members and that every now and then something new seems to emerge. This led Darwin to speculate that, in the struggle, some types or forms are likely to prove more successful than others, simply because these types or forms will help their possessors against others. Given enough time, these types will spread through the group and eventually there will be full-blooded change:

Can the principle of selection, which we have seen is so potent in the hands of man, apply in nature? I think we shall see that it can act most effectually. Let it be borne in mind in what an endless number of strange peculiarities our domestic productions, and, in a lesser degree, those under nature, vary; and how strong the hereditary tendency is. Under domestication, it may be truly said that the whole organisation becomes in some degree plastic. Let it be borne in mind how infinitely complex and close-fitting are the mutual relations of all organic beings to each other and to their physical conditions of life. Can it, then, be thought improbable, seeing that variations useful to man have undoubtedly occurred, that other variations useful in some way to each being in the great and complex battle of life, should sometimes occur in the course of thousands of generations? If such do occur, can we doubt (remembering that many more individuals are born than can possibly survive) that individuals having any advantage, however slight, over others, would have the best chance of surviving and of procreating their kind? On the other hand, we may feel sure that any variation in the least degree injurious would be rigidly destroyed. This preservation of favourable variations and the rejection of injurious variations, I call Natural Selection. (C. Darwin 1859, 80–81)

Note that Darwin's natural selection does not just bring about change; it brings about change of a particular kind. Organisms are adapted—they have adaptations, features that aid them in the struggle to survive and reproduce. Eyes, noses, teeth, hands, penises, vaginas, bark, leaves, flowers, seeds, and more—these are things that are "as if" designed, that is to say that they are put together in order to help their possessors.

In the *Origin*, Darwin said very little about humans. This was deliberate. He wanted to get the case for the basic theory out and on the table before it got swamped (as he knew it would be and as indeed happened) by the clamor over our own species. But he was no coward. He did not want to conceal his personal conviction that we are part of the natural world and evolved by the same means as all other organisms. At the end of the book, in what must be the most understated paragraph of the nineteenth century, he wrote: "In the distant future I see open fields for far more important researches. Psychology will be based on a new foundation, that of the necessary acquirement of each mental power and capacity by gradation. Light will be thrown on the origin of man and his history" (C. Darwin 1859, 488).

As is well known, Darwin had formulated his basic ideas some twenty years before he broke into print on them. From the first, he had been stone-cold certain about our natural status—most probably because of his firsthand experience with the natives at the bottom of South America (the Tierra del Fuegians) and his conviction that even the most civilized of us is but a hair's breadth from the apes. Indeed, the first real indication in his private notebooks that he had hit upon natural selection was a passage applying it to humans—and not just to humans generally but to human intelligence. "An habitual action must some way affect the brain in a manner which can be transmitted.—this is analogous to a blacksmith having children with strong arms.—The other principle of those children, which *chance?* produced with strong arms, outliving the weaker ones, may be applicable to the formation of instincts, independently of habits" (C. Darwin 1838e, N 42). (Note incidentally the reference to the supposed inheritance of acquired characteristics, what today is known as "Lamarckism," although it was in fact but a minor part of Lamarck's theory and not original with him. Darwin always accepted this as a secondary mechanism.)

After the *Origin* was published, Darwin set about writing works to deal at length in turn with each of his chapters, although he tended to be distracted by other beckoning projects. He wrote, for instance, a little book on orchids. As it happens, this work elaborates on how Darwin thought that the chance variations that make the building blocks of evolution can be taken and used by natural selection, although it probably says little directly about how he thought humans might have evolved (Beatty 2006a). I doubt, in fact, that Darwin would have turned full-time to humankind and thought hard about our evolution had the man who co-discovered natural selection (and whose work prompted Darwin to write the *Origin*), Alfred Russel Wallace, not become a spiritualist and started arguing that human origins had to be nonnatural (Schwartz 1984; Ruse 1999).[2] Dropping everything, Darwin plunged into the topic of our ancestry, and the *Descent of Man and Selection in Relation to Sex* was published in 1871. Much of it is fairly standard and without great surprise, as Darwin argues for an African origin to our species and tries to show how both physically and mentally we have been shaped by natural selection. What is surprising, however, is that much of the book is given over to a discussion of Darwin's secondary mechanism of sexual selection, involving competition for mates. Darwin had always subscribed to this mechanism from virtually the first thoughts about natural selection, but until now (as in the *Origin*) it had been merely mentioned (C. Darwin 1859, 89). The reason for this newfound prominence was that Darwin intended to use sexual selection to counter Wallace, who argued that much about human nature—our intelligence and our hairlessness, to take two examples—cannot be explained by natural selection. Agreeing with Wallace that natural selection cannot always do the job but disagreeing with Wallace about the need to seek out a nonnatural cause, Darwin invoked sexual selection instead. Particularly, he thought it played a major role in racial differences.

> Kafirs, who differ much from negroes, "the skin, except among the tribes near Delagoa Bay, is not usually black, the prevailing colour being a mixture of black and red, the most common shade being chocolate. Dark complexions, as being most common are naturally held in the highest esteem. To be told that he is light-coloured, or like a white man, would be deemed a very poor compliment by a Kafir. I have heard of one unfortunate man who was so very fair that no girl would marry

him." One of the titles of the Zulu king is "You who are black." Mr. Galton, in speaking to me about the natives of S. Africa, remarked that their
ideas of beauty seem very different from ours; for in one tribe two slim,
slight, and pretty girls were not admired by the natives.

Turning to other quarters of the world; in Java, a yellow, not a white
girl, is considered, according to Madame Pfeiffer, a beauty. A man of
Cochin-China "spoke with contempt of the wife of the English Ambas-
sador, that she had white teeth like a dog, and a rosy colour like that of
potato-flowers." We have seen that the Chinese dislike our white skin,
and that the N. Americans admire "a tawny hide." In S. America, the
Yura-caras, who inhabit the wooded, damp slopes of the eastern Cordil-
lera, are remarkably pale-coloured, as their name in their own language
expresses; nevertheless they consider European women as very inferior
to their own. (C. Darwin 1871, 2:347)[3]

Passages like this jolt us back to our main interest, the possible direction of
evolutionary change. A couple of points are almost self-evident. First, Darwin
believed absolutely and completely in social progress. Apart from the Darwin
family tradition going back to Erasmus, Charles Darwin and his wife (his first
cousin) were children of the Industrial Revolution. Their shared grandfather
was Josiah Wedgwood the potter, a man who made huge amounts of money
out of the factory system and whose wealth carried on to the third generation, freeing Darwin from the necessity of ever making a living for himself.
(Making a salaried living, that is. He was incredibly hard-working and very
much respected as a professional scientist.) There was no doubt about social
progress and the desirability of the actual form and course it had taken. "In
all civilised countries man accumulates property and bequeaths it to his children. So that the children in the same country do not by any means start fair
in the race for success. But this is far from an unmixed evil; for without the
accumulation of capital the arts could not progress; and it is chiefly through
their power that the civilised races have extended, and are now everywhere
extending, their range, so as to take the place of the lower races" (C. Darwin
1871, 1:169). Second, Darwin believed absolutely and completely in biological
progress, leading up to humankind. We have won the race. "From the war of
nature, from famine and death, the most exalted object which we are capable
of conceiving, namely, the production of the higher animals, directly follows.
There is a grandeur in this view of life, with its several powers, having been
originally breathed into a few forms or into one; and that, whilst this planet

has gone cycling on according to the fixed law of gravity, from so simple a beginning endless forms most beautiful and most wonderful have been, and are being, evolved" (1859, 490).

More than this, as the condescending passage about the differences between races shows only too clearly, certain parts of the species merit bigger prizes than the rest. Europeans are above other races, and the denizens of a small island off the mainland are the supreme winners. And, incidentally, we are not now referring to the Irish. In the *Descent*, Darwin worried that the Irish had bigger families than the Scots and hence were beating them in the struggle for life. But then he consoled himself that although the Irish had big families, they did not care for them very well, whereas the Scots did a grand job and surely on balance beat out the Irish (C. Darwin 1871, 1:174–75). All of this not much more than twenty years after the Great Famine.

NATURAL SELECTION AND BIOLOGICAL PROGRESS

So how then did Darwin spoil things? At one level, he certainly showed that the link between evolution and beliefs in social progress can be broken. Before the *Origin*, it seems fair to say, evolutionary theorizing existed on the back of views about social progress. Those who endorsed evolutionary theorizing endorsed biological progress, and they did this because of their prior commitments to social progress. In the *Origin*, Darwin made the case for the fact of evolution without any recourse to commitments to progress in the social and intellectual world. He did it rather on the information from studies of instinct, from paleontology, from organic geographical distributions, from morphology, systematics, embryology, and more. Whatever his own beliefs, he showed it was possible to be an evolutionist without any beliefs in improvement.

Perhaps this is not spoiling things. Perhaps it is rather liberating. But at another level, Darwin's effect on the question was truly corrosive. This, as he saw himself, was because of the mechanism of natural selection. It is not, as later critics have sometimes claimed, a tautology—those who survive and reproduce are those who survive and reproduce. However, it is relativistic. That means that what may well lead to success in one situation may well not lead to success in another situation. And this applies to human beings and their intelligence. Having big brains requires lots of protein and hence probably, since big protein supplies generally come packaged in the form of animals, lots of effort to get it. It does not necessarily follow that the effort is worth it. If there is a lot of cheap fodder around like grass, you might be better off going vegetarian and spending your time grazing and munching. In the immortal words of the late paleontologist Jack Sepkoski: "I see intelligence as just one of a variety of adaptations among tetrapods for survival. Running fast in a herd while being as dumb as shit, I think, is a very good adaptation for survival" (Ruse 1996, 486). Darwin would have agreed with the sentiment, if not the language. Moreover, although he was pretty uncertain about the nature and causes of heredity, Darwin saw full well that new variations, the building blocks of evolution (what today we call "mutations") are random not in the sense of uncaused but in the sense of not appearing according to the needs of their possessors. No direction is going to come from them, as was supposed by those who, like Darwin's great American supporter Asa Gray, thought that it is here that God steps in and gives the evolutionary process a shove along the right path—that is, toward humans (Ruse 2003).

So there is no guaranteed progress to humans. But as you can garner from the quote given at the head of this essay, taken from the scribbling of Darwin on his copy of *Vestiges*, this did not mean that he gave up and went home. He thought he could get progress from natural selection. In the *Origin* he rather assumes that it will happen almost automatically. "The inhabitants of each successive period in the world's history have beaten their predecessors in the race for life, and are, in so far, higher in the scale of nature; and this may account for that vague yet ill-defined sentiment, felt by many palæontologists, that organisation on the whole has progressed" (C. Darwin 1859, 345). By the end of the book, it is clear that in the mind of Charles Darwin, the sentiment is not so very vague. "And as natural selection works solely by and for the good of each being, all corporeal and mental endowments will tend to progress towards perfection" (489). But then, through later editions of the *Origin*, Darwin started to see that more was needed. It was a matter first of definition—he needed something that humans would have in abundance, greater than other organisms—and then, second, a matter of seeing that selection could deliver the goods, if not absolutely then at least very probably.

As far as the first problem was concerned, relying on the best recent German biology, Darwin (writing in the third edition of the *Origin* in 1861) opted for some kind of organization and differentiation and specialization.

> Natural selection acts, as we have seen, exclusively by the preservation and accumulation of variations, which are beneficial under the organic and inorganic conditions of life to which each creature is at each successive period exposed. The ultimate result will be that each creature will tend to become more and more improved in relation to its conditions of life. This improvement will, I think, inevitably lead to the gradual

advancement of the organisation of the greater number of living beings throughout the world. But here we enter on a very intricate subject, for naturalists have not defined to each other's satisfaction what is meant by an advance in organisation. Amongst the vertebrata the degree of intellect and an approach in structure to man clearly come into play. It might be thought that the amount of change which the various parts and organs undergo in their development from the embryo to maturity would suffice as a standard of comparison; but there are cases, as with certain parasitic crustaceans, in which several parts of the structure become less perfect, so that the mature animal cannot be called higher than its larva. Von Baer's standard seems the most widely applicable and the best, namely, the amount of differentiation of the different parts (in the adult state, as I should be inclined to add) and their specialisation for different functions; or, as Milne Edwards would express it, the completeness of the division of physiological labour. (C. Darwin 1861, 133)

Then, with this criterion in hand, Darwin argued that overall selection would promote excellence, that is to say, the organisms that better show functioning differentiation or an efficient division of labor. "If we look at the differentiation and specialisation of the several organs of each being when adult (and this will include the advancement of the brain for intellectual purposes) as the best standard of highness of organisation, natural selection clearly leads towards highness; for all physiologists admit that the specialisation of organs, inasmuch as they perform in this state their functions better, is an advantage to each being; and hence the accumulation of variations tending towards specialisation is within the scope of natural selection" (C. Darwin 1861, 134)

ANTI-PROGRESS

Was Darwin just whistling in the wind?[4] Some thirty years ago, the late Stephen Jay Gould, paleontologist and popular science writer, began fulminating against biological progress. In earlier writings he had not been entirely unsympathetic to the idea. But then he swung around and castigated it bitterly. Sensitized by the controversy over human sociobiology—the attempt to apply Darwinian principles to human social behavior—Gould thought that, as an idea, biological progress was racist and sexist and much more. Particularly (and at an obvious personal level), he disliked the way progressivist ideas had been used to belittle the status and talents of Jews (Gould 1981). He spoke of biological progress as "a noxious, culturally embedded, untestable, non-operational, intractable idea that must be replaced if we wish to understand the patterns of history" (1988, 319). He argued that there is nothing inevitable about the emergence of humans. Jokingly referring to the asteroid that hit the Earth 65 million years ago and wiped out the dinosaurs, making possible the age of mammals, he wrote: "Since dinosaurs were not moving toward markedly larger brains, and since such a prospect may lie outside the capabilities of reptilian design . . . , we must assume that consciousness would not have evolved on our planet if a cosmic catastrophe had not claimed the dinosaurs as victims. In an entirely literal sense, we owe our existence, as large and reasoning mammals, to our lucky stars" (1989, 318).

In fact, as we shall see, Gould's thinking was rather more complex than this. Almost paradoxically, he was not that opposed to the nigh-inevitable appearance of humans, or at least humanoids (meaning humanlike beings), if not here on Earth then elsewhere in the universe, nor was he against our having some level of complexity above all others. What he really objected to was the idea—an idea that we have just seen in Darwin—that humans emerged because they were biologically better in life's struggles. In that way lies racism, and, as we have seen when looking at Darwin, Gould had a bit of a point.[5] Almost paradoxically, therefore, here we had someone who believed in social progress but who also believed that in order to achieve such social progress, it was necessary to deny biological progress, at least as brought about in the Darwinian struggle!

We shall return to Gould. For now, let us stay with those who put a more positive spin on matters having to do with selection and improvement.

ARMS RACES

Natural selection was never a great success in Darwin's lifetime, or indeed for many years after that. It was not until the 1930s that, with the development of an adequate theory of heredity—then Mendelian, now molecular—selection could really come into its own and be recognized for the power that it has, a recognition that is as strong today as it has ever been (Provine 1971). But from Darwin on, there has been a continuous thread of thinkers who have agreed with the founder that natural selection can bring about some kind of improvement, ending with humans, simply because the later forms have beaten out the earlier forms. Even back then we find the language of "race" being used, as one kind of organism competes against another, but in the twentieth century, with the development in military circles of thought about competition—specifically about "arms races"—we have a refinement of Darwin's thinking, with extensive use of the military metaphor as transferred to biology.

It was Julian Huxley, the biologist grandson of Darwin's great supporter Thomas Henry Huxley and older brother of the novelist Aldous Huxley, who first and most explicitly spelled out this kind of thinking. It comes in a little book he published in 1912. He gave a graphic description of an arms race, couched in terms of the then-state-of-the-art, naval, military technology. "The leaden plum-puddings were not unfairly matched against the wooden walls of Nelson's day." Now, however, "though our guns can hurl a third of a ton of sharp-nosed steel with dynamite entrails for a dozen miles, yet they are confronted with twelve-inch armor of backed and hardened steel, water-tight compartments, and targets moving thirty miles an hour. Each advance in attack has brought forth, as if by magic, a corresponding advance in defence." Likewise in nature, "if one species happens to vary in the direction of greater independence, the inter-related equilibrium is upset, and cannot be restored until a number of competing species have either given way to the increased pressure and become extinct, or else have answered pressure with pressure, and kept the first species in its place by themselves too discovering means of adding to their independence" (J. S. Huxley 1912, 115–16). Eventually, "it comes to pass that the continuous change which is passing through the organic world appears as a succession of phases of equilibrium, each one on a higher average plane of independence than the one before, and each inevitably calling up and giving place to one still higher" (J. S. Huxley 1912, 116).

More recently, it has been the well-known popular science writer Richard Dawkins who has declared for biological progress—"Directionalist common sense surely wins on the very long time scale: once there was only blue-green slime and now there are sharp-eyed metazoa" (Dawkins and Krebs 1979, 508)—and arms races are the key. Speaking in today's language, he points out that, more and more, today's arms races rely on computer technology rather than brute power, and—in the animal world—Dawkins translates this into bigger and bigger brains. There are winners and losers. Referring to a notion known as an animal's EQ, standing for "encephalization quotient" (Jerison 1973)—a kind of cross-species measure of IQ that factors out the amount of brain power needed simply to get an organism to function (whales require much bigger brains than shrews because they need more computing power to get their bigger bodies to function) and that then scales according to the surplus left over—Dawkins writes: "The fact that humans have an EQ of 7 and hippos an EQ of 0.3 may not literally mean that humans are 23 times as clever as hippos! But the EQ as measured is probably telling us *something* about how much 'computing power' an animal probably has in its head, over and above the irreducible amount of computing power needed for the routine running of its large or small body" (Dawkins 1986, 189). That "something" presumably is that which tells of the ability to write best-selling books against God, promoting atheism.

ECOLOGICAL NICHES

Are the Darwinians right in thinking that natural selection unaided leads to progress? Obviously, it is hard to say. We humans have emerged thanks to natural selection. Given that apparently we are the only beings like us ever to emerge, it is difficult to calculate how likely it was to happen. Arms races have their supporters, it is true. They also have their detractors, people who think their importance has been overrated. The classic case of an arms race is between chasing predator and fleeing prey. Apparently the fossil record does not always show that predator-prey interactions lead, as one might expect, to ever-faster animals (Bakker 1983). At best, the skeptic might return the Scottish verdict of "not proven." All of which prompts one to wonder if there are other approaches to the question of improvement. Other than arms races, might one expect natural selection to lead to some kind of progress, with the necessary emergence of humans or humanlike creatures?

Gould thought so. Think about ecological niches—such as the shelter provided by trees or the safety of burrows underground. One might say that thanks to natural selection, organisms are constantly looking and prodding to find new niches, and when they do, they enter and exploit them. The first animals to get up to and survive on dry land found a whole new area in which they and their descendants could live and thrive. Culture can be described as another class of niche that humans have found and entered. If not us, was it not predictable that someone on Earth would find and enter the niche? Why should this not happen repeatedly, if not here on Earth (because you might say that we have now blocked the entrance), then elsewhere in the universe? Note that selection here is doing the work, but it is not doing it by one organism beating out another and thus proving its superiority. The superiority, if such there be, comes from the niche itself, which confers it on its inhabitants. So there are no longer the obvious opportunities for racism and sexism.

In an essay on the search for extraterrestrial life, Gould wrote: "I can present a good argument from 'evolutionary theory' against the repetition of anything like a human body elsewhere; I cannot extend it to the general proposition that intelligence in some form might pervade the universe" (1985a, 407). He then went on to quote the leading twentieth-century evolutionist Theodosius Dobzhansky, writing in a textbook with other major evolutionists: "Granting that the possibility of obtaining a man-like creature is vanishingly small even given an astronomical number of attempts . . . there is still some small possibility that another intelligent species has arisen, one that is capable of achieving a technological civilization" (411). About this passage, Gould commented: "I am not convinced that the possibility is so small" (411). He then went on to give an argument that evolutionary convergence (where two different lines evolve essentially similar adaptations to survive and reproduce) suggests that even though major intelligence has arisen but once on this earth—one takes it that "major intelligence" includes all hominins and perhaps even the great apes (all hominids)—it is quite possible that elsewhere in the universe it has arisen quite independently. "Conscious intelligence has evolved only once on earth and presents no real prospect for reemergence should we choose to use our gift for destruction. But does intelligence lie within the class of phenomena too complex and historically conditioned for repetition? I do not think that its uniqueness on earth specifies such a conclusion. Perhaps, in another form on another world, intelligence would be as easy to evolve as flight on ours" (412).

This argument has been picked up (apparently independently) and promoted by the paleontologist Simon Conway Morris, who (as a Christian) is very keen to argue for the inevitability of the appearance of humans.[6] He argues that only certain areas of what we might call "morphological space" are welcoming to life forms (the center of the sun would not be, for instance) and that this constrains the course of evolution. Again and again, as Gould argued, organisms take the same route into a preexisting niche. The saber-toothed, tigerlike organisms are a nice example, where the North American placental mammals (real cats) were matched right down the line by South American marsupials (thylacosmilids). There existed a niche for organisms that were predators, with catlike abilities and shearinglike or stabbinglike weapons. Darwinian selection found more than one way to enter it—from the placental side and from the marsupial side. It was not a question of beating out others but of finding pathways that others had not found.

Conway Morris argues that, given the ubiquity of convergence, we must allow that the historical course of nature is not random but strongly selection-constrained along certain pathways and to certain destinations. Most particularly, some kind of intelligent being was bound to emerge. After all, our own very existence shows that a kind of cultural adaptive niche exists—a niche that prizes intelligence and social abilities.

If brains can get big independently and provide a neural machine capable of handling a highly complex environment, then perhaps there are other parallels, other convergences that drive some groups towards complexity. Could the story of sensory perception be one clue that, given time, evolution will inevitably lead not only to the emergence of such properties as intelligence, but also to other complexities, such as, say, agriculture and culture, that we tend to regard as the prerogative of the human? We may be unique, but paradoxically those properties that define our uniqueness can still be inherent in the evolutionary process. In other words, if we humans had not evolved then something more-or-less identical would have emerged sooner or later. (Conway Morris 2003, 196)

SPENCERIAN THEMES

Our first two approaches to trying to generate a natural explanation of biological progress have both made natural selection the prime mover. Are there other plausible options? What about complexity or some such thing, something that one associates with humankind, coming about by chance or essentially nonselective reasons? Interestingly, Darwin himself in his notebooks (twenty years before the *Origin*) toys with this kind of thinking.

The enormous *number* of animals in the world depends [on] their varied structure & complexity.—hence as the forms became complicated, they opened fresh means of adding to their complexity.—but yet there is no *necessary* tendency in the simple animals to become complicated although all perhaps will have done so from the new relations caused by the advancing complexity of others.—It may be said, why should there not be at any time as many species tending to dis-development (some probably always have done so, as the simplest fish), my answer is because, if we begin with the simplest forms & suppose them to have changed, their very changes . . . tend to give rise to others. (C. Darwin 1838c, E 95–96)

Darwin certainly thought of the complexity as, ultimately, adaptive. Immediately after the just-quoted passage, he added: "it is quite clear that a large part of the complexity of structure is adaptation" (C. Darwin 1838c, E 96–97). But even if you think that the complexity is adaptive, and that in the end the complexity leads to the ultimate adaptation—humanlike intelligence—you might still think that selection did not play a role or at least not a major role. This seems to have been the position of Darwin's fellow-Englishman contemporary, the polymath Herbert Spencer. Like Darwin in being influenced by Germanic thinking—although turning more to the romantic philosopher Friedrich Schelling than to the embryologist Karl Ernst von Baer—Spencer likewise adopted a criterion of progress that involved division and specialization, or, as he called it, a move from the homogeneous to the heterogeneous:

> Now we propose in the first place to show, that this law of organic progress is the law of all progress. Whether it be in the development of the Earth, in the development of Life upon its surface, in the development of Society, of Government, of Manufactures, of Commerce, of Language, Literature, Science, Art, this same evolution of the simple into the complex, through successive differentiations, holds throughout. From the earliest traceable cosmical changes down to the latest results of civilization, we shall find that the transformation of the homogeneous into the heterogeneous is that in which Progress essentially consists. (Spencer 1857, 446–47)

For Spencer, everything obeys this law. With respect to other animals, humans are more complex or heterogeneous; with respect to (what he would have thought of as) savages, Europeans more complex or heterogeneous; and with respect to the tongues of other peoples, the English language more complex or heterogeneous. Unlike Darwin, Spencer saw a certain inevitability to the evolutionary process and felt no need to fit his thinking to more selection-based notions like the division of labor. Causes just keep multiplying effects, and this leads to ever-greater complexity, and somehow value or worth gets pulled in along the way. This all seems to occur in waves. Something disturbs the natural balance, forces work to reachieve the balance, but when this occurs, we have moved a stage higher. An overall process of "dynamic equilibrium."

I do not want to postulate direct links between Herbert Spencer and Stephen Jay Gould, although it is tempting to speculate on the influences on someone whose main theory was called "punctuated equilibrium" (Ruse 2013). However, the later Gould did see the course of evolution as having distinctive Spencerian characteristics, with complexity emerging almost inevitably and with no selective guidance (Gould 1996). He thought that life is a bit asymmetrical. Necessarily, it started simple. Necessarily, it cannot get less simple. However, it can get more complex. Not through any guiding power, but because this is the way things are. It is rather like the old tale of the drunkard and the sidewalk. On one side, the sidewalk is bounded by a wall, and on the other side lies the gutter. It may take a long time, but in the end the drunkard will wind up in the gutter. This is not through any conscious choice, but because the drunkard can fall off the sidewalk and he cannot walk through the wall. His random staggering will eventually lead to the gutter. So it is with evolution. There is no progress in nature, but there is direction. Apparently this can end up with humanlike beings, although whether this is because of niche existence or complexification is not always easy to tell.

Explicitly Spencerian is the recent thinking of paleontologist Daniel McShea and philosopher Robert Brandon (2010).[7] They see a kind of non-Darwinian upward momentum to life's history. Introducing what they call the "zero-force evolutionary law" (ZFEL for short) they write: "In any evolutionary system in which there is variation and heredity, in the absence of natural selection, other forces, and constraints acting on diversity or complexity, diversity and complexity will increase on average" (McShea and Brandon 2010, 3). It seems that for them (as was the case for Spencer) things just naturally keep complexifying—one cause leads to several effects and these in turn multiply. They are not committed to the kind of surging view that characterizes dynamic equilibrium—although Gould's non-Darwinian punctuated equilibrium surely echoes—but they do see an upward drive as part of the ontology of the universe. And, while they stress that they are not into the adaptation business (complexity is complexity in its own right and could be nonadaptive) and while they are carefully silent on the question of human emergence—by cloaking themselves in the mantle of Spencer as well as by expressing a sympathy for such older mechanisms as orthogenesis, which saw a momentum in evolution and certainly thought it led to humans—the reader would not be unduly distorting were he or she to conclude that McShea and Brandon themselves would be happy if their thinking could be shown applicable to the human case.

CONCLUSION

Today's practicing evolutionary biologists, who are almost to a person Darwinian to a greater or lesser extent, tend to be a bit cagey on the overall course of evolution, whether it shows a trend and where humans fit into the picture. None would feel comfortable with saying that humans absolutely and necessarily had to emerge. Relativistic natural selection and unguided mutations put paid to that fantasy. I suspect, however, that few would think the emergence of humans was entirely a matter of chance. Or if they would, then like Gould they would not care to deny that somewhere, somehow in the vast universe, humanlike beings have emerged or will emerge. So to answer the question in my title: "Probably not, but I wouldn't want to bet on it unless I could afford to lose the wager." One presumes that this is at least part of the reason why the US government, in flush times, was prepared to support the search for extraterrestrial intelligence (SETI) but in recent more impoverished times has been considerably less forthcoming with support.

NOTES

1. I take it that this is why this essay ties in with and is relevant to the whole theme of this volume. If there is progress in evolution, with humans or humanlike beings destined to appear, then in some sense chance is being overridden. This is not to say that chance may not be involved, for instance in the appearance and nature of new variations, but that ultimately something—for the Darwinian, presumably natural selection—is going to work so that the expected or desired end is going to appear, no matter when or what the variations.

2. Embarrassingly groundless is the claim that it was the antislavery commitment of Darwin and his family that guided his thinking about human evolution. See Adrian Desmond and James Moore, *Darwin's Sacred Cause: How a Hatred of Slavery Shaped Darwin's Views on Human Evolution* (2009) for the pro side to this claim and the devastating review by Robert J. Richards (2009) for the con side to this claim.

3. In the first paragraph, Darwin is quoting from *Mungo Park's Travels in Africa*, published in 1816. In the second paragraph, from the third edition of J. C. Pritchard, *Researches into the Physical History of Mankind*, published between 1836 and 1847. In the *Descent of Man*, far more than in the *Origin of Species*, Darwin is given to quoting from the work of others.

4. I am focusing in this discussion on Darwin's claims about progress and more generally about direction in the history of organisms. I see this as part of a general question about Darwin's thinking on end-directed processes, what Aristotle called the problem of "final causes," language that Darwin used also and that today is often spoken of as "teleological." There is also the question of final cause in individual organisms, such as the purpose of the hand or the eye, the problem known to today's evolutionists as the problem of "adaptation." This was the problem for Aristotle, who, like most Greeks, did not have a sense of history that would have made issues of progress very meaningful. How far Darwin's thinking on the final-cause problem made him a full-blown teleologist is much debated. Ghiselin (1994) thinks not and Lennox (1993, 1994) thinks he was in a sense a teleologist. I agree with Lennox (Ruse 1996, 2003), as does Beatty (2006a). We don't think he was an Intelligent Designer, at least not in his science, although when writing the *Origin*, Darwin definitely believed

God designed at a distance (by the time of the *Descent* he was an agnostic), and he did not believe in vital forces directing nature. But he certainly viewed organisms through the metaphor of design and understood what was happening in terms of what he thought would be future effects.

5. Ameliorating Darwin's position a little bit, in the *Descent* Darwin did argue that a major reason why Europeans usually beat natives in their own lands is not because of superior strength or intelligence but because they bring diseases that the natives cannot withstand. Although it did not much if at all influence his science, Desmond and Moore (2009) are completely right in pointing out that Darwin, like his family, especially the Wedgwoods, was always strongly against slavery and, atypically for the British—the cotton industry was heavily dependent on the plantations in the American South—was a fervent supporter of the North in the Civil War.

6. I have had people (the Chicago evolutionist Jerry Coyne for one) simply refuse to accept that Gould anticipated Conway Morris, given that the latter has spent so much time criticizing Gould's (1989) view of the Burgess Shale and, even more, Gould's view of Conway Morris on the Burgess Shale (Conway Morris 1998). But he did. As I have intimated earlier, Gould on progress is complex; indeed, an unkind commentator might say "confused." Based on a twenty-year friendship, I think Gould always had a hankering for biological progress. Most evolutionists do, whatever they may say publicly (Ruse 1996). But as I have explained, he was wary of the idea because of what he took to be potential racial extrapolations. Reading Darwin and many others, Gould had a point.

7. McShea (2005) acknowledges a debt to Gould.

* 2 *

Chance in the Processes of Evolution

CHAPTER 6

The Reference Class Problem
in Evolutionary Biology:
Distinguishing Selection from Drift

Michael Strevens

1. THE REFERENCE CLASS PROBLEM

"The reference class problem is your problem too," writes Alan Hájek in a penetrating paper of the same name—regardless, he argues, of your interpretation of probability (2007). He is right, and the problem is a particular difficulty for the foundations of evolutionary biology, as it threatens to undermine the distinction between natural selection and drift. This chapter outlines the problem and presents a solution. At the chapter's end, many important questions about the proper definition of drift will remain unresolved, but I hope to have provided at least a robust criterion for individuating the probabilities that so many writers invoke in attempting to give those questions answers.

In its broadest formulation, the problem of the reference class as follows: in determining the value of the physical probability of an outcome, what factors should be taken into account? And why? The problem is especially salient in those domains of scientific theory and everyday thinking in which the probabilities customarily assigned to certain outcomes take into account some causal factors relevant to the outcomes but ignore others. In determining the probability of heads on a coin toss, for example, we take into account the physical symmetry of the coin, ascribing a probability of one-half to heads on a toss of a fair coin but not on a toss of an unbalanced coin. Yet we ignore other initial conditions of the toss, such as its initial translational and angular velocity: since these are more or less sufficient to determine the outcome, were we to take them into account we would say that the probability of heads is always either one or zero, depending on the initial conditions. (There is good reason to think that coin tosses are effectively deterministic.)

We make a distinction, then, between two kinds of physical quantities relevant to the outcome in question. Some are taken into account in determining physical probabilities. Call them *parameters*; as the parameters change, then, the probabilities change. The rest—call them *variables*—are not taken into account; differences in their values make no difference to the probability assigned to an outcome. It is as though all processes with the same values for their parameters are considered as a unified class, in spite of the differences in the values of their variables—differences that have just as big an impact on the outcome as differences in parameters. In the frequentist tradition, such classes are called *reference classes*; the problem of the reference class, then, is the problem of deciding how to individuate classes, or in other words (when dealing with causal processes), the problem of deciding whether a causal factor is to be treated as a variable or a parameter.

The physical probabilities at the heart of both classical statistical mechanics and evolutionary biology are like the probability of heads in this crucial respect: some causally relevant quantities are treated as parameters and some as variables. Some causally relevant quantities, then, are ignored when fixing probabilities. On what basis? If the question matters to you at all, then the reference class problem is your problem too.

The case of evolutionary biology is especially interesting for reference class theorists because it presents good reason to think that the distinction between parameters and variables has objective significance. It is not just a matter of what we know, or of how we have decided to organize inquiry; rather, it reflects deep differences in the way the world works. I say this because it seems, if the majority of philosophers of biology writing on the topic are correct, that the distinction between natural selection and drift is only as real as the distinction between parameters and variables. But biologists regard selection and drift as qualitatively quite different kinds of explanation for evolutionary change. Such an attitude can be maintained only if the reference class problem can be solved—for biological probabilities at least.

Two caveats before I go on to explain the problem and to develop a solution. First, I will not supply my own characterization of drift or seek to judge between characterizations that others have offered. Rather, I hope to provide something that (almost) all participants in these debates are in need of—something that is foundational, and therefore crucial to the debates, but that does not dictate any particular solution or support any particular side.

Among the many questions on which I will do my best to remain neutral are the following: Are selection and drift distinct kinds of processes, or are they different aspects of the same process (Millstein 2002; Matthen and Ariew

2002; Brandon 2005)? If they are aspects of the same process, is that process causal (M. J. S. Hodge 1987; Walsh, Lewens, and Ariew 2002)? Is drift constitutive of population equilibrium, or is it a force disturbing equilibrium (Brandon 2006)? Can the contribution of drift to a change in relative frequency be quantified precisely (Beatty 1984; Millstein 2008)? Indeed, this chapter does not even resolve all issues that might be grouped under the heading of "finding the right reference class" (I am thinking of the questions raised by Abrams [2009] in particular).

It is not that the problem of the reference class is completely independent of these other problems. Some arguments for the "statisticalist" position that neither selection nor drift is a causal process invoke the presumed arbitrariness of reference classes (Walsh 2007; Matthen 2009). But I will make no attempt here to untangle the often very complex connections involved.

The second caveat is that my resources for dealing with the reference class problem are drawn from the general philosophy of science: I will argue that the parameter-variable distinction can be understood in the same way in biological systems as in simple gambling devices, in the systems treated by statistical physics, and so on. This is not an essay, then, for those who say, "Philosophy of biology solutions for philosophy of biology problems" (as one anonymous referee made very clear). At the same time, I want to emphasize that what I say here is only a small part of a complete treatment of the nature of drift—a complete treatment that most certainly will derive much of its content from what is special and distinctive about the philosophy of biology.

2. PARAMETERS, VARIABLES, AND DRIFT

Many philosophers have recognized that the problem of the reference class poses a prima facie challenge to the selection-drift distinction and thus arguably to the thesis that fitness and selection are real and explanatorily potent: Sober (1984, 129–34), Rosenberg (1994), Brandon (2005), and Matthen (2009), among others. Let me begin by presenting what I take to be the standard understanding of the challenge.

Begin with a paradigm example of drift: neutral evolution. Neutral evolution occurs when frequencies (of genes, traits, or whatever) change without selective pressure; although no particular gene or trait has a systematic advantage, their relative numbers change simply "by chance."

The clearest cases of neutral evolution are those in which variations in the structure of a gene have no phenotypic consequences. One variant may be reproduced at a greater rate than another, but this cannot be due to some feature of the variant that makes it more likely to be reproduced; the difference in outcomes is therefore due to "random genetic drift."

Perhaps, for example, the gene in question is wholly devoted to determining its organism's pattern of camouflage. Suppose that, because the gene's two variants are phenotypically equivalent, the organisms in which they appear have identical camouflage patterns (something of an idealization, to be sure). We may distinguish two classes of organisms, then, differing only with respect to which of the two genetic variants they possess (and ignoring the complications brought on by diploidy, gene linkage, and so on): call them α and β. There is variation within each class, but it is the same variation. Intuitively, then, the two classes should be, on the whole, equally susceptible to predation.[1] But it might be that over some period of interest, simply by chance, αs are eaten less frequently than βs. Then the proportion of αs, and so the proportion of the underlying genetic variant, will increase. This increase is due to drift.

If, by contrast, the α allele makes a difference to the camouflage pattern, and this difference results in a lesser susceptibility to predation, the decrease in the β population would be due to natural selection. These seem to be very different explanations of the relative increase in αs. The challenge to the philosopher of biology is to spell out the qualitative difference between the two explanations—or to argue that, contrary to appearances, there is no difference.

Traditionally, drift has been distinguished from natural selection using the figure of "sampling error." Fill an urn with balls, half red and half black. Conduct one hundred draws from the urn (replacing the drawn ball each time). You will very likely draw red about one-half of the time, reflecting the one-half probability of such a draw. But most likely, you will not draw red exactly fifty times. The difference between the actual frequency of red and the probability of red is called sampling error. The idea is then generalized: sampling error occurs when the frequency of an outcome differs from its probability.

An evolutionary outcome is (on the traditional conception) attributed to drift when it is due to sampling error. That such attributions can be made is clearest by far in cases of neutral evolution (hence my beginning with the neutral case). If all variants have the same probability of experiencing some evolutionarily significant outcome, such as successful mating or death, then any relative change in their numbers resulting from outcomes of that sort must be due to a deviation of frequencies from probabilities.

In the scenario sketched above, the probabilities of all evolutionarily significant outcomes are the same. In particular, the probabilities of predation, the kind of event responsible for the relative increase of αs, are the same. So that increase, due entirely to the deviation of predation frequencies from predation probabilities, is attributed to drift.

Enter the problem of the reference class. There is more than one way to determine predation probability, and different ways may yield different answers to the question of whether the probabilities of predation are the same or different for the two variants α and β, and so different answers to the question of whether α's relative success is due to drift or to selection.

To make the worry more precise, let me make an assumption for purely strategic reasons: I will suppose that the processes underlying evolution are fully deterministic. The precept of determinism makes it easier to press home a serious version of the reference class problem, so increasing the challenge of constructing a response. A solution to the problem developed under these conditions is, as a consequence, all the stronger.

Given determinism, an event in which a specimen of α is eaten but a nearby specimen of β is not must differ in some causally relevant way from an otherwise similar event in which it is the β that is devoured and the α that escapes to live, love, and leave its genes to the next generation. Perhaps two such events differ only in that the positions of the α and the β are switched, with the α occupying the fatally conspicuous spot in the first event and the β in the second.

Conditional on these and other details, the probability that the α is captured by the predator is much higher in the first kind of scenario than in the second, and vice versa for the β.[2] In other words, declare the details of position to be parameters rather than variables—adopt a very fine-grained individuation of predation probabilities—and the probabilities will vary a great deal from episode to episode.

In my evolutionary scenario, more βs are eaten than αs, presumably just because of unlucky positioning. This fact will be reflected in the fine-grained probabilities: the probability that a β will be eaten in any particular situation will be higher on average than the predation probability for an α.

But in that case, the relative increase in the number of αs caused by predation is not (principally) due to sampling error. It is not that αs and βs are equally likely to be eaten, but that simply by chance (that is, sampling error), more βs are eaten. Rather, βs are on the whole more likely to be eaten than αs, and this difference in probabilities is reflected in the outcomes. That is natural selection.

Use coarse-grained predation probabilities, by contrast, and, as I wrote above, the predation probabilities are equal and so the success of the αs is entirely due to sampling error. That is drift.

In short, whether the αs' increase in numbers should be explained by natural selection or by drift appears to depend on which causes of predation are treated as variables and which as parameters. Treat fine-grained matters of positioning as variables and you have drift; treat them as parameters and you have selection. It follows—so you might think—that there is no objective distinction between explanations from drift and explanations from selection. Yet many biologists, and philosophers of biology, feel the distinction in their bones. As Millstein (2002) points out, some of the most important debates in evolutionary biology since the modern synthesis turn upon it. We ought to look harder to see on what foundations it might rest.

Before beginning the search for an objective criterion to distinguish parameters and variables in the probabilities of evolutionary biology, however, I need to take a closer look at the existing literature on drift. Not every writer on drift characterizes it in the traditional manner reflected in my presentation above; some have proposed solutions to the reference class problem that ought to be considered; and some dispense with a probabilistic understanding of drift altogether.

3. THE PROCESS APPROACH TO DRIFT

To say, along with the textbooks and the traditionalists, that evolutionary change is due to drift rather than natural selection insofar as it constitutes sampling error, can seem rather unhelpful, even obscure. I toss a coin ten times and obtain six heads. If had obtained only five heads, the frequency of heads would have matched the probability of heads and so there would have been no sampling error. Is one of the six tosses that landed heads, then, responsible for the error? If so, which one? The question seems misconceived. There has been sampling error, but no discrete part of the process causes that error. To attribute an outcome to drift, then, cannot coherently be understood as attributing the outcome to some feature of the process that produced it—in which case drift cannot explain outcomes. What has gone wrong?

According to what I will call the process-driven account of drift, the problem is a definition of drift that focuses on outcomes—frequencies that diverge from probabilities—rather than the processes that produce the outcomes. (In discussing the process account, I do not assume that either the account or even its diagnosis of the problem is correct; I just want to cover all the bases, showing that almost every definition of drift has a reference class problem to wrangle about.)

The first step in the process account is to divide the causal totality driving evolutionary change in a population into discrete causal mechanisms, each potentially producing an evolutionarily significant outcome such as death. Some such mechanisms will be those that determine the consequences of encounters with predators; some will be those underlying other dangerous events such as electrical storms and forest fires; and so on. The division need not be completely determinate, just as the distinction between selection and drift might admit of some fuzziness. Mechanisms might run in parallel—there can be predation during storms. Some mechanisms, such as predation, will run to completion over short periods of time; some, such as gestation or drought, will unfold over months.

Once the division is made, the real business begins: classifying each mechanism as a discriminate or an indiscriminate sampler. The net evolutionary change in the population—the net change in the proportion of each variant—will be determined by the various outcomes of the many evolutionary episodes in which these mechanisms operate. Part of the net change will be explained by episodes of indiscriminate sampling, and part by episodes of discriminate sampling. The first part—the part due to indiscriminate sampling—is, according to the process-driven approach, explained by drift, while the second part is explained by natural selection. An explanatory attribution to drift just is, then, an explanatory attribution to indiscriminate sampling. (When all change is due to indiscriminate mechanisms, you have neutral evolution.)

Or at least, that is the basic idea. Many sophistications are possible. On Gildenhuys's (2009) account, for example, indiscriminateness of sampling is necessary but not sufficient for a mechanism to qualify as a drift-type explainer: the mechanism must also be "non-interactive" and "non-pervasive." I hope I will be forgiven for not going into further details about the taxonomy of process-driven accounts; there is just one feature of these accounts on which I think I need to focus, namely, the definition of indiscriminate sampling.

Traditionally, an evolutionary mechanism is said to sample indiscriminately if all variants are equally likely to experience each of its possible outcomes: "A sampling process is indiscriminate if and only if each entity in the pool to be sampled has an equal probability of being chosen" (Brandon 2005, 156). This is Gildenhuys's sense of indiscriminate sampling: "A cause [e.g., predation, lightning strike] is indiscriminate if its probability of influencing the reproduction of population members, in any given generation or time slice, is independent of the type of those members" (Gildenhuys 2009, 540), since probabilities that are independent of type are the same for all types, all other things being equal.

On this traditional conception of what constitutes indiscriminateness, the traditional problem of the reference class presented in section 1 plainly has its bite: probabilities that are equal when causally relevant factors such as position are treated as variables become unequal when the same factors are treated as parameters. Any notion of drift based on the equality of the probabilities of evolutionarily significant outcomes such as mating and death must provide or presume some solution, then, to the reference class problem.

There is, however, another notion of indiscriminate sampling in the literature on process accounts of drift. It is found in Beatty, who writes that indiscriminate sampling occurs when "physical differences . . . between the entities in question are irrelevant to whether or not they are sampled" (Beatty 1984, 189), and Millstein, who defines an indiscriminate sampling process as "a process where heritable physical differences between entities (e.g., organisms, gametes) are causally irrelevant to differences in reproductive success" (Millstein 2008, 353). I will call this the modern, as opposed to the traditional, view of indiscriminate sampling.

On the face of it, the modern view makes no appeal to probability at all. But I think that both Beatty and Millstein understand causal relevance in terms of probabilistic relevance: a factor is causally irrelevant to an outcome (equivalently, to an entity's being sampled) just in case its presence makes no (causal) difference to the outcome's probability. Does that mean that the reference class problem affects even the modern view of indiscriminateness?

Not necessarily. On one plausible understanding of difference-making, indiscriminateness in the modern sense does not require equality of probabilities, or indeed any fixed facts about probabilities at all. Consider, for example, the scenario introduced at the beginning of the chapter. Variants with the α camouflage gene get eaten less often than variants with the β camouflage gene, even though the two alleles are functionally identical and so result in exactly the same camouflage pattern. Specimens of β just happened to be in the wrong place at the wrong time more often than specimens of α. On a fine-grained individuation of the probabilities, details of positioning are treated as parameters, and so on average the αs' probability of death by predation is less then the βs' probability of the same. That makes sampling discriminate on the traditional view. On the modern view, however, it is indiscriminate: the probabilities may differ, but not in virtue of physical differences (let alone heritable physical differences) between the two variants.

More generally, however the probabilities are individuated, however you draw the boundary between variables and parameters: neither the genetic difference nor (arguably) any other intrinsic physical difference between the variants will affect the probability of death. So even without there being a fact of the matter about the probabilities, there can be a fact of the matter about whether physical differences affect the probabilities (by supervaluation, if you like). In this case, they have no such effect, and so the increase in αs can be attributed, objectively, to drift.

Other cases, however, pose greater problems for the modern view. Many organisms appear to make foraging decisions at random, as though they conduct a kind of internal mental coin toss to decide where to seek out food. Imagine a population of organisms that function in just this way: when confronted with the sort of choice that would defeat Buridan's ass, they toss some neurological equivalent of a coin, and then act accordingly: heads they go left; tails they go right. Or at least, some of them—the αs—work this way. In the others, the βs, it is the other way around: heads they go right; tails they go left. (Although this is a ridiculously simple toy example, it seems plausible that there are many real-life, though of course more complicated, analogues.)

In some situations, if you go left you get a nice lunch, whereas if you go right you are lunch; or in other words, the outcome of the mental coin toss can make a difference between life and death. Over a period of time, suppose, many of the βs get unlucky in just these circumstances and so get eaten: consulting their mental coin toss, they happen to go in precisely the wrong direction. The αs are on the whole more fortunate, and so the proportion of αs in the population increases. Selection or drift? On the process-driven approach, the question becomes: discriminate or indiscriminate sampling?

If we are allowed to construct probabilities in a fine-grained way, then we can take both the position of the hungry predator—whether it is on the left or on the right—and the outcome of an organism's mental coin toss to be parameters. Such probabilities will vary from encounter to encounter, but in general, βs will have a higher probability of being eaten than αs. Now the modern test for indiscriminateness: does the physical difference between the two variants make a difference to the probabilities? Clearly it does: had the βs followed the α rule, they would (usually) have fared much better.[3]

Relative to the fine-grained probabilities, then, the sampling counts as discriminate. But relative to more coarse-grained probabilities that treat predator position and so on as variables, the difference between the variants will make no difference to the probability of death (which will be equal for both). So the sampling will count as indiscriminate. In the absence of a solution to the reference class problem, the modern view therefore fails to deliver a determinate judgment about this scenario. But it is clearly a case of drift.

This is not an argument against the modern view of indiscriminate sampling or more generally against the process-driven account of drift. It is rather an argument that these approaches to thinking about drift, like all others that depend on probabilities, require some solution to the reference class problem, some criterion for distinguishing variables from parameters—a criterion that is as robust and objective as the distinction between selection and drift itself.

Do we need probabilities to draw a line between selection and drift? As intimated above, the modern notion of indiscriminateness might be rendered nonprobabilistic if it were supplied with a suitable nonprobabilistic criterion for causal relevance—a criterion that sometimes counts even decisive difference-makers like position relative to a predator as irrelevant. But I know of no such criterion.

I do know of a characterization of episodes of drift that plainly makes no appeal to probabilities, that suggested by Godfrey-Smith (2009). It is not clear to me, however, that Godfrey-Smith's criterion succeeds in distinguishing episodes of drift from episodes of selection. His criterion for a drift-type episode—roughly, that it involve external causes and that the outcome of the episode be sensitive to small changes in initial conditions—seems to apply equally well to certain selection-type episodes, such as those involving predation in which small differences in position make a difference to whether the prey is or is not spotted.

These brief and far from decisive considerations hardly close the door, however, on nonprobabilistic approaches; they should be regarded only as giving my own reasons for not pursuing this path.

4. SOLVING THE REFERENCE CLASS PROBLEM

If drift and the reference class cannot be separated, then what to do about the reference class?

You might opt for a brute force or trivializing solution: declare that every factor that plays a role in causing an outcome should be treated as a parameter for the purposes of determining the outcome's probability. In a deterministic world, then, all probabilities are zero or one. This is, in effect, Rosenberg's (1994) solution; he concludes that drift (or the concept thereof) is more an instrumental aid to theorizing than an objectively valid explanatory construct.

Another option is to relativize the distinction between variables and parameters, and so (possibly) between drift and selection. Matthen (2009) cogently explores two forms of relativization.

The first form is epistemic. Perhaps information about parameters is easy to come by, whereas information about variables is often unknown and expensive to acquire. (Certainly, that is true of the coin toss.) Reichenbach's solution to the reference class problem is to "use the narrowest . . . class for which reliable statistics are available" (Reichenbach 1948, 374); what makes a causal factor a parameter, on this approach, is that we are in a position to learn its value and then to use that value to reliably determine a probability (in Reichenbach's frequentist interpretation of probability, by consulting statistics). The resulting doctrine on drift implies, like Rosenberg's, that the concept of drift is useful only because we are imperfect knowers of causes.

Matthen's second form of relativization is institutional: on this approach, the distinction between variables and parameters is dictated by a theoretical framework, with different theories or domains prescribing different distinctions. It is part of the rulebook for doing statistical physics, for example, that you take into account the temperature of a gas but not the velocities of particular gas molecules. It is part of the rulebook for doing the science of games of chance that you take into account a coin's physical symmetry but not its initial spin speed. And so on.

One rationale for establishing demarcations of this sort is epistemic; the epistemic and institutional forms of relativization are not, then, mutually exclusive. Another rationale, which presents a genuine alternative to the epistemic approach, is the efficient organization of research: just as a factory assembly line assigns different squads of workers to different tasks, so science tells evolutionary biologists to pay attention to some variables but not to others—with responsibility for the others being assigned to lower-level sciences such as physiology, molecular biology, and physics. What count as variables in one science, then, might be perfectly knowable, but pursuit of such knowledge is for reasons of economy assigned to a different, complementary branch of inquiry. Whereas on Reichenbach's epistemic approach, the parameter-variable distinction will dance around as our epistemic situation changes, on the economic approach it is sociologically fixed: not immovable, then, but slow to change.

Matthen argues that physical probabilities built on a parameter-variable distinction dictated by a socially mandated division of labor are genuinely explanatory.[4] Thus drift is genuinely explanatory—though this explanatoriness must surely be relativized to the sanctioning institutional structure.

The importance and (relativized) explanatory significance of divisions of cognitive labor are important themes in my own work—see Strevens (2008) and especially Strevens (forthcoming). Nevertheless, I will not take the institutional approach in the present chapter. Something better is possible: an entirely nonrelativistic conception of drift's explanatory power.[5]

The objective explanatory validity of drift must rest, I believe, on an unrelativized, objective criterion for marking the boundary between parameters and variables. Seeking to provide an objective foundation for the probabilities in his statistical-relevance account of explanation, Salmon (1970) developed just such a criterion. More recently, Brandon (2005) has suggested that Salmon's criterion can be put to use to ground the objective explanatory status of drift.

Salmon's proposal turns on what he called the homogeneity of reference classes. Formulated in terms of the parameter-variable distinction, the homogeneity criterion stipulates that a causal factor should be counted as a variable rather than a parameter just in case conditionalizing on the value of the variable does not affect the probability of the outcome. Mechanisms that differ only with respect to quantities satisfying this condition form what Salmon calls a homogeneous class (with respect to the outcome in question).[6]

The homogeneity criterion is suitable for some purposes, but it does not draw the line between parameters and variables in the places needed to make sense of biologists' judgments about what is selection and what is drift. The problem is that any causal factor that makes a difference to whether an outcome occurs will count, according to the homogeneity criterion, as a parameter for the purposes of determining the probability of that outcome. But biologists treat many such causal difference-makers as variables.

Consider, for example, the scenario from section 2 in which variants α and β have identical camouflage patterns, but in which β, over some period of interest, suffers relative to α because its specimens more often, simply by chance, find themselves in the "locus of doom," where they will be picked out and picked off by a predator. We want to count this as a case of drift, which means—on the traditional approach—assigning to both variants equal probabilities of death by predation.[7] The homogeneity criterion does not oblige. It requires that a specimen's being located within the locus of doom, since it makes a big difference to the probability of death, be treated as a parameter. The probability of death must take position relative to predators into account, then, and so will vary from specimen to specimen (and from moment to moment) and will be higher on average for βs. In short, the homogeneity criterion, applied to this scenario, judges that the proportion of βs decreases not because of drift but because of selection against bad timing and unhappy placement.

Indeed, in a deterministic world, the probabilities delivered by the homogeneity criterion are invariably equal to zero or one; it finds stochasticity only where there is indeterminism. A parameter-variable distinction based on homogeneity, then, is none other than Rosenberg's trivializing distinction. Some other approach to the reference class problem is required.

We want a parameter-variable distinction that provides probabilities suitable for objectively valid explanation. I propose that we work backward from explanatory concerns: what distinguishes parameters and variables, I suggest, is that parameters play an explanatory role that variables do not, somehow connected to their counting as probabilistically relevant to evolutionary outcomes in a way that variables are not. This explanationist approach promises not only to draw a line between parameters and variables, but also to tell us why we ought to care about that line: it demarcates factors that are explanatorily relevant to evolution, neutral or selective, from those that are not.

What is explanatory relevance? According to Salmon (1984), Woodward (2003), and many other writers, it is causal relevance. On such a view, the explanationist approach to the parameter-variable distinction seems to be entirely unhelpful. Initial conditions such as position, which I have taken to be paradigms of the sort of quantities that we want to count as variables rather than parameters, are, as I have emphasized throughout, highly causally relevant to the outcomes in question: an organism's position may play as important a causal role as any fact about it and its situation in determining whether or not it gets eaten. Surely, then, position is as good an explainer as camouflage or predator sensory prowess or any other paradigmatic parameter?

If it is individual outcomes that are to be explained, this is true: a concern with the explanation of some particular death provides no reason to favor, say, facts about camouflage patterns over facts about position. But what matters in evolutionary explanation are the frequencies of outcomes, not the outcomes themselves. The evolutionary biologist cares, and attempts to explain, why the relative frequency of αs has increased at the expense of the βs (or why the frequencies of some larger range of variants have changed) but does not care at all which particular αs and βs reproduced and which did not.

Some initial conditions that are causally relevant to individual outcomes are explanatorily irrelevant to frequencies because they do not make a difference to frequencies, I will argue in section 5—and it is precisely what biologists typically take to be the variables that are irrelevant and the parameters that are relevant. The probabilities of evolutionary models are constructed, then, to take into consideration just what matters in explaining frequencies, and so just what matters in accounting for the standard explananda of evolutionary theory—namely, relative changes in the numbers of genes, traits, organisms, and other biological protagonists.[8] The distinction between selection and drift mirrors this explanatory divide.

5. THE PROBABILITIES THAT EXPLAIN FREQUENCIES

5.1. A Wheel of Fortune

Simple stochastic models such as the drawing of balls from an urn have often served as toy examples for thinking about drift. Let me follow suit.

Begin with an especially simple probabilistic setup that has long been used to understand physical probabilities in deterministic systems, the wheel of fortune (Poincaré 1896). The wheel is a disc with alternately red and black sections, mounted on an axis around which it turns freely. In addition to the wheel itself, there is a fixed pointer. To conduct a trial on the device, the wheel is given a vigorous spin. Eventually it comes to rest; the outcome of the trial, *red* or *black*, is then determined by the color of the section indicated by the pointer.

Such a wheel might have 100 sections, 50 red and 50 black, all of equal size. Or it might have the same 100 sections but with each of the black sections slightly wider than the red sections, so that the wheel is only 40 percent red.

Spin the wheel many times, and as everyone knows, the frequency of *red* in the resulting sequence of outcomes will tend to equal the proportion of the wheel that is painted red: you will get around 50 percent *red* from the first wheel described above and about 40 percent *red* from the second wheel. Why? A schematic answer: First, because of the wheel's construction, the physical probability of obtaining *red* on any given spin is equal to the proportion of red. Second, because of the law of large numbers, the frequency of *red* will, over many spins, tend to equal the probability of *red*. What explains the frequency, then, is whatever explains the probability.

Now intuitively, what explains the probability are properties of the wheel such as its characteristic paint scheme, and in particular the ratio of the width of the red sections to the width of the black sections; the fact that the wheel rotates smoothly around its axis; and the fact that the wheel's spin speeds are sufficiently fast and variable. As explainers of the probability, these are also explainers of the frequency.

Certain facts that are causally relevant to individual outcomes—facts that explain why some particular spin yielded *red* rather than *black*—do not help to explain the probability: the speed of any particular spin, the position of the pointer (provided that it has some position or other), the diameter of the wheel. The position of the pointer, for example, makes a difference to whether any particular spin gives you *red* or *black* but makes no difference to the probability of *red*: put the pointer anywhere you like (or even vary its position between spins) and the probability of *red* remains the same, equal to the ratio of red to black. Because these things do not explain the probability of *red*, they do not explain the frequency of *red*—in spite of the fact that they help to cause and therefore to explain the individual occurrences of *red* and *black* that jointly determine the frequency.

The same line of thought applies to the biological case. Imagine two variants α and β, differing perhaps only in their camouflage, with α being slightly less conspicuous to predators. Their population dynamics may be modeled as follows. Each variant has its own wheel of fortune. Once a month each organism spins the applicable wheel. If the outcome is *black*, it survives predation for that month. If the outcome is *red*, it dies. Because α has the more effective camouflage scheme, it has a little more black and a little less red on its wheel than β. Over time, then, it will tend to suffer less from predation, and—all other things being equal—it will more likely than not take over the population. (Ignore drift for now.)

Suppose that this is indeed what happens: α goes to fixation at the expense of β. The fixation of α, and the extinction of β, is of course to be explained by selection, which means that it is to be explained by a difference in the frequency of (in this case) predation, which is explained in turn by a corresponding difference in the probabilities of predation: β's more bloody wheel renders it more likely to be picked off. What explains selection, then, are those differences between α and β that contribute to the probability difference. In the model, the relevant difference is the proportion of red on the wheel; pointer position and the magnitudes of individual spins are by contrast probabilistically irrelevant. In the real biological world, the relevant difference is something about the camouflage scheme; the locations of particular specimens during predator-prey encounters and various fixed features of the environment (the analogues of pointer position) are probabilistically irrelevant.

5.2. The Skeptical Response

I have elicited your intuitions about what is relevant to the probability, and so to the matching frequency, of *red* on the wheel of fortune, but I have not given those intuitions an objective vindication. Counting the red to black ratio but not the pointer position as relevant to the probability of *red* seems right, seems natural—but so what? Skeptics such as Rosenberg and relativists such as Matthen will put this down to nothing more than prejudice, context, convention.

Let me run the argument undermining the parameter-variable distinction in more detail. Looking at the wheel of fortune, the skeptic says: What we want to explain is the fact that the frequency of *red* outcomes on some particular long series of spins is close to one-half (supposing in this case that the wheel in question is half red and half black). What determines the facts about the frequency? The outcomes of individual spins, of course. What determines the outcomes of individual spins? A number of things: the paint scheme, the pointer position, the initial speed of particular spins, and so on. All these quantities are bona fide causes of, and explainers of, outcomes, thus they are causes and explainers of any facts determined by outcomes—including the frequency of *red*. If we are more inclined to attribute the frequency to the paint scheme than to the pointer position, it is only because the role of the paint scheme is more salient, both in its spatial extent and because of the striking mathematical correspondence between the proportion of red paint and the frequency of *red*.

How to answer the skeptic? Intuitions about probabilistic relevance are not enough—they can be explained away. Some physical and philosophical foundation for the intuitions must be found.

In what follows, I examine the physical structure of the wheel of fortune to find a difference between what we instinctively take to be the parameters of the probability of *red* and what we take to be mere variables. My principal move is, however, philosophical rather than physical: on a conception of explanation as difference-making, I will argue, what explains the low-level facts that determine a high-level fact such as frequency may not explain that high-level fact.

5.3. Microconstancy and Smoothness

Take a closer look at the wheel of fortune. Suppose that the outcome of a spin on the wheel is entirely determined by the initial spin speed v and the fixed properties of the setup (thus, fluctuations in atmospheric pressure, vibrations from passing trucks, and quantum probabilities are assumed to have no impact on the wheel's final state). Then a function can be defined specifying the outcome obtained for any value of the spin speed—a function mapping values of v to either one, representing the outcome *red*, or zero, representing *black*. I call this the wheel's *evolution function*. (Here "evolution" refers to the time evolution of dynamic systems in general, not to biological evolution.)

The form of the evolution function (for a plausible though idealized physics of the wheel's operation) is shown, for a wheel with equal amounts of red and black, in figure 6.1. The gray regions mark values of v that produce *red* outcomes; the white regions values that produce *black* outcomes.

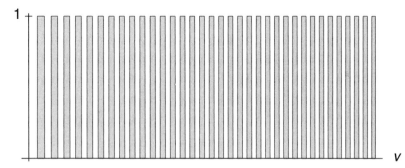

FIGURE 6.1. Evolution function for the idealized wheel of fortune.

Since physical probability muscled its way into the sciences in the later nineteenth century, philosophers, mathematicians, and scientists have perceived in the evolution function of the idealized wheel of fortune a clue to the existence (or apparent existence) of probability in deterministic systems (von Kries 1886; Poincaré 1896; Reichenbach 2008; Hopf 1934; Strevens 2003; Abrams 2012).[9] What is striking about the evolution function is that, given almost any probability distribution over the spin speed, the probability of *red* will be the same, equal to the ratio of red to black paint on the wheel: in this case, one-half.

As an informal visual demonstration of this claim, consider figure 6.2. The figure shows three possible probability densities for spin speed superimposed on the evolution function. The probability of *red* is equal to the proportion of the density that coincides with gray parts of the evolution function. Mash, stretch, knead the density almost any way you like, and the shaded area will persist in filling about half of the total, implying a probability for *red* of about one-half.

Almost any way, but not any way at all: there are some probability distributions over spin speed that will induce a probability for *red* other than one-half. The distribution in figure 6.3, for example, results in a probability for *red* that is much greater than one-half (since considerably more than one-half of the density is shaded, implying that values of the spin speed *v* that result in *red* are considerably more probable than values resulting in *black*). A sufficient condition for the probability's being one-half is that the probability distribution is relatively smooth, meaning that the corresponding density changes only slowly over small intervals—ruling out the rapid oscillations shown in figure 6.3. From here on, the term *smooth* should be understood in this technical sense.[10]

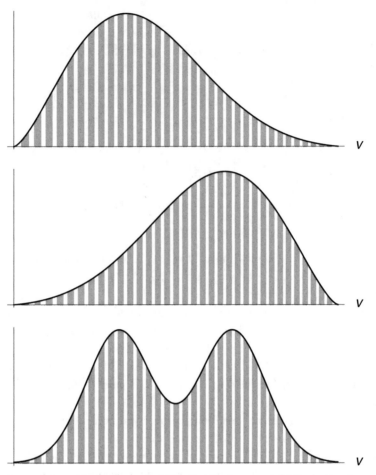

FIGURE 6.2. Almost any probability distribution over initial spin speed v results in a one-half probability for red.

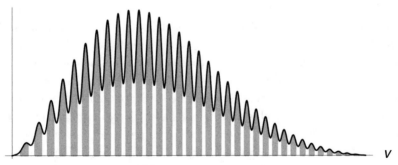

FIGURE 6.3. A rapidly oscillating probability density induces a probability for red much greater than one-half.

Let me state a more general result. The distinctive property of the wheel of fortune's evolution function I call *microconstancy*. Thus, a deterministic system is microconstant with respect to a given outcome if its space of initial conditions can be divided into many small ("microsized") contiguous regions, in each of which the proportion of initial conditions producing the outcome is the same. I call this proportion the *strike ratio* for the outcome. As you can see by inspecting the wheel of fortune's evolution function for *red* (fig. 6.1), the spin speed v can be divided into many small intervals in each of which the proportion of speeds producing *red* is one-half, so the wheel of fortune is microconstant with respect to *red* with a strike ratio of one-half.

The following is true: if a system is microconstant with respect to a given outcome with a strike ratio of p, and its initial condition distribution is smooth in the sense characterized above (with both microconstancy and smoothness assessed using the same standard of what is microsized), then the probability of the outcome is approximately equal to p.[11]

Some writers have attempted to build a metaphysics of physical probability in deterministic systems upon this result, arguing that there is something inherently stochastic about a microconstant dynamics (Reichenbach 2008; Rosenthal 2010; Strevens 2011; Abrams 2012). In the remainder of section 5, by contrast, I will not argue that the microconstancy-based probability has a special ontological status, but rather that it has a special explanatory status: it is the probability that best explains an approximately one-half frequency of *red*.

5.4. Explanation as Difference-Making

A long series of spins on the wheel of fortune yields *red* with a frequency of approximately one-half; how should that frequency be explained? Intuitively, it is the microconstancy of the wheel's dynamics, and in particular its strike ratio for *red* of one-half, together with the smoothness of the initial condition distribution, that provides the best explanation of the frequency. Citing less information (omitting the value of the strike ratio, for example) would leave the frequency a mystery, while citing more information—say, the position of the pointer, or the diameter of the wheel, or the exact speed of the nineteenth spin—would add nothing explanatorily enlightening.

Against this line of thought, the skeptic argues in this way. Pointer position, wheel diameter, and the speeds of particular spins are uncontroversially both causes and explainers of individual outcomes. Frequencies are nothing over and above patterns in individual outcomes. Thus, pointer position, wheel diameter, and individual spin speeds are causes and explainers of frequencies.

Where has the skeptic gone wrong? The principle of explanatory transitivity implicit in the skeptical argument does not, I submit, hold. Often, what explains a low-level state of affairs is irrelevant to the explanation of a high-level state of affairs that is realized in part by the low-level state, because the explainer is responsible for some aspect of the low-level state of affairs that makes no difference to the obtaining or otherwise of the high-level state of affairs. The seam on a baseball that breaks a window may explain why some particular shard of glass spins away during the breaking in the way that it does, but it is irrelevant to the breaking itself: with or without the seam, the window would break.

What is driving these judgments of relevance is the notion that, in order to explain an event or state of affairs, a causal factor must make a difference to whether or not the event occurs or the state of affairs obtains. How can something be a causal factor but not a difference-maker? As follows: its causal contribution to the high-level explanandum event is to make a difference to *how* the event occurs without making a difference to the fact *that* it occurs—it makes a difference to how the event is realized without making a difference to the fact that it is realized. The baseball's seam, for example, causally influences the window's breaking by helping to determine the way that the breaking is realized, namely, in part by some particular shard's spinning in some particular way. Difference-making in the "how" sense is quite consistent with the seam's not being a difference-maker in the "that" sense.

That "that" difference-making is necessary for explanatory relevance is an old idea in the philosophy of explanation (Salmon 1970, 1997; Garfinkel 1981; Lewis 1986; Woodward 2003); I will not defend it here. Nor will I advance a specific account of difference-making, though elsewhere I have argued for my own "kairetic" account over the counterfactual and probabilistic approaches offered by other writers (Strevens 2008). For the purposes of this chapter, the choice does not matter greatly. In what follows, I will rely on a simple counterfactual criterion for difference-making: a causal factor makes a difference to an explanandum just in case, had the factor been absent, the explanandum would almost surely not have occurred. Although the counterfactual criterion is universally acknowledged to have limited validity, it delivers reliable judgments for principled reasons in the cases I will discuss (Strevens 2008, secs. 2.4, 3.81).[12]

Consider, then, pointer position and wheel diameter. Had the pointer position been moved prior to spinning the wheel, the sequence of outcomes would almost surely have been different, but the frequency of *red* would almost surely have been the same—around one-half. Likewise, had the wheel been slightly larger or slightly smaller (while retaining the same symmetry and paint scheme), particular outcomes would have differed but the frequency would have been roughly the same. Thus, though position and diameter make a difference to how the one-half frequency of *red* is realized, they do not make a difference to the fact that that approximate frequency, and not some other, is realized.[13]

What about spin speeds? They are not entirely irrelevant: they must be smoothly distributed. But given smoothness, particular values of actual spin speeds are irrelevant: had the nineteenth spin had a speed of 7.8 rather than 8.7 radians per second, for example, the frequency of *red* would have been much the same.

To make sense of these claims, it must be borne in mind that it is not things but the properties of things that are difference-makers (although in some cases, no property of a thing makes a difference, in which case there is no harm in saying that the thing itself is not a difference-maker). It is not quite correct, then, to say that the spin speeds were not difference-makers for the frequency. A certain property of the speeds (their smooth distribution) was a difference-maker, and certain other properties (their exact values) were not.

5.5. The Reference Class Problem Solved

The story so far: certain features of a wheel of fortune's physics and paint scheme and the smoothness of its initial condition distribution are relevant to explaining why the frequency of *red* outcomes in some long sequence of spins is approximately one-half; the precise values of the spin speeds and the position of the pointer are, by contrast, explanatorily irrelevant.

This shows why a one-half frequency of *red* can be explained by the microconstancy-based one-half probability of *red*: the facts that determine, and so constitute, that probability are identical to the explanatorily relevant facts.

It also shows why the frequency cannot be explained by a set of more specific probabilities that conditionalize on precise spin speeds and such matters as the pointer position. Suppose, for the sake of the argument, that such fine-grained probabilities exist, say, probabilities of nearly one for *red* on about half of the spins and probabilities of nearly zero on the rest. Such probabilities together make it very likely that the frequency of *red* will be approximately one-half. But they are not properly explanatory, because they are constituted in part by information that is irrelevant to the explanation of the frequency. The problem with an explanation that deploys the probabilities, then, is not that it leaves something out but that it contains *too much* information, thus explanatorily irrelevant information.

In short, for explanatory purposes, you should cite a probability for *red* that takes into account the structure of the wheel's paint scheme but not its pointer position or any of the speeds with which it is initially spun, which is to say that when individuating explanatory probabilities, you should use a reference class constructed using the former but not the latter information. In other words, you should treat the paint scheme as a parameter and the pointer position and spin speeds as variables (in my technical sense, of course—the pointer position does not actually vary).

We have a solution, then, to the reference class problem. Treat as parameters those features of a system that determine microconstancy, strike ratio, and smoothness of the initial conditions; treat all other causally relevant features as variables.

This prescription is limited in two ways. First, the correctness of a reference class is relative to an explanatory task—to an explanandum. I have not said that there is a uniquely correct, objective probability for *red*, only that there is a uniquely correct probability of *red* for the purpose of explaining frequencies of *red* on long series of spins that approximately match *red*'s strike ratio. Second, the solution as framed requires a microconstant dynamics (although as Strevens [2008, sec. 10.3] shows, there is some scope for generalization).

6. BIOLOGICAL PROBABILITY AND DRIFT

6.1. Individuating Probability for Selection

The ecological dynamics that produce evolutionarily significant outcomes such as feeding and dying are microconstant with respect to those outcomes. Or at least, they are microconstant when the processes run for reasonably long times: the evolution function for a snail's being eaten by a song thrush over the course of a month is microconstant, though the evolution function for its being eaten over the course of the next three minutes most likely is not.

Further, the same is true for less significant outcomes such as being in such and such a physical state or even having such and such a position in the habitat. Thus, provided that your waiting period is long enough (a month rather than three minutes), and given a smooth distribution over initial conditions, there is a probability of a snail's being in some particular place rather than some other place that is entirely determined by the factors that determine the strike ratio of the microconstant evolution function for the outcome of *being in position x at the end of time period t.* Or as I will say, there is a microconstant probability distribution over position at the end of *t*—for any sufficiently long *t*. (Whereas in the case of the wheel of fortune, there is just one varying initial condition, spin speed, in the ecological case there will be a huge number of such conditions.)

The justification of these claims about pervasive ecological microconstancy is, as you might expect, a complicated matter; I make my best attempt in *Bigger than Chaos* (Strevens 2003). In this chapter, I will simply assume microconstancy without further argument.

Biological outcomes that have microconstant probabilities will tend to occur with frequencies equal to strike ratios. When explaining such frequencies—not just the frequencies of evolutionarily significant outcomes such as births and deaths, but even the frequencies with which snails are found in this position, as opposed to that position, on the old log at the back of the garden—you have a rationale for making the distinction between parameters and variables: what is explanatorily relevant to the frequencies, aside from the smoothness of the initial conditions, is just the set of properties that determine the strike ratio for the outcome in question; these are the properties, then, that should be used to individuate the probabilities used to explain the frequencies, solving the reference class problem.

The properties relevant to the strike ratio are more or less the properties you would expect: a snail's camouflage is relevant to death by predator, and the distribution of nutritious moss is relevant to the snail's location on the log, but the snail's exact starting position at the beginning of the month (or other long time period) is, like initial spin speed on a wheel of fortune, relevant to neither.

The facts about the microconstancy of biological dynamics are reflected, then, in our intuitions about probabilistic relevance in biology. Why are we so expert in these matters? This is a question I answer in *Tychomancy* (Strevens 2013), but not here.

To summarize the story so far: microconstancy provides an objective individuation of probabilities for an array of outcomes—ranging from position to personal extinction—of relatively long-term biological processes, provided that those probabilities are used to explain corresponding strike-ratio-matching frequencies.

When do frequencies that match probabilities appear in evolutionary explanation? Most saliently, they appear in explanations that attribute changes in the makeup of a population to selection. Suppose, for example, that variant α has replaced β in the population because of its superior camouflage scheme. In virtue of the difference in camouflage (compare to the difference in the wheels' paint schemes in section 5.1), α's probability of going for a month without being eaten is somewhat higher than β's probability of the same. Typically, the frequencies will roughly match the probabilities: α will evade predation more often than β. All other things being equal, the relative frequency of α will therefore continue to increase, month after month, until β is crowded out of the habitat. Because such an explanation turns entirely on probability-matching frequencies, it should be conducted entirely in terms of information relevant to explaining such frequencies. Probabilities individuated as I have recommended are the right tools for the job.

So much for explanation by natural selection. How should the parameter-variable distinction be drawn for probabilities invoked in explanation by drift? We should use the same strategy as in the case of selection, counting as parameters just those factors that affect evolutionarily significant outcomes' strike ratios—so I claim.

It looks to be an easy claim to sustain on the traditional view of drift as sampling error, since precisely the same sorts of probabilities feature in both cases: the probability of being eaten in the course of month and so on. But that is too quick: not only the probabilities but the explananda must be of the same type. You might wonder whether the explananda targeted by drift explanations are frequencies. If not, there is trouble, since it is only with respect to frequencies that microconstancy draws an explanatory distinction between variables and parameters. This issue is explored in the next section, section 6.2.

What about alternatives to the traditional view of drift? I will consider only the process-driven view (section 3). The probabilities that matter in the process view are those invoked in the characterization of indiscriminate sampling. On both the traditional and modern notions of indiscriminateness, many of these probabilities are attached to processes unfolding over relatively brief periods of time, such as the time taken by a typical predator-prey encounter. The dynamics mapping the initial conditions of such processes to evolutionarily significant outcomes are likely not microconstant, so the process view's drift-defining probabilities are likely not microconstant—in which case, the individuation criterion proposed above does not apply. This objection is confronted in section 6.3.

6.2. Drift Explanation Is Frequency Explanation

Two identical twins are strolling along a ridge top; one is killed by lightning, but the other survives. Why? You might be tempted to reply "Drift." In this case, drift explains a single evolutionarily significant event, rather than a frequency with which such events occur. The explanation is equivalent, in effect, to saying that the outcome occurred "by chance."

The great majority of serious evolutionary explanations citing drift are not, however, mere attributions of arbitrary deaths and wonky statistics to chance. They rather use mathematical models of evolutionary processes to make predictions about differential reproduction.

Consider, for example, the explanation of homozygosity due to a population bottleneck. A study of northern elephant seals, to take a superb example of the genre, found population-wide homozygosity at a large number of genetic loci. This is explained by the seals' having been hunted almost to extinction in the 1890s (with perhaps only twenty specimens at one point surviving). As the population recovered from this low ebb, it was highly likely that any particular allele would drift to extinction, even given modest fitness advantages to heterozygotes. The vast majority of alleles of the genes in question evidently did go extinct, leaving only one survivor at each locus (Bonnell and Selander 1974).

The probabilistic reasoning here is as follows. Any allele can go extinct if has a run of bad luck (if, for example, its sole possessors, through no fault of the allele, fail to contribute offspring to the next generation—quite probable in elephant seals, since a few males do all the reproduction—or if the offspring they contribute by chance contain only its rival allele). In extremely small populations, this run need not be very long; if the population remains small for a while, then, so that there are many chances for any given allele to suffer such a run, there is a high probability that most alleles go to extinction. Hence wholesale homozygosity.

Consider the external ecological outcomes that play a role in allele extinction (putting aside, then, the vicissitudes of meiosis, the fertilization process, and so on). The relevant outcomes are those that enable or frustrate an allele's transmission, thus the same outcomes by which selection occurs, such as birth and death. Compared to explanation by selection, there is a complication: we are interested in explaining not frequencies of individual births and deaths, but rather frequencies of "bad runs"—the evolutionary equivalent of explaining the frequency with which you see, on a wheel of fortune, ten *red* outcomes in a row. The same probabilities that explain frequencies of individual outcomes, however, explain frequencies of runs. A probability for *red* of one-half, for example, explains both why the number of *red* outcomes in 1,000 trials on a wheel of fortune is about 500, and also why the number of sequences of five or more *red* outcomes in a row in the same number of trials is about 16.[14]

The probabilities that explain evolution by drift, then—as opposed to one-off incidents—have the same sort of explananda as the probabilities that explain evolution by natural selection: frequencies of evolutionarily significant events.

6.3. Short-Period Probabilities

To formulate the problem of probabilities attached to short-period ecological processes that lack microconstancy, consider again an encounter between a song thrush and a snail. Divide the initial conditions for the encounter into two sets, central and peripheral. In the central set, the snail is near the center of the thrush's visual field, while in the peripheral set, it is nearer an edge. Consequently, more initial conditions within the central set than the peripheral set result in snail capture and death. Define a binary variable, *centrality*, to represent whether the initial conditions of a particular thrush-snail encounter are central or peripheral. (Perhaps centrality takes the value one when the conditions are central, zero otherwise.)

Now consider the short-period probability of a snail's surviving a thrush encounter—the sort of probability that figures explicitly in the process approach's traditional definition of indiscriminate sampling and that is also, I have argued, presumed by the modern definition. For the purpose of applying these definitions, relative to what reference class ought the probability to be defined? Should we distinguish two "narrow probabilities," the probability of death given that centrality is equal to one and the probability given that centrality is zero? Or should we use a single "broad probability," a combination of these two probabilities weighted by the probability distribution over centrality? In short: should we treat centrality as a parameter or as a variable?

If the dynamics underlying each of the narrower probabilities were microconstant, with a higher strike ratio for death when centrality is equal to one, a straightforward answer could be given: centrality affects the strike ratios, so it should count as a parameter. Suppose, however, that the dynamics is not microconstant. Can any objective, substantive parameter-variable distinction be sustained? Yes—and in seeing how, you will see also that the "straightforward answer" is in fact incorrect.

Consider a wheel of fortune analogy. You generate a series of outcomes, *red* and *black*, using the following procedure: first, you spin a wheel that is half red and half black—call it the *even wheel*. Then, depending on the outcome, *red* or *black*, you spin one of two other wheels; call them the red and black wheels. The red wheel is 75 percent red and so produces *red* with a three-quarters probability; the black wheel is 75 percent black and so produces *red* with a one-quarter probability. (Think of these two wheels as representing the two kinds of thrush-snail encounter, and think of the outcome that determines which of the two wheels to spin as representing the value of centrality. In the wheel of fortune case, the red and black wheels have a microconstant dynamics, whereas in the biological case, by assumption the analogous dynamics is nonmicroconstant, but I promise not to appeal to microconstancy; this will enable me to move faster by not having to describe a new, nonmicroconstant probabilistic setup.)

Now suppose that you conduct 100 trials, generating 100 outcomes. (You will make 200 spins, then, since each trial consists of two spins: a spin on the even wheel to determine whether to use the red or the black wheel, and then a spin on the indicated wheel to determine the final outcome.) About one-half of the outcomes are *red*. How to explain this probability-matching frequency?

Here are two strategies. First, you could note that the probability of *red* on any particular trial (consisting of two spins) is one-half, and that the frequency reflects this probability. Second, you could record which of the red and black wheels was spun on each trial, noting that (as I will presume) in about one-half of the trials the red wheel was spun and in the other half the black, and that the frequency of *red* resulting from the spins of the red wheel was about three-quarters and that the frequency of *red* resulting from the spins of the black wheel was about one-quarter, for a net frequency of about one-half.

The first of these explanatory strategies appeals only to the coarse-grained or broad probability of *red*, a composite of the probabilities attached to each of the three wheels, which is the same in each trial; the second appeals to the fine-grained or narrow probability of *red*, which depends on whether the red or the black wheel is spun and which therefore varies from trial to trial. Both strategies predict what is to be explained, the one-half frequency of *red*, by citing factors causally involved in the production of that frequency. But the first explanation does so with strictly less information—it does not specify which of the red and black wheels was used for each trial, and so it does not specify the narrow probabilities at work in each trial. That the prediction is possible using broad probabilities alone shows that the additional information needed to deploy narrow probabilities does not make a difference to the frequency of *red*, any more than does the exact sequence of *red* and *black* outcomes that determines the frequency.

I want to say the same thing about the explanatory relevance of information about centrality in the thrush-snail encounters, namely, that centrality is explanatorily irrelevant to the frequency with which thrushes eat snails. If it is these frequencies you are trying to explain, then—as is the case when invoking the mathematics of drift to explain why, say, the ratio of two equally effective snail camouflage schemes in a small population exhibits a certain pattern of variation over time—the snails' centrality or otherwise in particular encounters should not be mentioned.

To individuate short-period probabilities, then, you should as a general rule put probability distributions over variables such as centrality, citing for explanatory purposes broad probabilities, rather than conditioning on specific values and thereby delineating and citing narrow probabilities. Of course, centrality is a contrived variable, an unsubtle coarse-graining of predator and prey position. The probability distribution should go over the predator and prey positions themselves, and over all other initial conditions causally relevant to the outcome: orientation, hunger, physical condition, and so on.

Where do these initial condition probabilities come from? With respect to what reference class should they be individuated? They are the long-term microconstant probabilities introduced at the beginning of section 6.1, individuated in the same way as all microconstant probabilities. To find the probability distribution over centrality, then, look to the microconstant probability distribution over the positions of snails and thrushes induced by the dynamics of the previous month (or any other appropriately long period—it will be the same distribution for any sensible choice).

The individuation rule for short-period probabilities advocated here—do not condition on actual values of initial conditions of short-period encounters, but rather put a long-term probability distribution over absolutely all of them—creates a pressure toward breadth in probability, that is, a pressure toward using the widest feasible reference class. Other principles of explanation will prevent things from going too far. Consider, for example, the following worry. Some snails are eaten by thrushes; others are struck by lightning. In treating these events, either to define drift using short-period probabilities or for any other purposes, you want to distinguish the probability of being eaten (when in close proximity to a thrush) from the probability of being struck by lightning (when caught in the open during an electrical storm). But you might think that the impetus toward breadth forces you to amalgamate the two, as follows. Consider as a "scenario type" a snail's either running into a thrush or being caught in a storm. (Both thrush and storm encounters fall under this disjunctive type, then.) Define the variable thrush-not-storm as follows: it has the value one for thrush encounters and zero for storm entrapments. Like the centrality variable, then, it divides a broader scenario into two narrower scenarios. Intuitively, the division is in this case a good one; it is a mistake to amalgamate for explanatory purposes the dynamics of predation and lightning into a single broad probability. But can you not put a probability distribution over the thrush-not-storm variable to define a broad probability of death-by-thrush-or-lightning-strike for snails that find themselves in a thrush-or-storm scenario? If such a distribution exists, which I will not question, you can. Why not use the broad probability then?

You should not use it because explanations should cite a single causal mechanism, not a disjunction of causal mechanisms (a consideration understood very well by process theorists of drift). When explaining a death, then, you should specify the particular mechanism at work: bird or bolt. Thus, you should not cite probabilities that fail to distinguish which of these mechanisms is at work.[15]

Let me finish there. Microconstancy, I have argued, provides the foundation for an objective individuation of the probabilities used by process theorists to characterize indiscriminate sampling, and so for an objective process-theoretical account of the distinction between selection and drift. This may not comprise a complete solution to evolutionary biology's reference class problems, but it is a good start.

ACKNOWLEDGMENTS

Thanks to anonymous referees and to the editors of the volume for helpful questions and guidance.

NOTES

1. Throughout this chapter, I consider only simple binary scenarios of evolutionary change, that is, scenarios in which the population is divided into two variants. Evolutionary biology is frequently concerned with scenarios in which there are many variants, even (notionally) continuously many, as in the case of traits such as height, weight, or leaf size. I am confident, though I will not make the case in these pages, that the problem of the reference class can be substantially posed and solved using only binary scenarios, although some discussion, omitted here for reasons of length, is needed to explain how to set up the treatment in sections 5 and 6 for the more complex cases.

2. Take into account enough information, and the probabilities are all ones and zeros, but there is no need to go this far; all that is dialectically necessary is information sufficient to generate a probabilistic inequality.

3. It is important for this conclusion that, as specified previously, the βs' bad luck consists largely in their choosing more often than not the direction in which the pred-

ator lies, and not in their running into such situations more often in the first place, or in their more often getting into situations in which there are predators on both sides.

4. Note that on Hempel's epistemic approach to statistical event explanation, the inductive-statistical model, probabilities built on a parameter-variable distinction dictated by epistemic limits are also genuinely explanatory (Hempel 1965).

5. A third explanation of institutional parameter-variable distinctions, also not to be pursued in this chapter, puts them down to historical contingencies in the development of the various disciplines that have no single explanation and perhaps no rational justification at all.

6. Salmon, at this stage of his thinking, avoids all mention of causation. Thus he does not restrict his potential probability-determiners—the factors that might count as parameters—to factors that play some causal role in producing the outcome. Rather, as Brandon explains, he restricts the potential probability-determiners to factors that are not characterized in terms that depend on the outcome of interest. This chapter has presumed from the very beginning that potential probability-determiners must be involved in causally producing the outcome; my formulation of Salmon's criterion imports that presumption—not altering Salmon's view, but rather applying it within the chapter's framework.

7. I omit here the parallel argument that the homogeneity criterion will not cooperate with a process-driven conception of drift driven by a "modern" definition of indiscriminate sampling (section 3).

8. The view, then, is the reverse of Matthen's, which is also in some sense explanationist: Matthen's institutional standards prescribe (relativized) explanatory standards, whereas my (objective) explanatory standards prescribe institutional standards.

9. The early history of what has been called "the method of arbitrary functions" is surveyed by von Plato (1983).

10. In more formal work I have called smoothness *macroperiodicity* (Strevens 2003) or *microequiprobability* (Strevens 2013).

11. A formal treatment is given in *Bigger than Chaos* (Strevens 2003).

12. The best-known failure of the counterfactual criterion occurs in scenarios where there is some backup to the actual difference-maker: because of the backup, even if the actual difference-maker had not been present, the explanandum would have occurred anyway.

13. Apparently crucial to the truth of these observations is the "almost surely" qualification in the counterfactual criterion for difference-making, concerning which, briefly, two points. First, the qualification bridges, in my own story about difference-making, a considerably more sophisticated story, developed in *Depth* (Strevens 2008), part 4. Second, since it is the objective validity of probabilities that is at stake in this chapter, you might wonder which probabilities are used to assess the "almost surely." If there is an objective probability distribution over the relevant initial conditions, then that one, of course. But I have assumed, for the sake of the argument, that the

world is a deterministic place. In that case, what is the source of the initial condition distribution function?

One approach to the question puts aside any need for probabilities over initial conditions. A sense is defined in which a large but finite set of actual spins can be smoothly distributed. This smoothness of the actual speeds implies, given microconstancy, a frequency approximately equal to the strike ratio (Strevens 2003, secs. 2.33, 2.72). Further, holding this smoothness fixed, changes to pointer position, wheel size, and so on determinately have no effect on the frequency. Thus, the "almost surely" qualification to the counterfactual criterion for difference-making can be dropped.

14. Not counting subsequences.

15. On the individuation of mechanisms for causal explanation, see Strevens 2008, sec. 3.6.

CHAPTER 7

Weak Randomness at the Origin of Biological Variation: The Case of Genetic Mutations

Francesca Merlin

Chance is a fundamental theme in biology, from molecular genetics to macro-evolutionary theories. Philosophers have focused their reflection on the ambiguous way biologists use this notion, in particular when they characterize a specific source of variation in living organisms, i.e., genetic mutation. The idea that all genetic mutations occur by "chance" with respect to adaptation is a cornerstone of the modern synthesis (1930s–1950s), and despite a number of controversies that have arisen since then, this is still the consensus view in contemporary biology (Ridley 2004). Nevertheless, the notions of chance that biologists invoke to talk about genetic mutations possessed an ambiguous character until very recently. Philosophers of biology have contributed to its clarification, suggesting precise definitions of chance from the point of view of evolution (Beatty 2006a, 2008; Millstein 2006, 2011; Merlin 2010; Ramsey and Pence, this volume, introduction; Plutynski et al., this volume, chap. 3). The result is a notion that focuses on the relation between mutation, selection, and adaptation, i.e., defining the "chancy" character of genetic mutations from the perspective of evolutionary biology. In other terms, the notion of "evolutionary chance," as I have called it (Merlin 2010), looks at genetic mutations from the point of view of their phenotypic result and refers to the fact that they are not specifically provoked with a view to the adaptation of the organisms concerned.

The majority of philosophical analyses of the notion of chance used to characterize genetic mutations are limited to an evolutionary perspective. However, the terms *chance, random, stochastic, probabilistic,* and so on are also used in biology to discuss mutations without making any reference to their utility from the evolutionary point of view (for instance, see Kacar, this volume, chap. 11). In particular, in classical genetics and in molecular genetics, they are used to characterize the distribution of a number of possible mutations along the DNA sequence and their distribution over time, and even to talk about each mutational event with respect to the physico-chemical processes at its origin.

The present chapter investigates the notion of chance when it is used in this context, i.e., to characterize mutational events at the level of the genome versus the phenotype. My aim is to identify the precise meaning and role of this notion. This will allow us to understand in which sense genetic mutations as changes of the genome sequence are a matter of "chance" (independently of their phenotypic effects and of their evolutionary consequences, and with regard to the physico-chemical processes causing them).

My overall objective is, on the one hand, to contribute to the philosophical debate on chance in biology by showing that, in the case of genetic mutation, the debate cannot be immediately reduced to the metaphysical alternative between "pure chance" as indeterminism and chance as ignorance (or no chance at all) in a deterministic framework. There's more to be said about the character of genetic mutations before dealing with such metaphysical alternatives, and this calls for a far-reaching study of the probabilistic physico-chemical processes producing mutations. On the other hand, the analysis provided here aims at having an impact on research directions in biology by pointing at the gap between the present state of knowledge about genetic mutations and the idealized character of models and experimental techniques that are still used by biologists to describe and quantify them.

Throughout this chapter, I maintain that the only way to explain the "chancy"[1] character of genetic mutations (e.g., a particular change of the DNA sequence, the distribution of mutations over a certain period of time or along the DNA sequence) is to study the characteristics of the physico-chemical (causal) process at their origin. So, first of all, let us distinguish the two following concepts: a chancy process and a chancy outcome (see Earman 1986).[2] In keeping with a visible trend across the scientific discourse about chance, I use the term *stochastic* to qualify chancy processes and the term *random* for chancy outcomes, and I define them from an epistemological point of view (i.e., referring to the models used to describe them and not, metaphysically speaking, looking at the structure of the real world).

On the one hand, a "stochastic" process is a process whose evolution can be described in accordance with probabilistic laws (also called stochastic laws). In other words, starting from a given (sub)set of initial conditions, the model used to describe it can evolve in several possible directions and so can produce many different results according to a certain probabilistic law. According to this definition, the stochastic character of some process does not imply any metaphysical commitment as to its indeterministic nature, i.e., it does not follow that, starting from the exact same starting point, the real process can actually have more than one possible evolution, and so many different possible results.

On the other hand, an outcome is "random" if it is characterized by irregularity, disorder; lack of structure, of pattern, or of law; etc., and so by unpredictability. Here I refer to the mathematical definitions of randomness, e.g., in terms of algorithmic or Kolmogorov complexity (Kolmogorov 1968; see also Delahaye 1999), the consensual definition of which today is Martin-Löf's concept of a random sequence (Martin-Löf 1966). Note that this way of defining random outcomes can be applied only to series of events, in particular to the result of a series of trials of a chance experiment (e.g., a numerical series generated by a certain number of independent rolls of a die). Now, some outcomes are single events because the initial conditions that bring them about cannot be repeated. Single genetic mutations in nature are an example of such unique events that are not characterized by properties such as irregularity, disorder, lack of structure, etc. So we actually can't declare that a single mutation is random without looking at the process that generated it: this is the only way to try to define and to account for its random character (i.e., as the result of a stochastic process; see the definition above).

It's also worth noticing that, if a result consisting in a sequence of events, such as the distribution of mutations over the genome or through time, is random, this does not necessarily imply that the process at its origin is indeterministic, metaphysically speaking. In fact, a given result can be due to a mechanism characterized by a fully deterministic internal structure, i.e., which is governed by nonprobabilistic laws, as well as to an indeterministic mechanism. For instance, random numerical series can be produced by a deterministic generator or by an indeterministic (quantum) generator (Glennan 1997).

As such, I maintain that the chancy (random) character of a mutational event (whether at a single time, like a point mutation, or extended over space or time, like the distribution of mutations along the genome) should be defined by looking at the chancy (stochastic) character of the process producing it. Again, revealing the origin of the event in question is indeed the only way to account for its alleged randomness.

The present chapter aims to identify the proper notion of chance needed to characterize genetic mutations at the genome level (for a useful distinction of five sense of chance, see Plutynski et al., this volume, chap. 3). It is structured as follows. First, I introduce two notions I label "strong randomness" and "weak randomness."[3] Then I analyze their respective status. The former ("strong randomness") reflects the idealized way molecular geneticists characterize genetic mutations when they statistically evaluate their rate of occurrence. On this basis, I show that the notion of "strong randomness" is not appropriate if our aim is to provide a realistic and precise account of genetic mutations. The latter ("weak randomness") corresponds to the molecular conception of mutations, considering all physico-chemical factors that are likely to influence the mutation process at the genome level. By introducing a certain number of mutational biases and some recent experimental studies on genetic mutations, I show finally that "weak randomness" is the appropriate notion for characterizing all genetic mutations from a molecular point of view. I conclude by replying to three possible objections that might be raised against my view.

"STRONG RANDOMNESS" AND "WEAK RANDOMNESS"

Strong randomness and weak randomness[4] rest upon the idea that a random event is the result of a stochastic process (see definition above). In other words, they don't define the random character of an event by providing a description of its properties as a result; rather, they look at the characteristics of the causal process producing it.

Let's introduce first the notion of "strong randomness." An event is "strongly random" if and only if it is the result of a stochastic (causal) process that fulfills the following two conditions:

C1: It is an *indiscriminate sampling* process[5] where no difference in the characteristics of the elements that can be sampled plays a causal role. Thus, the probability of being sampled is the same for all elements.

C2: It is a process *invariant over time*. The probability of an element's being sampled does not change over time; thus, such probability is independent with respect to other occurring events in the sense that no event is going to affect it in a differential way over time.

To put it differently, a "strongly random" event is the result of an indiscriminate sampling process with replacement, i.e., a process analogous to a series of draws from an urn where every ball has the same probability of being sampled and, when it is drawn, is systematically placed back into the urn.

The general notion of "strong randomness" can be applied to the phenomenon of genetic mutation as follows. A mutation is "strongly random" if and only if it is the result of a stochastic physico-chemical process that fulfills the two following conditions:

C1: It is an indiscriminate sampling process with regard to differences along the DNA sequence, i.e., a process where differences in the physico-chemical characteristics of the nucleotide sites do not play a causal role in the production of mutations (i.e., the fact that some sites mutate and others do not) and so in the distribution of mutations along the DNA sequence. The probability of a mutation's occurring is the same at all sites of the sequence.

C2: It is a process that is invariant over time (e.g., over a cellular generation), i.e., an indiscriminate sampling process with regard to differences in the physico-chemical conditions, inside and outside the cell, over the time period under consideration. In other words, it is a process where these differences do not play a causal role in the way mutations are distributed over time. The probability of a mutation's occurring at a particular site (or in the entire DNA sequence) is the same at each moment of the time period under consideration; thus, it is independent with respect to other occurring mutational and (broadly speaking) environmental events in the sense that no such events are going to affect it in a differential way over time.

In this case, the mutational process is analogous to an indiscriminate sampling process with replacement (see above).

Let us now introduce the notion of "weak randomness." An event is "weakly random" if and only if it is the result of a stochastic (causal) process that does *not* fulfill at least one of the two conditions required to be a "strongly random" event. Consider three weakly random scenarios:

S1: It is a *discriminate sampling* process *invariant over time*. More explicitly, it is a process where differences in the characteristics of the elements that can be sampled do play a causal role: the probability of being sampled is not the same for all elements and does not change over time (it is independent from other occurring events). In other words, it is a process analogous to a series of draws from an urn where some balls have a larger probability of being sampled than others in virtue of their physical properties (for instance, because they are sticky while the others are slippery)[6] and the drawn balls are systematically placed back into the urn.

S2: It is an *indiscriminate sampling* process *variant over time*. More explicitly, it is a process where no difference in the characteristics of the elements that can be sampled plays a causal role (thus, all elements have the same probability of being sampled), but this probability changes over time (it is not independent from other occurring events, which are likely to affect it in a differential way over time). Using the analogy, it is an indiscriminate sampling process without replacement, i.e., a process analogous to a series of draws from an urn where every ball has the same probability of being sampled and, when it is drawn, is not systematically placed back into the urn.

S3: It is a *discriminate sampling* process *variant over time*. Thus, it is a process analogous to a series of draws from an urn where some balls (e.g., the sticky ones) have a greater probability of being sampled than others (e.g., the slippery ones) and the drawn ball is not systematically placed back into the urn.

As in the case of "strong randomness," the general notion of "weak randomness" can be applied to the phenomenon of genetic mutation as follows. A mutation is "weakly random" if and only if it is the result of a stochastic physico-chemical process that does not fulfill at least one of the two conditions required to be a "strongly random" event, i.e., in the three following situations:

S1: It is a discriminate sampling process with regard to differences along the DNA sequence and invariant over time. More explicitly, it is a process where differences in the physico-chemical characteristics of the nucleotide sites do play a causal role in the mutational process and so in the distribution of mutations along the DNA sequence. Thus, the probability of a mutation's occurring is not the same at all sites of the sequence, but it does not change over the time period under consideration (e.g., a cellular generation). In this case, the mutation process is analogous to a discriminate sampling process with replacement (see above).

S2: It is an indiscriminate sampling process with regard to differences along the DNA sequence and variant over time. More explicitly, it is a process where differences in the physico-chemical characteristics of the nucleotide sites do not play a causal role in the mutation process; nevertheless, differences in intracellular and extracellular conditions during the considered time period do. Thus, the probability of a mutation's occurring is the same at all sites of the DNA sequence but changes over time (other mutational and environmental [broadly speaking] events are likely to affect it in a differential way over time). In this case, the process of mutation is analogous to an indiscriminate sampling process without replacement (see above).

S3: It is a discriminate sampling process with regard to differences along the DNA sequence and variant over time. So, in this case, the mutation process is analogous to a discriminate sampling process without replacement (see above).

The two notions of "strong randomness" and "weak randomness" can be formulated by looking at the mutational process at different levels, for instance at the level of a single nucleotide or of a nucleotide sequence of a certain length, or by considering the entire genome. Let us consider, for instance, a short nucleotide sequence composed of four nucleotide bases (ATTC: one adenine, two thymines, and one cytosine) and a certain period of time from t_0 to t_n corresponding to a cellular generation (see fig. 7.1). I maintain that if some mutation occurs at a certain moment between t_0 and t_n (for instance, as represented in the figure, a point mutation transforming the adenine into a guanine at t_1), and such mutation is due to a physico-chemical process that favors neither any possible alteration over the others nor any particular instant during the cellular generation considered, then the mutation in question is a

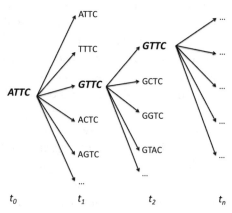

FIGURE 7.1.

"strongly random" event: it has the same probability of occurring as all other possible mutations and its probability does not change over time. On the contrary, if the process at the causal origin of the (point) mutation (from adenine to guanine) favors one or more possible changes over the others and/or one or more particular instants of the cellular generation considered, such a mutation is a "weakly random" event. To sum up, a random mutation is "weakly random" if it is not a "strongly random" event.

NO GENETIC MUTATION IS A "STRONGLY RANDOM" EVENT

Let us examine the notion of "strong randomness." At first glance, it does not look like an appropriate way, from the empirical point of view, to characterize mutations at the molecular level (without any reference to the adaptive value of their phenotypic effects). In fact, in the light of research advances about mutations (see the next section), it seems difficult to imagine a situation where no factor affects the mutational process in a differential way along the DNA sequence and/or over time. Yet, even though no one explicitly advocates the idea that all genetic mutations are strongly random, molecular geneticists continue to use this notion, in particular when they perform the fluctuation test to evaluate the mutation rate. The fluctuation test is a statistical method introduced by Luria and Delbrück at the beginning of the 1940s.[7] In 1949 Lea and Coulson expanded and generalized it (which is why it is now called the Lea-Coulson model). Today, despite some of its limitations (see, e.g., Stewart, Gordon, and Levin 1990; Rosche and Foster 2000; Foster 2006), it is the most-used theoretical and experimental model for estimating mutation rates.

To show where and how the notion of "strong randomness" intervenes in this statistical model, let us consider some of its constitutive hypotheses. In particular, let us focus on what the Lea-Coulson model assumes about the mutational process, a mutation being defined in this context as a change of the DNA nucleotide sequence.

According to the model's first hypothesis, the probability of a mutation's occurring (e.g., a spontaneous point mutation) is independent of previous mutational events. Second, such probability is constant over the cellular life cycle. Third, the proportion of mutants in a population of clonal cells is very low. Fourth, reverse mutations, bringing a mutant back to the wild type, are negligible. Fifth, when the population is placed in a selective milieu, no mutations occur. Sixth, the probability of a mutation's occurring per each cellular life cycle is constant during the growth of the cellular culture.[8] An additional hypothesis is introduced to calculate the mutation rate per gene (or per nucleotide) per generation based on the mutation rate per genome per generation: a mutation has the same probability of occurring over a generation at each gene (or at each nucleotide) of the genome considered. As such, the mutation rate per gene (or per nucleotide) per generation is an average value, i.e., the ratio of the mutation rate per genome per generation over the total number of genes (or nucleotides). Thus, the fact that the mutation rate per generation can be different depending on the gene (or the nucleotide) considered is neglected in this context. It is also important to note that mutation rates evaluated using the Lea-Coulson model are then introduced, as parameters, in population genetics models to take into account the role of mutations in the evolution of natural populations (see, e.g., J. H. Gillespie 2004).

It is clear that the notion of "strong randomness" characterizes a certain number of hypotheses at the core of the Lea-Coulson model, in particular the hypotheses of the equal probability of mutations along the DNA sequence, of the invariance of the probability of a mutation over time, and of its independence from other mutational events. Nevertheless, all these hypotheses are simplifications (and biologists know that!) with respect to the real characteristics of the mutational process—more precisely, they are idealizations with regard to the state of knowledge about genetic mutations during the first half of the twentieth century and, more importantly, with regard to current research in molecular genetics on genetic mutation. I refer to the set of factors I call "mutational bias," which are likely to affect, in different degrees and ways, the site, type, moment, and rate of mutational events.

Although molecular geneticists acknowledge that at the molecular level every mutation is inevitably biased by a variety of physico-chemical factors, when using the Lea-Coulson model they *choose to ignore* these phenomena, taking its hypotheses as reasonable idealizations of the real process of mutation. Their objective is not to provide the most realistic description possible of what is going on at the molecular level when a change of the DNA sequence takes place. By making all these assumptions, allowing them to estimate average values, molecular geneticists rather aim at simplifying the estimation of mutation rate: their objective is to have a both general and precise model, i.e., a model they can use to estimate the mutation rates in a variety of different organisms.[9]

First of all, one should ask whether, from this point of view, the notion of "strong randomness" is actually suited for characterizing genetic mutations, even though it corresponds to an idealization with regard to what effectively happens at the molecular level. Do the simplified hypotheses of the Lea-Coulson model really have no impact either on the evaluation of the mutation rate or on the predictions obtained by evolutionary models of which the mutation rate is a parameter? The answer to this question would require a comparative statistical analysis of how alternative assumptions would change all these values, and I'm aware of none for the moment.

But then, even if "strong randomness" were a reasonable idealization of the real mutational process, I would still argue that if our aim is to characterize genetic mutations at the molecular level, i.e., as changes in the genome sequence and with regard to the physico-chemical processes at their origin, we should take into account the fact that the most recent results of research about mutations in molecular genetics significantly move away from the simplified hypotheses of the Lea-Coulson model. From this perspective—in other words, if we want to provide the most realistic (accurate, in Foster's terms) and precise description of the mutational process in order to characterize the random character of genetic mutations—I maintain that the notion of "strong randomness" invoked by the Lea-Coulson model is not empirically suited: it is too idealized and is far removed from what we know now about the mutational process.[10] As I will say later, better theoretical and experimental methods are now available to obtain data about the mutational process, in particular, to estimate the mutation rates (e.g., sequencing methods [Alphey 1997] improved by the use of new computing techniques that reduce the potential sources of error in the estimation; in situ real-time detection methods [Elez et al. 2010]).

To be clear, let me be precise that I don't intend to categorically exclude the (physical) possibility that a "strongly random" mutation could occur. Nevertheless, I want to stress again that the probability of such a change would be extremely low because of the variety and the abundance of factors that are likely to bias the mutational process.

NEW MUTATIONAL BIAS

Let us examine the notion of "weak randomness" to show that it is the appropriate notion, from the empirical point of view, for characterizing all genetic mutations at the molecular level. To this end, an analysis of the present state of research about the mutational process, in particular in molecular genetics, is needed.

But first, it is important to recall what biologists, in particular biologists of the modern synthesis, have already acknowledged about genetic mutation, during the first half of the twentieth century, on the basis of research results in classical genetics, and then, starting from the 1950s, further to the reformulation of evolutionary theory in molecular terms. My objective is to highlight that the way they conceived genetic mutations as random has not changed and corresponds to what I call the notion of "weak randomness."

Before the discovery of the molecular structure of the genetic material (DNA) in the 1950s, biologists of the modern synthesis described the causal sources of genetic variation, i.e., the processes of mutation and recombination, using the vocabulary of classical genetics, in terms of the modification of genes and chromosomes whose molecular structure was unknown at that time. They acknowledged that mutations are not equally probable across the entire genome: some genes are more mutable than others. In the 1950s and 1960s, the discovery of so-called hot-spots of mutation corroborated the fact that the probability of a mutation occurring is higher in some genomic regions because of the physico-chemical characteristics of the DNA sequence (Benzer and Freese 1958; Benzer 1959, 1961, 1962). Moreover, before the molecular turn in biology, biologists of the modern synthesis acknowledged that different types of mutations do not have the same probability of occurring, that a variety of physico-chemical agents (e.g., ultraviolet rays, mustard gas, etc.) can increase the mutation rate, and that the latter is under genetic control and can be enhanced by the presence of specific genes called "modificators."

After the discovery of the structure of DNA in 1953 and in the light of the first results of research in emerging molecular genetics, biologists of the modern synthesis reformulated in molecular terms all the physico-chemical mutational biases they had already admitted. As such, they acknowledged that some sites of the genome are more mutable than others due to the physico-chemical characteristics of the nucleotide sequence (for instance, in the case of mutational "hot-spots," the mutation rate is higher at some sites because of nucleotide repetitions inducing errors in the replication enzymes). The double-helix structure of DNA is also at the origin of differences in the mutation rate across the genome. They also admitted that some physico-chemical agents tend to produce mutations at the level of specific nucleotide sequences and that the presence of modifications in some genes involved in the production of specific enzymes triggers a global increase in the mutation rate. In the 1960s, they also started to admit that mutations can occur both during growth and stationary phases and that different rates of mutation characterize different steps of the developmental process. Finally, in the 1990s, many biologists started to argue that the discovery of mutator mechanisms that are both specific and adaptive (i.e., inducing a local increase in mutation rate in response to environmental stress) did not challenge the modern synthesis paradigm and, in particular, did not alter the chancy character of genetic mutation from the evolutionary point of view (see Merlin 2010).

Other biases concerning the origin, the rate, and the distribution of mutations have been observed only very recently (the first articles date back to the beginning of the 1990s; in particular, see Ninio 1991). Since the turn of the century, such biases have been studied in detail and have been the object of several experimental corroborations (see Drake 2007a, 2007b; Garcia-Villada and Drake 2010). Here we see the discovery of statistical correlations, probably due to causal links, between genetic mutations affecting the same DNA sequence: these correlations involve both the time and the site of the nucleotide sequence where the changes occur. Such biases certainly represent a significant advance over the way genetic mutations were conceived by the first geneticists, by the founders of population genetics and of the modern synthesis, and, more generally, by biologists just after the molecular turn. They also bring in new elements challenging the underlying hypotheses of the statistical methods used in molecular genetics to calculate the mutation rate (in particular, the hypotheses of the Lea-Coulson model). Let us take a look at these developments.

The history of their discovery begins in 1991, when Ninio observed microbial populations characterized by a temporary decrease in DNA replication fidelity, and so a temporary increase of mutants with two or more mutations (what Ninio called "double" or "multiple" mutations; Ninio 1991, 961). To account for his observations, Ninio suggested the following scenario: the hypothesis of "transient" or "phenotypic" mutators according to which double or multiple mutations are due to the temporary hypermutability of a microbial subpopulation. Such an explanation is plausible and still advanced by other biologists today; however, at the beginning of the 1990s it was rather speculative because it was not based on any experimental proof.

Some years later, Ninio (1996) noticed that multiple mutations (he called them "complex mutational events") can also occur in multicellular organisms (from *Drosophila* to humans). For these cases, he suggested the following two explanatory models. According to the first, multiple mutations are independent from each other and occur simultaneously, as in the case of "transient" or "phenotypic" mutations in bacteria. However, this first scenario implies too high a level of deleterious mutations that would not allow the organisms concerned to survive. According to the second explanatory model, preferred by Ninio, multiple mutations are the result of successive events that are correlated with both the time and the site of the DNA sequence at which they occur. In the context of this second model, Ninio proposed two hypotheses about the mechanisms at the origin of multiple mutations. According to the first hypothesis, an epigenetic mark (e.g., a methyl group) forms following a mutational event and then other mutations occur around the genomic site in question because of the presence of the epigenetic mark. According to the hypothesis preferred by Ninio today, multiple mutations occur successively at the level of "hot-spots" of recombination, due to the local heterogeneity of homologous chromosomes. In other words, multiple mutations occur during the process of "genetic conversion"—the alteration of a DNA strand that makes it perfectly complementary to the strand it is paired to. Even though Ninio's models and explanatory hypotheses are interesting and foreshadow recent experimental results on double or multiple mutations, no experiment was performed at that time (in the 1990s) to test them. So they could be considered simple theoretical speculation.

Ten years later, Drake analyzed several mutants generated in vitro by DNA polymerase errors in RB69 bacteriophages. The number of mutants with double or multiple mutations he observed was significantly higher than the number expected according to the classical vision of mutation distribution as following a Poisson law (both at the populational level and at the level of the genome sequence). Drake waited some years, then submitted the article "Too Many Mutants with Multiple Mutations," which was rejected because of its title. But in 2005, Drake succeeded with a new title: "Clusters of Mutations from Transient Hypermutability" (Drake et al. 2005).[11] Like Ninio, he suggested explaining the high number of double or multiple mutations by the temporary hypermutability of a subset of the cellular population.

During the past ten years, mutants with multiple mutations have been observed in both RNA and DNA viruses, in prokaryotes, in eukaryotes like yeast, and in the somatic cells of many multicellular eukaryotes (see Drake 2007b for further references). The transgenic mouse called Big Blue is a paradigmatic example of these studies (Buettner et al. 2000; Wang et al. 2007). In many mutants of this organism, two or more mutations occur at very close genomic sites: this is why molecular geneticists have started using the expression mutational "agglomerates" or "isolates" to talk about them. More precisely, they observed Big Blue mutants with multiple mutations (doublets, triplets, etc.) that are densely distributed (one mutation every three kilobases) in a specific and circumscribed genomic region (the *lacI* region and its surroundings within thirty kilobases).[12] They started talking about "mutational showers" when referring to this kind of mutational pattern: in contrast with the classical hypothesis of a Poisson distribution, genetic mutations are distributed in populations and in the genomic sequence according to an exponential law.

The molecular mechanisms at the origin of multiple mutations have not been precisely characterized yet. However, with regard to the distribution of mutations along the DNA sequence, molecular geneticists consider less plausible the hypothesis that such mutations are due to several mutational events occurring during successive replications of the same DNA molecule. Why is this the case? Because this hypothesis would imply a too-high mutation rate, which would be deleterious for the organisms concerned. For the same reason, molecular geneticists set aside the hypothesis that these mutations could occur independently from one another during the same replication process. They prefer the idea that multiple mutations occur synchronically (i.e., during the same replication) and that they are causally linked to each other. As for the distribution of mutations at the population level, they believe that the high number of mutants with multiple close mutations is probably *not* due to the activity of constitutive mutator mechanisms; rather, it is due to the fact that a subpopulation is temporarily hypermutable (see Ninio's hypothesis of "transient" or "phenotypic" mutants, which he formulated in 1991).

What do these new observations and the corroboration of the hypotheses suggested for accounting for them imply, particularly in reference to the way molecular geneticists currently describe and predict mutational events by using the Lea-Coulson model? And what is their impact on the way genetic mutations are conceived as random? Molecular geneticists should now acknowledge, first, that the probability of a mutation's occurring[13] can change, not just with regard to the physico-chemical characteristics of each particular site and its nucleotide context, but also because of the presence of other modifications of the DNA sequence in the immediate vicinity that turn out to be part of the causal story. Second, they should admit that such probability is not constant over time; on the contrary, it can change depending on the presence of a variety of mutator agents, on the particular stage of the developmental process, and also because of the temporary activation of some mutagen mechanism and of the presence of changes that have affected the same DNA sequence before. Thus, the discovery of all these new mutational biases expands the list of the physico-chemical factors that can affect the mutational process and provides one more reason to consider genetic mutations as "weakly random" events.

EVERY GENETIC MUTATION IS A "WEAKLY RANDOM" EVENT

In the light of current research in molecular genetics, I conclude that, from the empirical point of view, the more appropriate notion of chance for characterizing most genetic mutation—or even all genetic mutations—is the notion of "weak randomness."

Three objections might be raised to this view. The first is methodological: how is the probability of a mutation's occurring along the DNA sequence and over time evaluated? In fact, depending on the level one chooses to focus on (e.g., the nucleotide level, the level of a certain portion of the DNA sequence, the gene[14] level, the chromosome level, or the entire genome level) and on the time span considered (e.g., one generation, many generations, an arbitrary period of time, etc.), it could happen that the same mutation comes out both strongly and weakly random, which is obviously contradictory. For instance, the probability of a point mutation's occurring can be the same at each nucleotide of a given DNA sequence, but this could lead to different values when evaluated at the gene level. This would be due to differences in gene size (i.e., the number of nucleotides composing a gene). Conversely, the probability of a point mutation's occurring could be the same at every gene of a given DNA sequence, but different when it is evaluated at the nucleotide level. In this case, even though the probability of a mutation is different depending on the nucleotide considered, the sum of these probabilities for each gene would be the same. Moreover, the probability of a point mutation could be constant or could change over time depending on the period of time considered.

My reply to this first possible objection is that the random character of genetic mutations is defined at the molecular level, i.e., it considers mutations as changes in the DNA nucleotide sequence and with regard to the physico-chemical processes at their origin. In other words, the question here concerns the probability of a mutation's occurring with reference to the smallest and best-defined biological scale, which is the nucleotide level. It is at this level that I suggest evaluating the probability of genetic mutations. It is true that, from the practical point of view, it is very difficult—and perhaps even impossible—to evaluate the specific probability of all possible changes at the nucleotide level in a precise and objective way. However, from the theoretical point of view, and, in particular, with a view to a relevant definition of randomness in this context, I suggest that we speak of the probability of a genetic change with reference to the molecular level of biological reality, i.e., considering, in principle, every property and every physico-chemical factor that could have a causal role in the mutational process.

As regards the probability of mutations over time, I suggest evaluating it per cellular generation because it is a relatively small and, from the biological point of view, well-defined period of time. It corresponds to the temporal process leading from birth to reproduction (duplication) of a cell in a given organism.

A second possible objection to my view concerns the extension of the notion of "weak randomness." In fact, this notion could apply to a very large set of physical and biological phenomena, and that could be considered a problem for its conceptual value and relevance, in particular in the characterization of genetic mutations.

In response to this objection, I maintain that a notion *can* apply to a large number of objects and, at the same time, be relevant for characterizing them. Moreover, if it is an explanatory notion, its role is not defined by the extent of its applicability. For instance, if the notion of "weak randomness" were appropriate for characterizing every biological phenomenon, this would not mean that my argument is incorrect or that I haven't provided a good definition of the notion. Rather, this would show that all biological phenomena are "weakly random" (and this seems very plausible to me), or that "weak randomness" corresponds to a very general and unifying notion that can be developed into several more precise and specific notions of randomness.[15]

With regard to the specific case of genetic mutations, the notion of "weak randomness" has been shown to be appropriate in characterizing all genetic mutations (or, at least, the majority of physically possible mutations) from the molecular point of view. Indeed, it accounts for every physico-chemical factor that is likely to influence the mutational process (i.e., the mutational biases). By contrast, this does not mean that it isn't possible to limit the extension of "weak randomness" and to make this notion more precise in the context of each particular study. On the contrary, in the light of experimental data that are available today (e.g., as to the probability of mutations over time and along the DNA sequence, and about the independence or dependence relationships between mutational events), we might be able to determine whether certain kinds of mutations under study fulfill the C_1 and C_2 conditions (see above).[16] On this basis, a narrower and more specific definition of "weak randomness" could be formulated and would be appropriate, from the empirical point of view, for characterizing the mutations under investigation. In sum, "weak randomness" allows both capturing aspects that are common to all genetic mutations and making comparisons between different kinds of mutations and their respective random character.

A third possible objection follows the second one. It is about the meaning of the notion of "weak randomness": it might be objected that to say that an event is "weakly random" simply means that it is due to a stochastic process (as defined above in epistemological terms) or, in other words, that it is an event to which we ascribe a certain probability of taking place. However, I think that the problematic aspect of my definition of "weak randomness" does not lie in the fact that it is defined by referring to the probabilities used to describe an event and its generating process (Plutynski et al., this volume, chap. 3, show that all the authors of the modern synthesis often talked about chance as proxy for probability). Rather, the major issue here (which concerns every notion defined in terms of probability) has to do with the meaning and the role of the notion of probability invoked in the definition of this notion. The question of the interpretation of probabilities used to describe and predict a given phenomenon (in this case, genetic mutation) is central here, and its resolution could allow us to grasp the nature (objective or subjective) of weak randomness.[17]

Another advantage of raising the question of probability interpretation is that this could also provide some hints to address the metaphysical issue of the very nature (deterministic or indeterministic) of the process producing the phenomenon under investigation (in this case, the mutational process). But, as suggested at the beginning of this chapter, I maintain that this issue should be the final step of any philosophical analysis about the meaning and role of chance in biology and, more generally, in the natural sciences. In fact, there's a lot to be said about the way scientists describe and investigate natural phenomena (using a plurality of theoretical models and experimental techniques), and no meaningful argument about the metaphysical alternative between determinism and indeterminism can be advanced without a firm anchoring in scientific practice.

CONCLUSION

In this chapter I have shown that, from the empirical point of view, what I call "weak randomness" is the most appropriate notion for characterizing all genetic mutations at the molecular level, i.e., looking at the physico-chemical processes at their origin rather than at the adaptive value of their results at the phenotypic level. Indeed, by contrast with the notion of "strong randomness," "weak randomness" takes into account all possible mutational biases that are likely to affect the mutational process. So if our aim is to provide a realistic and precise description of the way genetic mutations occur, we should invoke this notion and specify it on a case-by-case basis.

My thesis has some important implications for the way genetic mutations are modeled. By pointing at the idealized character of the most-used theoretical and experimental models for estimating mutation rates, it suggests that they be reconceived in the light of the present state of knowledge about genetic mutations. In particular, molecular techniques of DNA sequencing are now available that are not based on the assumptions of the Lea-Coulson model and allow making good estimations of mutation rates by comparing sequence data over many generations. Moreover, these techniques have been improved very recently by the use of a new software called "DeNovoGear" (Ramu et al. 2013). It consists in a probabilistic algorithm that calculates the likelihood of mutation at each site of the genome based on estimations of mutation rates, sequencing error rate, and the initial genetic variation present in the population under study. This new method allows avoiding a number of potential sources of error in the estimation of mutation rates by DNA sequencing, including insufficient sampling of the genome, mistakes in the gene sequencing process, and errors of alignment between sequences.

Another alternative method for estimating mutation rates has been developed very recently. It aims at facing some limits of DNA sequencing techniques, in particular the fact that they are expensive and have limited ability to determine when mutations occur (Elez et al. 2010). This method uses a fluorescent fusion protein (MutL-GFP) to detect mutations as they emerge in living cells, independently of their potential phenotype. In fact, this protein both repairs DNA mismatches and forms green spots, visible with the microscope, when the mistakes are not repaired and are going to convert into mutations. So, the number of spots makes it possible to estimate the mutation rate in real time. Even though this new method has some limits too (see S. M. Rosenberg 2010), it's innovative because it allows calculating mutation rates before any selective process likely to bias the estimations and, above all, because it does not assume that genetic mutations are "strongly random" events.

NOTES

1. I use the terms *chancy* and *chance* to refer to the general idea of chance, including in its scope more mathematically connoted notions like "stochasticity" and "randomness."

2. See Millstein 2002 for an elaborated argument showing the relevance of the distinction between process and outcome. Specifically, she argues that this distinction is the key to conceptually distinguish natural selection and random drift, which are distinct as processes but not as outcomes.

3. As I said above, *chance* here is used as a general term, and the term *random* refers to chancy outcomes. Now, genetic mutations are "random" outcomes of the physico-chemical processes at their origin.

4. Another reason why I use the term *randomness* here is to distinguish the chancy character of genetic mutations at the molecular level ("randomness") and at the evolutionary level ("evolutionary chance"). For a detailed analysis of this last notion, see Merlin 2010.

5. See Beatty 1984 for an influential analysis of the importance of indiscriminate sampling processes, such as drift, in biology. See also Millstein 2002 for further developments about the distinction between indiscriminate and discriminate sampling processes in evolutionary biology.

6. I borrow the example of sticky and slippery balls from Brandon and Carson (1996), who use it to illustrate the phenomenon of genetic drift.

7. The fluctuation test consists in performing a statistical analysis of the distribution of mutants in parallel cultures to evaluate their mutation rate. For more details, see Luria and Delbrück 1943; Lea and Coulson 1949.

8. The other hypotheses of the Lea-Coulson model are as follows: cells grow exponentially; mutants and nonmutants have the same growth rate; cellular death is negligible; all mutants are detected; and the initial number of cells is negligible with regard to the final number of cells.

9. In Foster's words: "The goal of a fluctuation assay is to maximize the precision with which the mutation rate is estimated. Precision is a measure of reproducibility, not accuracy (accuracy is how well the resulting estimate reflects the actual mutation rate, and that will depend on how well the underlying assumptions reflect reality)" (Foster 2006, 200).

10. Levins (1966) argues that a model should aim at maximizing the three following desiderata: generality, realism, and precision. Nevertheless, this cannot be done, because these three desiderata don't have identical goals. Thus, he maintains that one of them must be sacrificed depending on the objective of the modeling process. I argue that, in the case of genetic mutations, we must sacrifice generality to realism and precision if our goal is to understand what happens at the molecular level when mutations occur and to accurately evaluate their chance.

11. Drake used the original title for another article he published in 2007 (Drake 2007b).

12. Similar multiple mutations have also been observed in *Escherichia coli* bacteria living in the gut of mice (see Hill et al. 2004).

13. Here I refer both to the probability of a mutation's occurring or not at a given site and to the probability of having a mutation at one site rather than another.

14. There is no precise and unique way to define what a gene is (see Keller 2000). In this chapter, I use the term *gene* to refer to a DNA segment situated at a precise place (locus) on the chromosome and producing a functional RNA molecule. Note, however, that this conception has been challenged by the discovery of genetic splicing in eukaryotic organisms, i.e., the process consisting in the elimination of portions of the nucleotide sequence (the introns) during the synthesis of RNA molecules.

15. In a paper published in 2011, Millstein identifies and defines what she calls the "Unified Concept of Chance" (UCC; Millstein 2011). As with the UCC, "weak randomness" is a very general notion that could seem too large to be interesting. With Millstein I claim that, on the contrary, such general notions are relevant because they

allow grasping what a plurality of chancy or random events have in common, and so allow for comparisons.

16. As explicitly stated by an anonymous referee, "weak randomness" comes in degrees. Thus, in the light of experimental data about genetic mutations, we might also be able to evaluate to what extent the mutational process under investigation fulfills these two conditions (C1 and C2).

17. The question at stake here is the following: does "weak randomness" refer to objective or subjective chance when it is used to characterize genetic mutations? In other words, does it refer to some property of the mutational process or, on the contrary, to our ignorance or partial knowledge of this process? I suggest that, in virtue of the privileged link between chance and probability, showing that the probabilities invoked in the definition of "weak randomness" are objective provides good reason to conclude that this is a notion of objective chance. For more details, see Merlin 2013.

CHAPTER 8

Parallel Evolution: What Does It (Not) Tell Us and Why Is It (Still) Interesting?

Thomas Lenormand, Luis-Miguel Chevin, and Thomas Bataillon

PARALLEL EVOLUTION: ONE PATTERN, MANY INTERPRETATIONS

Since the modern synthesis, the topic of chance in evolution has mostly focused on the role of genetic drift as a source of randomness in evolutionary trajectories, as opposed to the more deterministic evolutionary force of natural selection (Plutynski et al., this volume, chap. 3). However, the stochastic interplay of selection and mutation is also of fundamental importance for understanding the chancy nature of evolution (Lenormand, Roze, and Rousset 2009). Indeed, the response to selection is ultimately contingent on the occurrence of favorable alleles by mutation. Recently, the importance of chance in the interplay of mutation and selection has come to center stage in debates about parallel evolution, which is the topic of this chapter.

Following Arendt and Reznick (2008), we use both parallel and convergent evolution throughout to refer to the repeated occurrence of similar phenotypic or genotypic features in independently evolving populations. It is a pervasive pattern found across many biological levels of organization. Interestingly, this pattern has attracted many contradictory interpretations and has been used to support radically different views of evolution.

Interpreting "coincidences" is often tricky, especially to infer common cause (Sober 2012), and surely the observation of parallel evolution has led to vastly different causal interpretations. Outside evolutionary biology, it was used in a theological context as a "proof" of the existence of a conscious design. This is a variant of Paley's (1802) argument, according to which parallel change would indicate the reuse of the same template, hence proving the existence of a design, and of a designer. This line of argument predates Darwinism (e.g., Newton 1950 [originally 1710]) and has been rebutted elsewhere (see, e.g., arguments against design in Sober 2004; Coyne 2009). Parallel evolution has later been used as a strong argument against Darwinism. For instance, Mivart used it to argue in favor of "some unknown internal laws which determine variation at special times and in special directions" (1871, 311). Similarly, Bergson used convergent evolution as a proof of his *"élan vital"* and concluded: "Convergence does not appear possible in the Darwinian, and especially the neo-Darwinian, theory of insensible accidental variations" (1907, 83). Even within modern evolutionary biology, which is our focus here, we will argue that two broad and somewhat antagonistic interpretations of parallel evolution reflect a deeper and long-lasting tension between what is often termed a selectionist and a mutationist view of evolution. Even if this is not always fully recognized since the global adhesion to the modern synthesis, this antagonism persists today and still fuels many debates within evolutionary biology, dating back to the beginning of genetics (Stoltzfus 2006).

According to the "selectionist" view, variation is barely limiting, and therefore natural (or sexual) selection is the main driver of evolution in a given environment, as long as it overcomes random genetic drift. The environment is metaphorically seen as a mold in which species are passively shaped by natural selection, and evolution can be interpreted as a process by which organisms store heritable information about their environment (Maynard Smith 2000). From an empirical perspective, understanding evolution at any biological level thus requires only identifying and quantifying selective pressures operating at this level. Without going into a detailed historical account, this view has been championed by Darwin ("omnifarious" variation upon which selection acts); Weismann (pan-selectionism); biometricians such as Weldon and Pearson, Fisher (1918, 1930; micromutationism, emphasizing natural selection acting upon large amounts of available genetic variation), and Dobzhansky (1970), who again emphasized that large amounts of heritable variation was available and that response to artificial selection was effective on most traits. Even Lewontin (1974)—who cannot be viewed as a naive adaptationist—largely contributed to showing that there was ample standing variation on allozymes, upon which selection could act. This viewpoint was crystallized in the modern evolutionary synthesis by the idea that selection was the major driver of evolution (Beatty 2010).

The "mutationist" view, in contrast, emphasizes that the process of mutation determines and orients evolution at least as much as natural selection does. In this view, identifying selection pressure(s) is not enough to understand evolution, because the genetic variation required to respond to these pressures may not exist, and mutations may be available only in some specific phenotypic directions. Historically, this view has been advocated by supporters of "orthogenesis" (e.g., Eimer 1898), many Mendelian geneticists (Bateson's discontinuous evolution, Punnett's saltationism, de Vries's "mutation theory," or Morgan, arguing that variation was limiting; see, e.g., T. H. Morgan 1916, chap. 4), Haldane (1932, 75: "natural selection can only act on the variations available, and these are not, as Darwin thought, in every direction"), and Gould (1989; importance of "constraints" and contingency in macroevolution). The issue of mutation is often entangled and mixed up with a discussion about "constraints," whose status in the modern synthesis has been widely debated (Maynard Smith et al. 1985). Emphasizing that constraints limit variation within specific phenotypic directions is, however, just another way to say that mutations are not occurring in other possible directions. Note that this does not undermine the central neo-Darwinian tenet that mutation is random with respect to adaptation (i.e., mutations are not "directed"; Lenski and Mittler 1993). The distinction between these different concepts of "random mutation" was made early on, and especially in reference to the issue of parallel evolution (Shull 1935). The role of constraints is currently strongly advocated in the field of evo-devo, where the processes at the origin of genetic novelty are thought to be of critical importance for understanding the evolutionary process (Wake 1991; Gilbert, Opitz, and Raff 1996; Stern 2000; West-Eberhard 2003; Brakefield 2006; Salazar-Ciudad and Marín-Riera 2013 for a recent example), as well as in the study of molecular pathways (Zuckerkandl 1994) or molecular evolution (Nei 2007, 2013; Takahata 2007; Stoltzfus and Yampolsky 2009).

We emphasize that these views are extremes within a continuum and that many current evolutionary biologists tend to accept arguments from both sides. However, we argue here that parallel evolution is one of the topics that currently most strongly reveals the tension between these two views. On the one hand, parallel evolution is often used as a proof of adaptation, confirming the prominent role played by natural selection in determining the course of evolution (Losos 2011), mostly at the phenotypic level (see below). On the other hand, it is also used as indirect evidence that the set of available adaptive mutations (or combinations thereof) is very limited in a given environment and genetic context, causing the same mutations to spread again and again under similar ecological conditions, which argues against the idea that natural selection is the sole driver of evolution. The selectionist/mutationist views are sometimes labeled differently, in terms of external versus internal laws (Metcalf 1913), or functionalism versus structuralism (Dwyer 1984; Wake 1991), respectively. We opt for a terminology that emphasizes the relative roles of the underlying processes (mutation, selection), because it directly relates both to evolutionary theory (see below) and to the broader debate about the major evolutionary forces in the modern synthesis (fig. 8.1).

One of the ultimate reasons why parallel evolution is a subject of such active research may be that it materializes the selectionist-mutationist controversy sharply, by generating opposing, testable predictions (either minimizing

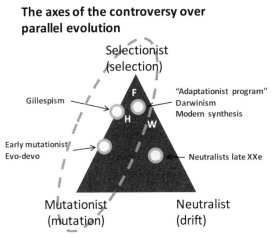

FIGURE 8.1. The "debate" triangle in evolutionary theory. Although the role of natural selection is widely acknowledged (all dots are toward the top corner), substantial differences persist among different views of evolution. The "positions" of the three founders of mathematical population genetics are indicated by letters: *F* (Fisher), *W* (Wright), and *H* (Haldane). Fisher advocated micromutationism, while Haldane emphasized the possible orienting effect of mutations (see references in text). The debate over parallel evolution often takes place along the left side of the triangle (the mutationist-selectionist axis). The role of drift in evolution is much more loosely linked to the topic (even though the role of biased mutation is becoming an important theme in neutral molecular evolution) and forms another important but distinct axis of controversy in evolutionary biology (Lenormand, Roze, and Rousset 2009; see also Plutynski et al., this volume, chap. 3).

or emphasizing the driving role of mutation in evolution). A more proximal/ mundane reason might be that experimental evolution (Garland and Rose 2009) combined with the development of genomics (Elmer and Meyer 2011) has recently provided the tools to experimentally investigate fine-grained patterns of parallel evolution at the genetic level. Parallel evolution is also often investigated for technical reasons, because it complicates phylogenetic reconstruction (the issue of homoplasy; T. H. Morgan 1923; Wake 1991; Wood, Burke, and Rieseberg 2005), but we will not address this issue further here. To argue our point, we will first expose the different patterns of parallel evolution. We will then focus on the different mechanisms at the origin of parallel evolution. We will also review the evidence in favor of selectionist and mutationist positions and discuss whether and how this antagonism can be resolved.

PATTERNS OF PARALLEL EVOLUTION

We first review the evidence for parallel evolution, categorizing it according to levels of biological organization.

Levels of Parallelism

Parallel or convergent evolution has been described at many levels of biological organization. In general, parallelism at lower levels (protein, gene, mutation) is thought to reveal constraints limiting the number of possible paths for the evolutionary process, consistent with the mutationist view, while parallelism at upper levels (organism, species, clade) is often taken as a proof of the power of natural selection (selectionist view).

   Genetic and genomic level. Many studies report evidence for independent fixation of the same mutation, or of another mutation at the same gene, in different populations (reviewed by Wood, Burke, and Rieseberg 2005; Gompel and Prud'homme 2009; Christin, Weinreich, and Besnard 2010; Stern 2013). While most of these cases are incidental, based on a few genes, populations, or species, recent experimental studies provide more general and quantitative results. Tenaillon et al. (2012) sequenced the full genome of one clone for each of 115 replicate *E. coli* populations adapting to high temperature and showed that over 20 percent of genes that fixed novel mutations were shared among replicate populations. Two meta-analyses of the literature (Conte et al. 2012; A. Martin and Orgogozo 2013) both concluded that gene reuse in different populations or species is a surprisingly common pattern of adaptive evolution.

Parallelism is also found in the origin of new genes. For instance, antifreeze glycoproteins of Antarctic fish all originate from a similar transformation of the tripsinogen gene, which happened independently in phylogenetically distant lineages leading to notothenioid fish and to Antarctic cods (L. Chen, DeVries, and Cheng 1997). Parallel whole genome duplications, amplification of gene families, and other modifications of the genome are also plausible. For instance, there is mounting evidence in model organisms, such as yeast, that adaptation in very diverged yeast lineages has entailed the selection of independent gene duplications and that gene duplications have occurred and subsequently have been retained independently in the same families in each of these two lineages to a far greater extent than is expected by chance alone (Hughes and Friedman 2003).

Protein and cellular function. Pelosi et al. (2006) showed parallel change in global protein expression profiles in *E. coli*, suggesting parallel evolution at genes specifically involved in the regulation of gene expression. In an experiment described above, Tenaillon et al. (2012) showed that ten broad gene classes with similar cellular functions (e.g., RNA polymerase complex, stress response mechanisms) contained more than 35 percent of all mutations, with substantial parallelism across lines for these units ($>30\%$). Roelants et al. (2010) showed the convergent evolution of the same toxic skin secretions in two distantly related frog lineages. Strikingly, this convergence involved different genes but parallel shifts in tissue specific expression in the skin.

Macroscopic phenotype. Parallelism and convergence at the level of the organismal phenotype include the most classic and well-documented cases. Historically, the case of phenotypic parallelism between mimics and models in butterflies was a source of dispute between a mutationist (Punnett 1915) and selectionist point of view (Fisher 1927). Recent and detailed examples include, for instance, ecomorphs of the island lizard *Anolis*, which exhibit convergent phenotypes between several islands (similar phenotypes are found in the same ecological niches), even though genetic distances are larger between islands within ecotypes than between ecotypes within islands (Losos et al. 1998). Other striking examples are found in sand lizards with convergent color evolution, involving mutations in the same gene (Rosenblum et al. 2010). Huey et al. (2000) showed that a latitudinal cline in wing length in *Drosophila subobscura* rapidly established after introduction to America, mirroring a cline that was previously reported in Europe. Travisano et al. (1995) showed, combining replicates starting from the same genotype of *E. coli* with replicates starting from distinct genotypes, that parallel evolution was observed for fitness and, to a lesser extent, for cell size, regardless of the starting conditions. An interesting and widespread example is given by the domestication of crops, which entails repeated phenotypic changes that are strongly favored under cultivation by man. These changes include loss of dispersal at maturity for seed crops or switches in apical dominance. Although these phenotypic changes can be achieved through different mechanisms at the genetic level, with selection of either a single major gene or several loci (and in fact this stereotypical variation was also interpreted in a mutationist perspective, as in Vavilov's "law of homologous series in variation"; Vavilov 1922), the phenotypic endpoint of evolution under domestication is strikingly repeatable (domestication syndrome; Ross-Ibarra, Morrell, and Gaut 2007).

Ecological function. Organismal phenotypes can be grouped into classes with similar function in terms of interaction with the environment and other organisms, and this may also be another level where parallelism occurs. This is particularly well known for the evolution of pollination syndromes in plants (S. D. Johnson and Steiner 2000) or parasitic morphology and strategies (Poulin 2011). This is also well documented for the evolution of ecotypic variation. For instance, McGee and Wainwright (2013) showed that benthic sticklebacks from different species evolved parallel ecological function (stronger suction, required to extract food from the benthos), but that this has led to diversification of the morphological traits involved.

Species formation. Similar environmental conditions may repeatedly lead to the origin of reproductive isolation, thereby causing parallel speciation. For example, benthic and limnetic species of sticklebacks found in sympatry in several lakes from northwestern America are thought to have originated independently in response to similar disruptive selection. This is supported by the fact that reproductive isolation is stronger between ecomorphs in a lake than between lakes for a given ecomorph, even though neither benthic nor limnetic ecomorphs are monophyletic (Rundle et al. 2000). Nosil, Crespi, and Sandoval (2002) showed that host plant adaptation led to the parallel evolution of reproductive isolation in *Timema* walking-stick insects. Parallel speciation was also documented recently in the cave-adapted fish *Astyanax* (Strecker, Hausdorf, and Wilkens 2012).

Macroevolutionary diversification. Adaptive radiation, the rapid expansion of lineages paired with phenotypic diversification, can be triggered in parallel in different clades by a similar increase in ecological opportunity or by a similar key innovation, which are thought to be prominent determinants of ecological speciation (Schluter 2000). For instance, Kozak, Mendyk, and Wiens (2009) showed that parallel adaptive radiation occurred in sympatry in two clades of salamanders in the North American Appalachian Mountains.

At larger time scales, Darwin used the convergent diversification of mammal phenotypes in Europe and America as evidence for natural selection.

Tendency across levels. An obvious, yet pervasive, consequence of nestedness in levels of biological organization is that parallel evolution may occur at a given level but not at levels below, because the mapping of one level to the next often entails some redundancy. For instance, parallel evolution of ecological function can occur without phenotypic parallelism, because of the many-to-one mapping between a phenotype and the function for which it is selected (Alfaro, Bolnick, and Wainwright 2005; McGee and Wainwright 2013). And phenotypic parallelism does not necessarily imply genetic parallelism: the cline in *Drosophila* wing size mentioned above was achieved through different genetic mechanisms in Europe and America (Huey et al. 2000). But parallel evolution at the genotypic level may be more likely between recently diverging populations/species with similar initial genotypes and phenotypes (Bollback and Huelsenbeck 2009). Tenaillon et al. (2012) quantitatively compared parallelism at several levels, confirming this effect of nestedness up to the level of cellular function in *E. coli* (fig. 8.2). However, to be more infor-

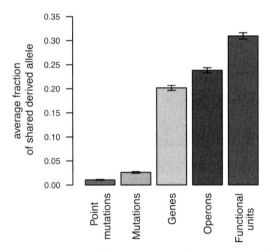

FIGURE 8.2. Amount of parallel evolution at different levels of biological organization, from mutation to functions. Parallel evolution may be more likely at higher levels of integration/variation. However, such a pattern would have to be compared to a neutral random baseline defining the set of possible variants, in order to establish whether the trend does not merely reflect the fact that there are fewer "units" at higher levels of integration (e.g., fewer operons than base pairs, and hence a higher chance to observe repeated evolution on operons). *Source*: Tenaillon et al. 2012, 460. Reprinted with permission from AAAS.

mative about the process underlying parallel evolution, such studies should correct for the number of possible units in each level, which determines the expected null level of parallelism (see below). Furthermore, the reverse pattern can also be envisioned for parallelism across biological levels: the same mutation fixing in different populations (genetic parallelism) may cause different phenotypic effects (no phenotypic parallelism), for instance because of epistatic interactions with other loci. We are not aware of clear examples of such a process.

Parallelism of State versus Parallelism of Process

While parallelism/convergence was initially investigated by considering the final states of evolved populations, which requires only comparative data, the advent of experimental approaches in evolution has led to an interest in parallelism in the evolutionary process itself. For instance in the above-mentioned case of sticklebacks, beyond the phenotypic similarity of benthic and limnetic species, what is striking is the parallel formation of new species under similar conditions, which is parallelism in process. Similarly, replicated laboratory experiments with extensive whole-genome sequencing of temporally spaced samples allow the exploration of parallelism in the very dynamics of selective sweeps (Lang et al. 2013). The dynamics of fitness increase over time in replicated *E. coli* adapting to glucose-limited medium is also very similar among replicates (Lenski and Travisano 1994), suggesting that that identical underlying processes are at work (see also Blount, this volume, chap. 10).

MECHANISMS OF PARALLEL EVOLUTION

We here review mechanisms that may cause parallel evolution, focusing on the genetic or phenotypic level.

Causes of Parallel Phenotypic Evolution

Over large evolutionary times, phenotypes generally undergo long stasis, suggesting that they are under stabilizing selection for an optimum set by the environment (Lande 1976; Kinnison and Hendry 2001; Estes and Arnold 2007; Gingerich 2009; Uyeda et al. 2011). Parallel evolution at the level of the organismal phenotype is thus often thought to result from exposure to similar environmental change causing similar shifts in the optimum. When parallel phenotypic evolution is found, it is used to argue for the simplicity of fitness surfaces, whereby only one or a few high fitness solutions exist within a given environment, even when many different phenotypes might be produced (degenerate phenotype-fitness map; Salazar-Ciudad and Marín-Riera 2013). If, in contrast, fitness peaks are separated by valleys of low fitness, different populations may evolve to different peaks by chance, that is, owing to genetic drift (Kirkpatrick 1982; Lande 1985). A mutation of major effect may also cause a population to shift from the domain of attraction of one peak to another (Agrawal, Brodie, and Rieseberg 2001), thus causing divergent phenotypic evolution in a common environment, which can be seen as a mutationist process. As mentioned above, phenotypic parallelism also depends on the functional redundancy among traits and on their hierarchy. More parallelism should be found for a trait that is the direct target of selection than for one of a number of underlying traits that affect another trait under direct selection (e.g., the various morphological features that affect benthic feeding in sticklebacks; McGee and Wainwright 2013).

Beyond selection toward a common optimum, parallelism in evolutionary trajectories may also occur if the combined effect of the direction of selection and the main axis of genetic variation (Lande 1979) are similar in different populations (Schluter et al. 2004). This process is often termed evolution "along genetic lines of least resistance" (Schluter 1996) and reflects the fact that phenotypic change over evolutionary time will happen as a compromise between selection and the available variation. Finally, extrinsic physical constraints may exist at large taxonomic scales on some traits, and these may be pervasive in shaping evolution of these traits even when considering distant species. For instance, metabolism and body size follow a power-law across all metazoans, partly (but not only) because metabolism is limited by diffusion in vascular systems (West, Brown, and Enquist 1997; Glazier 2010). Whether this regularity results from pervasive selective optimization (selectionist view) or the absence of variation (mutationist view) is difficult to judge, especially in the absence of any organism exhibiting large deviations from this pattern. Determining whether variation through new mutations does not occur at all, or does occur but cannot be observed because these mutations incur large *pleiotropic* deleterious fitness effects, is often very difficult. One important approach to distinguish between the two hypotheses is to perform artificial selection (Brakefield 2006).

Pleiotropic effects of mutations play a central role in both selectionist and mutationist viewpoints. According to the Fisherian/selectionist view, variation is not much constrained in its phenotypic direction, but it is constrained in phenotypic magnitude. Large phenotypic mutations are thought to be most deleterious due to their deleterious pleiotropic effects (as in Fisher's geometric model of adaptation; Fisher 1930). In contrast, in the mutationist view, variation is limited in phenotypic direction, but much less so in magnitude. In particular, mutations with large phenotypic effects are not necessarily associated with large deleterious pleiotropic effects (for instance, because the phenotype is modular, such that mutations impact only a small subset of traits, or are expressed in only a few organs). These views have long been strongly opposed (see, e.g., the debate over Goldschmidt's "hopeful monsters"; Dietrich 2003). We will return below in more detail to the issue of pleiotropy in the context of parallel evolution.

Causes of Parallel Genetic Evolution

Genetic parallel evolution can occur for a number of reasons, but in contrast to phenotypic parallel evolution, it is most often seen as a consequence of limited or stereotyped variation. Tinkering in evolution (Jacob 1977) proceeds by reuse and modification of existing genes and sequences. The reuse of the same building blocks increases the chance that similar genetic changes will occur independently. Furthermore, some mutations are more likely to occur than others. For instance, gene duplications are more frequent than point mutations increasing promoter strength. Hence, under a selective pressure for a higher dose of a specific protein, similar gene duplications (or amplifications) are likely to occur repeatedly in different populations (e.g., Raymond et al. 1998). Similarly, many mutational processes are biased (e.g., there is an almost universal bias toward AT mutation while gene conversion tends to also be biased in favor of GC nucleotides), and this bias will sometimes conflict with the direction of evolution driven by natural selection (Yampolsky and Stoltzfus 2001; Charlesworth 2013).

Next, the mutation rate (u) may be variable, and more mutable targets may be more likely to evolve repeatedly and cause parallel evolution (see also Merlin, this volume, chap. 7). Different loci can have different u for different reasons. If we consider that loci are genes, then they can differ in sequence length and therefore constitute mutational targets with different sizes (the same applies to quantitative traits; see, e.g., Houle 1998). Genomic location may also influence the mutation rate for double strand breaks, insertion/deletion, transposable elements, duplications, and repeat variation (D. N. Cooper et al. 2011; Hodgkinson and Eyre-Walker 2011), although perhaps less for point mutations (Kumar and Subramanian 2002).

Beyond these stereotyped mutational variations and heterogeneity in u, the pool of possible mutations available to change particular traits in a specific direction may be very limited. Parallel evolution is very likely to occur if there are only a few genetic changes (or even only one) that can confer a given phenotypic change. For instance, a single amino acid change in the acetylcholinesterase gene seems to confer high resistance to organophosphorous insecticides, as indicated by the repeated occurrence of this mutation in different genera of mosquitoes. Even more strikingly, only the species with codons allowing for the acquisition of this amino acid in a single point mutation evolved this resistance mechanism (Weill et al. 2003). With respect to variation in mutation effects on traits, this situation corresponds to restricted pleiotropy: there are directions in phenotypic space that are accessible only if mutation occurs at a narrow set of genes. A corollary is that the number of phenotypic directions available by mutation at each gene is low compared to the total number of traits under selection. As we will see below, it is thus possible to define an effective number of loci that can respond to selection in any given environment (Chevin, Martin, and Lenormand 2010), and the lower this number, the higher the chance of parallel evolution at the gene level.

This phenomenon can extend beyond one mutational step, raising the issues of epistasis among mutations fixed at different steps and mutation order effects (Mani and Clarke 1990). Whether or not epistasis favors parallel evolution is complicated by several factors and leads to contradictory views (Achaz et al. 2014). In many cases, the substitution of a first mutation will influence which mutation is likely to fix next, thus increasing or decreasing the likelihood of subsequent changes (positive and negative epistasis, respectively). The first substitution can then lead to divergence in adaptive trajectories. This process can occur with multiple adaptive peaks or simply by compensatory evolution with a single phenotypic peak (mutations correct the pleiotropic deleterious effects of previous substitutions in an "amelioration" process; Cohan, King, and Zawadzki 1994). It causes "historical contingency" (Merlin, this volume, chap. 7), which can be viewed as playing against the repeatability of evolution and lowering the amount of parallelism (Travisano et al. 1995; Wichman et al. 1999; Blount, Borland, and Lenski 2008; Kolbe et al. 2012; Bedhomme, Lafforgue, and Elena 2013). Indeed, in this situation, particular mutations are beneficial only in a specific genetic background and would be observed only after a series of specific previous substitutions had occurred. On the other hand, subsequent substitutions are not independent of each other due to epistasis, which greatly reduces the number of selectively favorable trajectories for a given set of possible substituted mutations and hence increases the chance of parallel evolution (Weinreich et al. 2006).

A convincing example of the role of epistasis as a constraint was investigated by Tenaillon et al. (2012) in reference to parallel evolution in the above-mentioned experiment on *E. coli* strains adapting to high temperature. They investigated the pattern of epistasis among multigenic functional groups ("units"). They found that mutations in these units clustered in few groups, exhibiting within-group positive epistasis and between-group negative epistasis, a pattern consistent with the existence of few competing, but historically contingent, adaptive trajectories. Within units, strains had many fewer multiple mutations than expected, indicating that the occurrence of a beneficial mutation prevented fixation of further mutation in that unit (negative epistasis). A similar pattern has been demonstrated for substitutions within genes by Woods et al. (2006). This is likely to occur, for instance, when selection favors a loss of metabolic function: the first mutation disrupting the metabolic pathway, regardless of where it occurs, is the only one that will be favorable, while all further mutations will be neutral (S. D. Smith and Rausher 2011). In all such situations, there will be much less parallel evolution among evolutionary replicates than expected from the number of beneficial mutations at the first step. In the extreme case of loss-of-function, the trajectory reduces to a single substitution step, annihilating all the parallelism that could have occurred in further steps. But the more general phenomenon is likely to be widespread, and it is crucial to understanding patterns of genotypic parallel evolution.

One last important consideration in the study of parallel genetic evolution is that different occurrences of a specific genetic change may not be due to independent mutation events. In particular, a given genetic variant may be maintained at low frequency in a "reservoir" population and could increase repeatedly in frequency in peripheral populations experiencing the same selection pressure. In this situation, gene flow is responsible for the spread of the same allele in different populations, which may prevent the fixation of other putative beneficial alleles not yet introduced by mutation (such as *eda1* in sticklebacks; Colosimo et al. 2005; Barrett and Schluter 2008; see also Bierne, Gagnaire, and David 2013). From a mutationist point of view, such a situation would not qualify as parallel evolution, since mutation occurred only once and spread by gene flow (it has even been argued recently that reduced gene flow, by favoring the spread of different equivalent alleles in different parts of the range, favors parallelism at the gene level; Ralph and Coop 2010). However, from a selectionist point of view, such a situation reflects parallel processes of natural selection acting independently, but in the same direction, in different peripheral populations. At the genetic level too, parallel evolution has a dual nature: it can be promoted both by the independent occurrence of the same genetic variant and/or by the independent occurrence of similar selective pressures (we will come back to this point when considering a more formal model for parallel evolution).

MODELING PARALLEL EVOLUTION

Mutation versus Selection as the Driver of Parallelism

A handful of models have investigated theoretically the drivers of parallel evolution (Orr 2005b; Chevin, Martin, and Lenormand 2010). These models are fairly simple and focus on the essential ingredients (mutation, selection), without dwelling on the complications introduced by some of the processes mentioned above (epistasis, gene flow). Although this simplicity may be seen as a limitation, it also makes it more straightforward to compare the roles played by mutation and selection in parallel evolution, and hence to evaluate the relative merits of mutationist and selectionist interpretations of this phenomenon.

Theory typically considers a collection of genetic variants (mutations at different loci, or alleles at the same locus) in order to make (probabilistic) predictions about which locus will contribute the next favorable mutant that will arise, escape stochastic loss, and sweep to fixation, thereby conferring further adaptation in the population. With little loss in generality, these models stipulate that mutations are the main source of variation and not other forms of inheritance (see Haig 2007 for a discussion of epigenetics). The underlying framework generally is the strong-selection-weak-mutation (SSWM) approximation by J. H. Gillespie (1983a, 1983b, 1984, 1991). In this limit, beneficial mutations are rare and selection is strong enough that a new beneficial mutation is either lost—because of demographic stochasticity—or undergoes fixation instantaneously. Thus there is no standing genetic variation in the population, and two beneficial mutations never compete (no selective interference), so the fate of a mutation depends only on its own selection coefficient and the effective population size. This considerably simplifies the mathematics of such models, which become a simple Markov chain, characterized by its transition probabilities and waiting times (time for establishment of a mutation escaping genetic drift). The analysis of this highly idealized condition yields simple analytical results and, more importantly, reveals some of the key factors underlying parallel evolution, with the caveat that the role of standing variation in driving patterns of parallelism is not addressed. Some models assume an arbitrary selection coefficient for the mutants, while others try to derive the distribution of selection coefficients from first principles. In the simplest approach, beneficial mutations are considered to occur in the right tail of the genomic distribution of mutation effects, which is assumed to follow some specified "extreme value" behavior (J. H. Gillespie 1983a). Properties of the extreme value distribution are thus used to find the probability that substitutions occurring in different populations involve mutations with the same fitness "rank" (best, second best, and so on; Orr 2005b). Such extreme value behavior can also be derived from explicit phenotypic landscape models (G. Martin and Lenormand 2008). Alternatively, and more directly focused on the question of phenotypic and genotypic parallel evolution, it is possible to investigate the question in a phenotypic landscape model with heterogeneity in mutations among genes. We develop here this more comprehensive theory.

Following Chevin, Martin, and Lenormand (2010), we consider an idealized genome with n_L loci (or sites) that can mutate at rates $u_1, u_2, \ldots, u_L$, with fitness effects $s_1, s_2, \ldots, s_L$. The latter are drawn from arbitrary and potentially locus-specific underlying distributions of fitness effects $f_L(s)$. Each locus then has its own "substitution potential" P_L, i.e., its probability of spawning the next substitution (unlike Chevin, Martin, and Lenormand 2010, we do not restrict our attention to beneficial mutations; for instance, neutral mutations can have a higher substitution potential than deleterious mutations). Denoting as $\pi(s)$ the fixation probability of a mutation with selection coefficient s (where s can be zero or even negative), the mean probability of fixation of a new mutation at locus L is

(1) $P_L = \int f_L(s)\, \pi(s)ds,$

where the integral is taken over appropriate bounds (i.e., the range of possible selective effects at locus L). Loci exhibiting a large fraction of beneficial mutations of large effects will have a high substitution potential.

We now consider patterns of parallel evolution between a pair of populations (although this can be generalized to a higher number of populations). Specifically, we focus on the probability (noted $p_{//}$) that the next substitution in the two populations diverging from a common ancestor occurs at the same locus. Assuming for simplicity that mutation rate and adaptive potential are not correlated among loci, we have (Chevin, Martin, and Lenormand 2010, eq. 8b)

(2) $P_{//} = \dfrac{\left(1+CV_g^2(P)\right)\left(1+CV_g^2(u)\right)}{n_L},$

where CV_g denotes coefficient of variation across loci in the genome. If all loci have the same mutation rate, $CV_g(u) = 0$, and substitution potential, $CV_g(P) = 0$, they are equivalent and parallel evolution is at its minimum ($p_{//} = 1/n_L$). This formula results from simple considerations about waiting times in Poisson-type processes, but a formal derivation is beyond the scope of this chapter (see Chevin, Martin, and Lenormand 2010 for more detail).

This simple expression (which has some connection to a decomposition of the "substitution spectrum" derived independently by Streisfeld and Rausher 2011) yields several heuristic insights. First it shows that the number of loci n_L will trivially always decrease $p_{//}$. Patterns of parallelism are thus expected to be strongly dependent on the number of sites in the genome that can yield beneficial mutations (or, more generally, the number of genetic variants with high substitution potential). In that respect, comparison of amounts of parallelism across different species might need some rescaling, especially if genome sizes are vastly different (say, e.g., virus versus bacteria versus eukaryotes). Similarly, pooling multiple loci into larger categories (i.e., nucleotides into genes, genes into gene units with similar function) automatically decreases the number of units (smaller n_L), thus necessarily inflating observed parallelism; if the entire genome was the unit of comparison, then parallelism would have to happen. To be more informative on causes of parallelism, studies such as that of Tenaillon et al. (2012) should therefore correct for the number of units at each level, so as to scale parallelism relative to a "neutral" benchmark. Given n_L mutable units (individual sites, genes, or group of genes), $1/n_L$ is a very simple baseline value, which corresponds to the expected probability of parallel evolution under the hypothesis that all units are equally likely to produce beneficial mutations. Qualitatively similar insights would be obtained when phenotypic rather than genetic parallelism is considered: n_L would reflect the level of "constraint" (how many physiological or developmental solutions there are to achieve the optimal phenotype), $CV(u)$ would indicate how heterogeneous these solutions are in terms of mutational occurrence, and $CV(P)$ how heterogeneous they are in terms of fitness.

Second, theory shows that under the SSWM selection regime (see the box for details), the likelihood of parallel evolution is *equally* increased by heterogeneity in mutation and substitution potential. In this respect, it plays the mutationist and selectionist views off against each other. The mutationist camp can invoke mutational heterogeneity as "sufficient" to explain patterns of parallel evolution, and similarly for the selectionists. Overall, both mutation and selection can cause parallel evolution (see Haldane 1932, 76, for the same insight about parallelism: "Related species will vary in similar directions and be subject to similar selective influence"). At the extreme, consider the scenario where all loci have the same adaptive potential (same s), but where a single loci is M times more mutable than the remaining $n_L - 1$ loci. The waiting time for the next beneficial mutation will be order on the order of 1 for this locus and M for the remaining loci. As M increases, $p_{//}$ will thus tend toward 1. A symmetric argument can be made with selection being stronger at a single locus. However, although mutation-rate heterogeneity matters as much as fitness-effect heterogeneity in the SSWM regime, it may have more limited influence under selection regimes characterized by greater levels of interference among mutations (i.e., more polymorphism, very low recombination). The intuition is that if beneficial mutations are typically introduced in appreciable numbers in every generation ($N u \, n_L \gg 1$), populations will harbor some amount of standing variation, and the probability that a focal beneficial mutation undergoes fixation is no longer merely a function of the average amount of genetic drift and its intrinsic effect but will be much more dependent on the presence of other beneficial mutations. Predicting amounts of parallelism is not trivial under those selection regimes, but selection is expected to play a greater role in conditions characterized by more competing beneficial mutations (see the box).

PREDICTING THE EXPECTED PROBABILITY OF PARALLELISM $\left(p_{//}\right)$ DURING ADAPTATION

Intrinsic factors (genetic architecture, amount of pleiotropy, etc.) underlying phenotypes under selection, together with extrinsic factors (type of environment, initial level of adaptation), determine the mutation rates and the underlying distribution of fitness effects at each locus (shaded box at top; each s_i is intended to represent a distribution of selection coefficients, rather than a single value). These can be predicted (upper single arrow) via extreme value heuristics (J. H. Gillespie 1991; Beisel et al. 2007; Joyce et al. 2008; G. Martin and Lenormand 2008), or using explicit fitness landscape models (G. Martin and Lenormand 2006; Chevin, Martin, and Lenormand 2010). Then the mutation supply (Nu) dictates what "selection regime" operates on the mutations. Three possible regimes are distinguished depending on the expected level of selective interference between mutations (depicted as unshaded boxes). Under each selection regime, expressions can be developed to predict the amount of heterogeneity among loci for their propensity to produce the next beneficial mutant to be substituted

(P_i's). These expressions are exact in the case of SSWM, and so far rely on heuristics confirmed by computer simulations in other regimes (Bataillon and Blanquart, unpublished data; Nagel et al. 2012). Altogether, these expressions reveal that the weight of selection in determining propensities (P_i's) increases with the supply of beneficial mutation, with the parameter γ controlling the bias towards selection as a determinant of parallelism. Based on these, we can predict (small arrows at bottom) the expected patterns of parallelism (box at bottom).

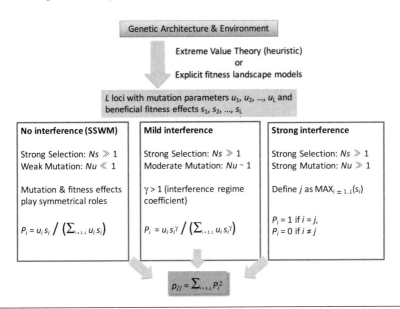

Third, and perhaps less appreciated, is the fact that even in the absence of mutational heterogeneity, observing higher than neutral parallel evolution does not guarantee that adaptation is taking place. The value of $p_{//}$ increases as soon as $CV_g(P) > 0$, which, in principle, can occur in the absence of beneficial mutations. For instance, if all loci produce only strongly deleterious mutations, except one that produces neutral mutations, there will be highly parallel evolution at this latter locus. Considering time-to-substitution, however, can distinguish the latter situation from a situation involving adaptation (neutral substitutions take a much longer time to fix; Kimura 1980).

In conclusion, the theory available so far indicates that the mere observation of a given pattern of parallel evolution cannot be used either as definite proof of adaptation or as a proof for mutational constraint. More information is required about variation in mutation rates and fitness effects across loci. For instance, Streisfeld and Rausher (2011) recently partitioned the contributions of mutation and of fixation probabilities to variation in substitution rates, in a data set on the anthocyanin pathway of flower color in plants and a larger data set on animals. Making sense of this information, especially with reference to a variable environment, is much facilitated by the use of fitness landscapes.

Phenotypic Mutation Landscape Models

Assuming a fixed and ad hoc distribution of fitness effects (DFE) fails to account for the fact that these DFEs—which may vary across loci—are themselves caused by various factors at the phenotypic level. Fitness effects often result from selection operating on phenotypic traits, and generally in an environment-dependent manner. Making this genotype-phenotype-fitness link is notoriously difficult, which limits our understanding of the relationship between parallel evolution at the genetic and phenotypic levels. One way to approach this question is to incorporate a specific and "constrained" mapping, using some informed mechanistic model of development (as is done in evo-devo approaches; see the introduction to this chapter). A more abstract and general alternative is to use phenotypic mutation landscape models, for instance assuming that fitness has a local maximum for an optimal combination of traits (Fisher 1930; Haldane 1932; Hartl and Taubes 1998; Orr 1998, 2005a, 2005c, 2006; Waxman and Welch 2005; G. Martin and Lenormand 2006, 2008; Gros, Nagard, and Tenaillon 2009; Chevin, Martin, and Lenormand 2010). This approach focuses our attention away from purely locus/allele-based models, because it relies explicitly on a set of causative traits for fitness. Such models can be used to make predictions at higher levels (i.e., phenotypes and fitness itself), and, in principle, such models could make joint predictions at all three levels (gene/locus, phenotype, fitness), which could be confronted quantitatively with empirical data stemming from, e.g., experimental evolution. Furthermore, this approach can yield important insights into the phenotypic origin of heterogeneities in adaptive potential, which can be parsed at the phenotypic level into *orientation, magnitude,* and *pleiotropy* heterogeneities (Chevin, Martin, and Lenormand 2010, fig. 1). It is important to note that Fisher's model, in its original formulation, predicts almost no parallel evolution at the genetic level, since all mutations are supposed to occur in any direction. Because parallel evolution at the genetic level has been known for a long time, this was a clear and strong argument against

Fisher's view of evolution (see, e.g., Shull 1935). In effect, to address the issue of parallel evolution, the original model needs to be modified.

For concreteness, assume that the phenotypic effects of mutations at locus L are drawn in a multivariate normal distribution with mean 0 and covariance matrix $\mathbf{M}_L$ and that selection can be described in this phenotypic space by a multivariate Gaussian fitness function (with covariance matrix $\mathbf{S}$). In this model, mutations may show preferred directions in phenotypic space: they are not random-in-direction. However they are not "directed": they do not point preferably toward the optimum (the mean phenotypic effect is zero and independent from the direction of the optimum). Hence, these two features alleviate the mutationist criticism against Fisher's model (that it totally ignores mutational constraint), while retaining the central Darwinian idea that mutations are random with respect to adaptation.

The fitness effect of mutations at locus L will depend first on the relative *orientation* of the $\mathbf{M}_L$ and $\mathbf{S}$ matrices. For instance, a first locus may exhibit mutations along phenotypic axes under strong selection, whereas another locus may mutate along axes with less intense selection. When adapting to a new optimum, the former locus is more likely to contribute to adaptation than the second. As intuitively expected, this effect of orientation is magnified when loci mutate only a small subset of traits (rank of $\mathbf{M}_L \ll$ rank of $\mathbf{S}$), because then only a small subset of loci can contribute to adaptation in a specific direction. Less intuitively, this effect of orientation is stronger closer to an optimum (Chevin, Martin, and Lenormand 2010), where the curvature of isofitness surfaces is stronger (far from optimum, all loci have 50% beneficial mutation, regardless of their orientation; Fisher 1930). Second, the *magnitude* of phenotypic effects may differ across loci. Different loci may mutate in the same phenotypic directions but cause smaller or larger phenotypic effects (which are measured by the trace of $\mathbf{M}_L$). This will largely impact the proportion and effects of beneficial mutations among loci. When adapting to a new optimum, loci with intermediate phenotypic effects overall will be more likely to contribute to adaptation (Kimura 1983). Hence heterogeneity in $\mathrm{Tr}(\mathbf{M}_L)$ will increase $p_{//}$. Third, different loci may exhibit different degrees of pleiotropy, i.e., some may affect few traits while others could affect many (this is measured by heterogeneity in the rank of $\mathbf{M}_L$). This latter form of heterogeneity can have complex effects on the outcome of parallel evolution. Indeed, a more pleiotropic locus is more likely to be able to mutate in the phenotypic direction favored by natural selection, but it is also more likely to produce mutations that "go wrong" because they explore wrong phenotypic directions (one of the costs of complexity of Fisher 1930; Orr 2000). Therefore, whether or not loci that are more pleiotropic contribute more to adaptation (higher P_L), thus increasing parallel evolution, is not obvious. Theoretical results indicate that for the first step of adaptation to a new environment, pleiotropy heterogeneity increases parallelism only modestly, except in situations where very little adaptation occurs (Chevin, Martin, and Lenormand 2010).

Extending the Model and Unanswered Questions

The theory presented above provides the minimal ingredients necessary for a comprehensive theory of parallel evolution. Many features could be added to build on this approach. We assumed for simplicity that two populations were diverging from an identical ancestor and evolving under identical conditions. This is the most favorable situation for parallel evolution, and it corresponds to the case of experimental evolution where replicate lines founded by the same genotype are evolving in a repeatable and identical abiotic environment. When considering sets of natural populations that have already diverged and that experience different environmental conditions, different sets of substitution potentials would be needed to compute the likelihood of parallel evolution, which will result in lower $p_{//}$. Selection regimes characterized by higher amounts of interference might be very hard to predict (but would not necessarily induce less parallelism; see the box). Indeed, as population mutation rates increase from $Nu \approx 1^{-10}$ to $Nu \gg 1$, the system moves from competition between single mutants to competition between genotypes that combine several mutations, for which pairwise or higher order epistasis needs to be considered. Similarly, predicting parallelism after more than one step of adaptive evolution implies deriving expectations for the magnitude of fitness effects in successive steps, which requires further assumptions, notably regarding epistasis (see above). Yet, considering this multistep scenario is an important extension, as parallel evolution is a central and recurrent theme in experimental evolution studies characterized by replicated "bouts" of adaptation, with identification of numerous substitutions. For instance, considering only Lenski's long-term experiment on *E. coli*, the issue of parallel evolution was raised in more than fifteen papers (Lenski and Travisano 1994; Travisano et al. 1995; Travisano and Lenski 1996; V. S. Cooper et al. 2001; T. F. Cooper, Rozen, and Lenski 2003; T. F. Cooper et al. 2008; Crozat et al. 2005; Crozat et al. 2010; Pelosi et al. 2006; Woods et al. 2006; Philippe et al. 2007; Philippe et al. 2009; Blount, Borland, and Lenski 2008; Ostrowski, Woods, and Lenski 2008; Barrick et al. 2009; Meyer et al. 2010). Blount (this volume, chap. 10) mentions several other cases. Finally, a nontrivial extension would be to revisit the above questions considering the joint roles of standing (i.e., preexisting) genetic variation and newly arising mutations, as well as obvious features that characterize evolution in nature, such as a changing environment and possible instances of frequency-dependent selection.

CONCLUSIONS AND OPEN QUESTIONS

The Fascination of Pattern and the Issue of Prediction

Patterns of parallel evolution are understandably fascinating because they can be seen as actual realizations of the "replaying life's tape" thought experiment (see Beatty 2006b), especially when replicated experiments can actually be performed on large temporal scales (e.g., on microbes), going back and forth in time (e.g., using frozen bacteria samples). In evolutionary biology, as in other historical sciences or retrospective thinking, one inevitably wonders whether "things would have been different" if a particular event had been changed in the past (as in the reverse evolution problem; Teotónio and Rose 2001). Debates about determinism are also pervasive in physics, where the question of time reversibility has been discussed extensively (e.g., Prigogine and Stengers 1984). To some extent, this may explain why parallel evolution is often studied in itself, for the sake of pattern. There is even now a strong tendency to use the topic of parallelism to justify the importance of studies carried out in genomic and experimental evolution research (see above). But parallelism is, after all, only a pattern, which is not that interesting if nothing is learned about the underlying processes (see also Rainey 2009, a news and views summary of Barrick et al. 2009).

Beyond the mere fascination of the pattern, parallel evolution indicates that some sort of regularity is at work. This regularity may be seen as a sign that there is an underlying law (or a combination of source and consequence laws; Stoltzfus 2006) that entails that evolution is repeatable and somewhat predictable. Given that evolutionary biology is mostly a historical and retrospective science, such a conclusion inevitably conveys the hope that developing an evolutionary "predictive science" is not beyond reach (even if such a view could, but need not, flirt with teleology; Conway Morris 2010). This issue is complicated by the fact that the different levels (phenotypic or genotypic) of parallel evolution tend to involve antagonistic processes in terms of prediction. For instance, the complete absence of genetic parallelism is compatible with precise prediction at the phenotypic level: the response to selection on a trait with polygenic inheritance will exhibit little genetic parallelism, yet could be nicely predicted at short to intermediate time scales. Along the same line, a "statistical" (as opposed to reductionist or mechanistic) treatment and calibration of mutational input, at least for fitness effects (G. Martin and Lenormand 2006), opens the possibility to build fairly general predictive evolutionary models (rates of adaptation, fitness trajectories) that ignore the genetic and functional basis of the phenotypic change. This prospect is, however, compromised by strong genetic parallelism. Indeed, if mutation effects are restricted to very few loci, a purely statistical modeling of mutation effects at a set of exchangeable loci will not likely be accurate enough to make a long-term prediction, and a more specific investigation of the nature of each locus and mutations involved becomes required. This may facilitate prediction of the functional and genetic basis of adaptation, but only after the process has been established and observed in some case study. This is exemplified in the candidate-gene approach, where, e.g., pesticide resistance alleles are successfully investigated across widely distant species. However, with a very limited set of possible mutants, the waiting time to the next adaptive substitution can be highly stochastic and difficult to predict over short to intermediate time scales. Overall, as we have seen, parallel evolution can have multiple causes and thus be unpredictable if the underlying processes are not properly identified and quantified. In contrast, the absence of parallel evolution can have a single cause that can be easily quantified and predicted. Overall, the issues of prediction and parallel evolution thus tend to come apart.

Several outstanding questions and prospects emerge regarding the causes of parallel evolution. The first is to be able to devise methods to determine the relative contribution of selection and mutation (see the interesting approach proposed by Streisfeld and Rausher 2011). In the case of genotypic parallelism, a second issue is to understand and describe the structure of pleiotropy (either universal or limited to "modules"; Klingenberg 2005; Wagner and Zhang 2011) and how such structure evolves (e.g., Jones, Arnold, and Bürger 2007). In the case of phenotypic parallelism, a third issue is to obtain precise information on the topography of phenotypic-fitness landscapes (single versus multiple peaks; Lobkovsky and Koonin 2012). These questions are best understood within a defined phenotypic space. Yet, such spaces may be difficult to circumscribe empirically, as the possible number of measurable traits is infinite in any circumstance. Whether these questions can be answered with either an arbitrary set of traits or only by statistically considering mutations and their fitness effects remains unclear. In any case, the probability of parallelism after one single adaptive substitution can be predicted from simple measurements of the mutability and underlying DFE of new mutations at each contributing locus. In principle, both mutability and DFE are quantities that can be estimated from direct experiment (although until recently this was experimentally prohibitive but for a few special cases) or inferred from long-term patterns of substitution (by making a number of assumptions about what drives substitution rates). Such investigations would be helpful to get a more balanced view of the respective roles of mutation and selection in driving parallel evolution.

Moving beyond Mutationism versus Selectionism

Selectionists usually emphasize phenotypic parallel evolution and the "mold" of the environment, while being puzzled by the observation that genotypic parallel evolution is also widespread. They tend to consider that mutations are random in phenotypic direction but limited in magnitude. They tend to conclude that the degree of parallel evolution decreases as we move from the level of fitness down to phenotype, genes, and finally particular base pairs. Mutationists, in contrast, emphasize genotypic parallelism (sometimes cascading up to the phenotype level), especially the fact that the number of possible mutations (or their combination or trajectory) is very limited, contributed by very rare events (duplications, etc.) in a given order, or that mutations occur on multiple "adaptive peaks" (although these peaks represent protein function more often than fitness itself). They tend to consider that mutations are restricted in few phenotypic directions but are not much limited in magnitude. They tend to consider evolution as fundamentally unpredictable and to be puzzled by strong phenotypic parallelism with no genetic parallelism.

The modern synthesis popularized the idea that evolution consists of changes in gene frequencies, with selection playing the primary role. This synthesis did not consider it essential to precisely incorporate and model the origin of variation and novelty. This is best seen in population genetics models that consider frequency change of given alleles in a "gene pool" but do not consider why these specific alleles are around, or in quantitative genetics models that consider selection on standing variation but do not explicitly account for a process at the origin of this variation (although the mutation process may be implicit in the maintenance of genetic variance). In both cases, mutations have a minimal role: they are assumed to be available and thus to not limit adaptation.

Some have advocated for a new "synthesis," taking into account the details of, e.g., development, especially with respect to morphology (Müller

2007; Pigliucci 2007; S. B. Carroll 2008). But a stronger emphasis on mutationist arguments goes beyond the question of the genetics of development, as it applies with equal force to any phenotype (e.g., metabolic, behavioral, etc.). This question also goes beyond reconciling population and quantitative genetics. The issue at stake is not whether trait variation is Mendelian or polygenic. It is to build models of evolution that incorporate both the process at the origin of variation, and natural selection sorting through that variation and generating adaptation as a by-product. Phenotypic landscape models can allow us to make significant progress in that direction, by providing a theoretical ground on which to build more complete models, thus patching the "blind spot" of the modern synthesis. In particular, such models may be very useful in understanding and quantifying the different processes at work behind patterns of parallel evolution. As such, they have the potential to put the evolutionary significance of parallel evolution in a much clearer and more quantitative context. More generally, better understanding and modeling of the stochastic interplay between selection and mutation—in addition to the better-studied process of random genetic drift—is key to determine and quantify how much evolution is chancy and contingent.

ACKNOWLEDGMENTS

We thank A. Stolzfus, J. Bridle, G. Martin, and an anonymous reviewer for insightful comments. We also thank C. Pence and G. Ramsey for their comments at various stages of the writing. This work was supported by ERC Grant QuantEvol to T. Lenormand, an FNU Grant and a visiting professor grant (Université Montpellier II) to T. Bataillon, and the ContempEvol grant from ANR to L.-M. Chevin.

* 3 *

*Chance and Contingency
in the History of Life*

CHAPTER 9

Contingent Evolution: Not by Chance Alone

Eric Desjardins

Darwinian evolution is commonly characterized as a historical process. This characterization may seem obvious if we think of evolution as continuous and long-lasting, involving the modification of life forms into different life forms. But extended time, continuity, and change are not the only properties that make evolution historical. Several biologists and philosophers have qualified evolution as historical in virtue of its *contingent* nature. For some, like Stephen Jay Gould, evolution is highly contingent, and contingency is the hallmark of any historical process: "Contingency is a license to participate in history" (Gould 1989, 285). As emphasized by Beatty (2006b), this type of claim can be vague because Gould (along with many others) uses the expression "contingency" to emphasize two aspects of evolution: unpredictability and/or causal dependence (for more meanings of "contingency," see also Erwin, this volume, chap. 12). The former is illustrated by a well-known thought experiment according to which the replay of evolution from an earlier stage would result in significantly different life forms. Causal dependence, on the other hand, means historicity, i.e., that history matters because the past is a difference maker. Gould was not the only biologist to insist on the historical and unpredictable qualities of evolution. François Jacob, a few years before Gould, had expressed similar views in a paper entitled "Evolution and Tinkering" (Jacob 1977). He expresses there his dismay about the general and popular belief that evolution was bound to produce life forms as we know them on our planet: "The surprising point here . . . is what is considered possible. It is the idea, more than a hundred years after Darwin, that, if life occurs anywhere, it is bound to produce animals not too different from terrestrial ones; and above all to evolve something like man" (1161). Jacob argues that Darwinian evolution invalidates this conception, which ignores the importance of history. "It is hard to realize that the living world as we know it is just one among many possibilities; that its actual structure results from the history of the earth. Yet living organisms are historical structures: literally creations of history. They represent, not a perfect product of engineering, but a patchwork of odd sets pieced together when and where opportunities arose. For the opportunism of natural selection is not simply a matter of indifference to the structure and operation of its products. It reflects the very nature of a historical process full of contingency" (Jacob 1977, 1166).

This quote nicely reinforces Beatty's point about the tendency of biologists to easily jump from one notion of contingency to another. In fact, we see not only two but three notions of contingency in these few sentences. First, the expression "just one among many possibilities" refers to contingency as not necessary. Second, in the expression "full of contingency," Jacob refers to contingency per se, or objective chance. Finally, when he says "results from the history of the earth," Jacob uses the causal notion of contingency. It is tempting to conclude from this passage that Jacob conflated these different notions. Perhaps, on this occasion, but a more charitable reading would suggest that he presents evolution as a combination of all these forms. He conveys the message that the present could have been otherwise because it causally depends on past contingencies, or, as he likes to say, historical "opportunities." Jacob, like Gould, looked at the imperfections of evolution as inevitable. From the strange "monsters" of evolution that often capture our imagination to the fact that a large proportion of embryos and fetuses in humans will spontaneously abort, life is depicted as the result of a historical process during which coincidences and opportunities arise and accumulate in an unpredictable and unique order. The whole fabric of evolution is formed from random parts that happen to be available at certain times and places (compare with Gould's "panda's thumb"; Gould 1980a). Using Simpson's (1952) terminology, Jacob describes a "net historical opportunity"—formed by the union of physical, ecological, constitutional, and genetic opportunities—"that mainly controls the direction and pace of adaptive evolution" (Jacob 1977, 1166). Evolution thus "guided" cannot but be unique: "Even if life in outer space uses the same material as on the earth, even if the environment is not too different from ours, even if the nature of life and of its chemistry strongly limits the way to fulfill certain functions, the sequence of historical opportunities there could not be the same as here. A different play had to be performed by different actors" (Jacob 1977, 1166). Aside from pointing to the fact that Jacob's views greatly resemble the ones later promoted by Gould—as if Jacob's paper murmured what Gould preached out loud in books and papers—these passages make a perfect starting point for the issues I discuss in this chapter. Like Jacob, Gould, and many others, I want to reflect on the *historical* and *contingent* aspects of evolution. But instead of focusing on their biological implications, I want to suggest an interpretation for their entanglement and highlight important differences both in their roles and in their conditions of realization. As suggested above, I see Jacob and Gould as presenting a view of evolution in which contingency and historicity happen in tandem. Yet, it is possible to isolate the one from the other, both conceptually and empirically. One of the main objectives of this chapter, then, is to show that historicity, i.e., the idea that historical "opportunities" matter in evolution, does not follow merely from the fact that evolution is a chancy process.

In order to make this point, I will build from my recent work (Desjardins 2011) and show that chance and dependence on the past can be integrated in the notion of path dependence (defined in the next section). My other objective is to go beyond the abstract criteria for path dependence and look at important instances of path dependence in evolution: phylogenetic constraints and generative entrenchment. The latter notion, championed by Bill Wimsatt and Jeffrey Schank, will help us to think more generally about the type of circumstances leading to historicity in macroevolutionary and macroecological processes.

WHEN HISTORY MATTERS

The following brief and rather abstract introduction to the notion of path dependence will serve as a conceptual basis for the rest of the chapter. As I argued earlier (Desjardins 2011), we can distinguish between two basic forms of historicity: (1) the commonly recognized phenomenon of dependence on initial conditions and (2) the less discussed (but more interesting) phenomenon of path dependence. I believe that the latter best captures what biologists mean when they claim that evolution is both contingent per se and historical. The focus will therefore be on this second notion.

When applied to personal individual histories, the idea of path dependence is fairly intuitive. It means that what/where a person has been during her life affects what/where she is now and will possibly be in the future. For example, had Willemby not decided to go for a run along the river after it rained a few Sundays ago, he would not have slipped on a wet rock and broken his arm, nor would he thus have a more fragile wrist for the rest of his life. These aspects of his life clearly depend on past contingencies about the weather and the decisions he made. Note, however, that not everything about him presents the same dependence on the past. The color of his eyes and the general structure of his body are traits that we could consider more resilient to the series of contingent events and decisions that followed his arrival in this world.

If we are to use this notion of historicity in the context of evolution, we have to look beyond personal individual histories and apply it to larger groups like populations and species. Nevertheless, the intuitive and more casual example mentioned above conveys two important ideas. First, it suggests that historicity comes in degrees. As subsequent chapters in this volume clearly indicate, (evolutionary) history can matter a lot or very little for the occurrence of a given outcome (see especially Erwin, this volume, chap. 12). This could be interpreted in at least two ways: either (1) different historical circumstances will lead to the evolution of little-to-very different outcomes, or (2) the probability of the occurrence of a given outcome will be little-to-very different depending on which historical circumstances take place. Second, this example entails that things could have been otherwise. This is because path dependence relies on a branching conception of history. Consider figure 9.1. Each tree represents multiple possible historical trajectories from a given initial state (S_0) of a system. Historical trajectories, or paths, are complete ordered series of events. The nodes in these trees represent the occurrence of particular events at a given time. Depending on the phenomenon investigated, events in a historical trajectory can be simpler (e.g., a point mutation) or more complex (e.g., the acquisition of a new developmental pathway by an evolving group of organisms). A path always encompasses *at least two* subsequent and causally connected events. This means that a single event is never a path in itself, and for an outcome to be *path* dependent, it has to be itself preceded by a series of at least two subsequent and causally connected events. In this chapter, I will consider the occurrence of state S_0 as the initial event,[1] whereas the last event of a historical trajectory is called the "outcome." The notions of "initial state" and "outcome" don't always apply to the absolute beginning and end of a processes. In fact, in most cases these terms will respectively designate the beginning and end of a scientific investigation on entities that can predate and/or continue to evolve after observation. This means that inferences about the path (in)dependence of outcomes will often be related to contextual and descriptive aspects of a situation. This is not to say that the events described are merely epistemic or that path dependence is not a real property. Although empirical inferences often have instrumental limitations, it does not mean that events are absolutely tied to their description or observation. As a last impor-

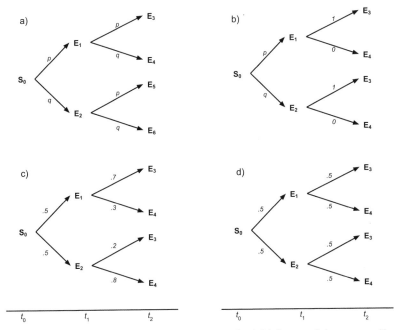

FIGURE 9.1. Branching trees. S_0 represents the initial state of the system. Each E_i represents the occurrence of an event. The p's and q's on the vertices stand for probabilities, where $(p + q = 1)$. The different t_n on the time line indicate instants. Scenarios: (*a*) path dependence, (*b*) path independence (condition 2 fails), (*c*) path dependence with partial convergence, (*d*) path independence (condition 3 fails). See text for detail.

tant point of clarification, the nodes in these graphs represent types of events. A given event can occur in more than one trajectory or be recurrent in a given historical trajectory. Nevertheless, we assume that time flows only in one direction and particular histories remain distinct once they have branched off one from another.

Let us now look at the conditions needed for path dependence to occur. First, there must be alternative paths from a given initial state. This criterion is always met if we are in a stochastic setting with extended historical trajectories. Second, there must be a genuine possibility to reach alternative outcomes. By "genuine possibility" I mean that there must be at least two biologically realistic outcomes with positive probability. In other words, it should make sense from a physical and biological point of view to say that "things could have been otherwise." So a closed system with a single strong point attractor will not meet this condition in the long run, although it can still be transiently path dependent if it is initially in a nonequilibrium state and there exist alternative routes to this outcome. Finally, at least one of the historical paths taken by a system must be a difference maker. This means that following different paths must affect the probability that a given outcome occurs. Path dependence can thus be summarized by these three conditions:

1. Alternative paths from an initial state,
2. Alternative outcomes, and
3. Probabilistic causal dependence between paths and at least one outcome.

Consider figure 9.1 again. Provided that p and q are greater than zero, all three conditions are met in scenario (a)—in fact, any situation where every path leads to a different outcome will be path dependent. In scenario (b), however, only one outcome, E_3, has a positive probability, thus violating condition 2. Looking at scenarios (c) and (d) now, we see that both meet conditions 1 and 2, but one of them is path dependent and the other path independent. Despite its partially convergent structure (i.e., the same outcomes appear in the final instant of alternative historical trajectories), scenario (c) is path dependent because the probability of the outcome E_3 changes as a function of the path taken, and ditto for E_4. Conversely, scenario (d) is path independent because it violates condition 3: the probability of occurrence of each possible outcome remains the same regardless of the path taken.

This last verdict proves that path dependence means more than contingency per se. In order to be able to say that a process is contingent in the latter sense, one only has to say/prove that "things could have been otherwise." But according to the definition adopted here, path dependence also requires that the path followed by a system is a difference maker and affects the probability of the outcome reached at a given instant. The next section will show more clearly why this is the case. In fact, an evolutionary process that would correspond to the one represented in scenario (d) could be qualified as random, because the probability of any admitted outcome is the same regardless of the path followed by the system. It would fail, however, to be historical in the stronger sense developed here.

MEASURING THE IMPORTANCE OF CHANCE AND EVOLUTIONARY HISTORY

Defining the conditions in which history matters is relatively easy compared to testing the hypothesis that it does so. Although we know many aspects of the evolutionary trajectory of several lineages, in most cases we can only guess at the trajectories that could have but did not occur, or did occur but did not leave enough traces for us to find out about them. Without knowledge of alternative evolutionary paths and outcomes, it is virtually impossible to decide whether history has been a difference maker. Moreover, even if we were able to know what these possible trajectories were, we would still have to assess different conditional probabilities of outcomes. All this information is obviously extremely difficult to obtain from natural evolution. So, discriminating empirically between the hypotheses of chance and historicity can be rather complicated.

The previous section showed that we can distinguish, at least conceptually, between scenarios where history is a difference maker and the merely random ones, where different paths leave the probability of a given outcome unchanged. Ideally, we would also be able to discriminate between the following two hypotheses on the basis of observational evidence:

H_c: Only chance plays a role in the evolution of these lineages.
H_h: History plays a role in the evolution of these lineages.

The problem is that these hypotheses are partially supported by the same kind of evidence and disconfirmed by the same evidence. Gould's "Replaying Life's Tape" thought experiment (1989) is perhaps the most commonly cited scenario for testing the chance hypothesis. In brief, Gould asks what would happen if we could rewind life's tape to a certain point in the past and let it run again. If evolution were to proceed in the same way, then evolution would not be contingent. The noncontingent scenario, sometimes called convergent, parallel, or deterministic evolution, is what Conway Morris (2003) and de Duve (1995) have been arguing for. For them, the evolution of life forms on this planet was meant to be roughly as we know it. This type of evidence disconfirms both H_c and H_h. If, on the other hand, we were to observe some divergence between the original run and the replay, then evolution would be contingent. Gould defended this position on many occasions, and as we saw, Jacob thought the same in 1977. The difficulty arises because we cannot distinguish between H_c and H_h on the sole basis of divergence. So the same type of evidential support seems to hold for H_c and H_h.

If the only method available to test these two hypotheses was the Replay experiment, then the discussion would be of very little interest to scientists. Fortunately, experimental evolution in highly controlled and tractable conditions can offer a solution (see also the following chapters in this volume).

One of the most impressive and fruitful such experiments is the Long-Term Experimental Evolution (LTEE) project directed by Richard Lenski at Michigan State University. In 1988, Lenski and his collaborators produced twelve cloned populations of a certain strain of *E. coli* and let them evolve independently in identical environmental conditions. No genetic engineering took place after t_0, so the only aspect that remains controlled is the environment in which these bacteria evolve (i.e., temperature and the nutrients composing the soup in which the microbes grow). All twelve lineages have completed more than fifty thousand generations since the beginning of the project (a summary of the experiment and links to many publications are provided at http://myxo.css.msu.edu/ecoli/; see also Blount, this volume, chap. 10).

It is common to read in the literature produced by this group that the LTEE project is creating the conditions imagined by Gould in his Replay thought experiment. Although I agree that there are similarities between the LTEE project and Gould's Replay experiment (for example, they both allow us to test whether evolution is affected by chance), I also think that there are significant differences. Leaving aside the qualms about the possibility of materializing exactly the conditions of the Replay thought experiment,[2] it is perhaps more accurate to say that the LTEE project creates twelve "plays" from the same evolutionary starting point. There have been moments when the evolution of a given lineage has been "replayed" from different stages (e.g., Blount, Borland, and Lenski 2008; more on this later), but these variations were not the motivation for the whole LTEE project. For present purposes, one of the most important differences between the Replay experiment and the LTEE project is that the latter created the conditions allowing for the discrimination of the chance and history hypotheses.[3] This new possibility comes essentially from the fact that the LTEE project does not only run twelve plays from the same starting point, but it also creates a library, by freezing a sample of each lineage every five hundred generations. So the researchers can actually return to, consult, and investigate the possibilities of various evolutionary pasts for all these lineages.

Given this experimental setting—initially nearly identical life forms evolving in nearly identical and controlled conditions—it makes sense to expect that all lineages should proceed and adapt to their environment in the same way. Therefore, the absence of chance and history could have been the most expected outcome at first. As of today, results show that some variables have evolved in parallel, whereas other have significantly diverged.[4] The fact that Lenski and his collaborators have observed some divergence confirms that chance or history plays a role in the evolution of certain aspects of these microbes. The question is, How can we distinguish empirically between the two types of phenomena?[5]

Although at a broad descriptive level, chance and historicity allow for the possibility of divergent evolutionary trajectories from the same starting place, at a finer descriptive level, they entail different patterns of divergence. In other words, the evidential support for chance and historicity are distinct enough to allow the formulation of different observational predictions. These predictions will take the following general forms:

Chance: If chance is an important factor in the evolution of these lineages, then we should observe divergence of type D_c.

History: If history is an important factor in the evolution of these lineages, then we should expect divergence of type D_h.

Before looking at how the LTEE group has been operationalizing this distinction, let us reflect for a moment on what has been said in the previous section about path dependence and try to specify in what ways D_c and D_h should differ. The only kind of case that will occupy us here should have multiple paths and multiple outcomes from a given initial state (like scenarios [c] and [d] in fig. 9.1). This entails that chance always plays a role—but history might not.

To put this in the context of evolution, let us suppose that the initial state for a set of populations is to have a trait or allele G_0. Now, imagine that some change in the environment occurs, such that two equally likely and immediately reachable mutations M_1 and M_2 could occur, both providing the same fitness advantage. In these circumstances, we would expect that about half the lineages would acquire M_1 and about the other half would acquire M_2. From these new genotypes or phenotypes, the lineages could either acquire mutation M_3 or M_4 with equal probability; i.e., M_3 and M_4 are equally likely to occur and they both provide the same fitness advantage. About half the population should therefore acquire state M_3 and about half M_4, regardless of which mutation they had evolved previously. Here, one would rightly conclude that chance plays a role, but not history. And because all outcomes are equally likely, we can say that this evolutionary process is purely random. (A graphical equivalent of this random scenario is represented in fig. 9.1d.)

Second, another type of divergent scenario, represented in figure 9.2a, involves chance alone but does not qualify as random per se, for one outcome may be more likely to occur than another one. Yet, this divergence would not be explainable in terms of path dependence; regardless of the path taken by a lineage, the probability of outcome M_3 is always the same, and ditto for M_4. In this scenario, some lineages take the evolutionary path P_1: G_0–M_1 and some take the path P_2: G_0–M_2. It does not matter whether one path is followed more often than the other (this is why I leave the probabilities unspecified for the first evolutionary stage). What matters for the investigator interested in deciding between H_c or H_h is the proportion of lineages evolving a given outcome M_i from P_1 compared with the proportion of lineages evolving *the same outcome* M_i from P_2. Because these proportions are the same, we disconfirm that history is a difference maker but not that chance plays a role in the evolution of the lineages. This presumes, of course, that we have the means to isolate and compare the two groups, i.e., that we can know which evolutionary path has been followed by which lineage. From there, we can understand why the scenario in figure 9.2b would allow our investigator to favor H_h over H_c. Just like the previous case, some lineages acquire M_3 via P_1 or P_1, while others acquire M_4 via the same P_1 or P_2. But unlike the cases of path independence,

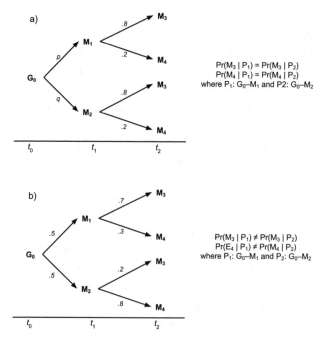

FIGURE 9.2. Branching trees representing divergent evolutionary scenarios: (*a*) chance only plays a role—revealed by equality in conditional probabilities for each outcome; (*b*) chance and history play a role—revealed by inequality in conditional probabilities for each outcome.

the proportion of lineages evolving a given M_i from path P_1 is significantly different than the proportion of lineages evolving the same M_i from path P_2.

Let us now look at one of the results obtained by the LTEE group, suggesting that a certain divergence in traits among the twelve lineages is possibly due to differences in history, and not a mere chance result. From Blount, Borland, and Lenski (2008), we learn that after approximately thirty-one thousand generations, one of the lineages evolved the capacity to metabolize citrate (hereafter Cit^+), which gave them an enormous boost in fitness and cell size. This result was surprising for different reasons. First, since citrate has been part of the bacteria's environment since the very beginning of the experiment (and a possible source of energy), one would have expected to see this advantageous trait appear earlier. Second, if one lineage did manage to evolve this trait, then it means that it was also among the possible traits for all lineages. Yet, only one of them acquired it. Blount, Borland, and Lenski (2008) reasoned that this could be explained by two hypotheses: chance or history. The chance hypothesis stipulates that this adaptation is extremely rare, but it does not require a particular evolutionary history to occur. The history hypothesis stipulates that this phenotype is rare and it requires a particular evolutionary history before becoming accessible. In other words, on the history hypothesis, the authors hypothesized that only this one lineage went through the right series of mutations in the right order, thus increasing the probability that the Cit^+ phenotype would occur.

The evidence sought by the group testing these hypotheses falls directly in line with H_c and H_h presented above. They created the conditions that would reveal whether something in the history of that lineage made the evolution of this new Cit^+ phenotype more likely. This is where the frozen library is of central importance. They took several samples of the lineage at different generations, some very early and others closer to the generation at which the Cit^+ phenotype manifested itself. Then they let these different temporal replicates evolve for several generations, thus creating multiple replays. The experiment did not produce only two groups, each with different evolutionary paths, but we can think of the setting as producing two types of groups: a first type replaying evolution from a distant past, and a second type replaying evolution from a recent past. If the historical hypothesis was true, then the distant replays would tend to remain Cit^-, whereas the recent replays would evolve Cit^+ much more frequently. Conversely, if the distant replays and the recent replays presented on average a similar number of Cit^+, then the historical hypothesis would be disconfirmed and the chance hypothesis favored.

The results obtained suggest that the historical hypothesis could be true.

The proportion of lineages evolving Cit$^+$ was higher in recent replays than in distant replays. They thus suggested that some mutations happened along the way (at ~20,000 generations) that provided some genetic background required for the subsequent evolution of the Cit$^+$ phenotype. However, as rightly recognized by the group, the alleged potentiating mutations remained unknown after this experiment. In order to further secure the historical hypothesis, the group needs to identify the potentiating mutations that allow the Cit$^+$ phenotype to go from potential to actual. So the results provided in Blount, Borland, and Lenski (2008) provide only partial evidence for path dependence. There is still a third possible scenario, only later discussed by the researchers, that would invalidate their conclusion that the Cit$^+$ phenotype requires a specific order of mutations.[6] Despite the collected evidence, it remains possible that the rare Cit$^+$ phenotype is a complex trait that requires numerous genetic mutations, but unlike the historical contingency hypothesis, the mutations can arise in any order. Consider the possibility that the Cit$^+$ phenotype and the original clones in the LTEE project were quite far apart in the sequence space of possible genomes. It may not matter which mutations happen first or second, but enough mutations are required to go from Cit$^-$ to Cit$^+$ that acquiring all of them takes a long time (~31,000 generations). Imagine that there are six mutations required and that a lineage acquires one of them (regardless of order) about every 5,000 generations. Then we reasonably expect the lineage at generation 20,000 to be more likely to evolve Cit$^+$ given another 10,000 generations of time than a sample at generation 5,000 would be. The probability of evolving Cit$^+$ in n new generations would increase linearly with the number of generations already elapsed. No path dependence is involved in this third scenario, but it is a biologically plausible explanation for the observed experimental "replay" data.

Putting this third scenario and the history one to the test requires that the team track more closely the mutations involved in the expression of the new trait. Blount et al. (2012) explain that this has been done by retracing the "phylogenetic history" of various Cit$^-$ and Cit$^+$ lineages. They discovered that there were at least three Cit$^+$ variants throughout the experiment, some being stronger than others. They also discovered that potentiation involves at least two mutations and that the expression of the new phenotype involves the amplification (duplication) of a genetic module (*rnk-citT*). In the 2012 paper, the authors distinguish between the history and the chance hypotheses thus:[7] "If expression of Cit$^+$ required earlier mutations, then the *rnk-citT* module should confer a weaker Cit$^+$ phenotype in a non-potentiated background than a potentiated background. Alternatively, if potentiation facilitated the amplification event itself, then that module should produce an equally strong Cit^+ phenotype in both potentiated and non-potentiated backgrounds" (Blount et al. 2012, 516). The first of these two alternatives implies path dependence, whereas the second alternative is equivalent to chance (for the expression of the complex phenotype is not sensitive to the presence of earlier mutations). Blount (this volume, chap. 10) provides a more detailed account of the experiment and of the possible mechanisms of potentiation. Here, I only wish to highlight that the evidence discussed in the 2012 paper suggests that the history hypothesis favored in the 2008 paper was most likely the correct one. In a nutshell, the attempt to introduce the *rnk-citT* module in the ancestral genome was not a great success, and the bacteria with the potentiated background had a greater increase in productivity on citrate than the ones that had not been potentiated.

HISTORICITY, CONSTRAINTS, AND GENERATIVE ENTRENCHMENT

The previous sections offered a general account of historicity on the basis of criteria and explained how it is possible to discriminate, conceptually and empirically, between chance and historicity in microevolutionary stochastic settings. In the remainder of this chapter, I will move to macroevolution and explore the relation between historicity, evolutionary constraints, and generative entrenchment. I will show that generative entrenchment is an important reason why evolution is contingent, but not by chance alone. Wimsatt (2001) makes a similar argument, but his analysis does not elaborate the connection with the notion of path dependence. In brief, I will argue that generative entrenchment, in virtue of being a source of phylogenetic constraints, generates the conditions for path dependence at the macroevolutionary level. Moreover, I will show that this interpretation can also be extended to macroecological processes.

Although the expression "constraint" typically carries negative or limiting connotations, I believe, as Gould (2002) and Amundson (1994) explain, that evolutionary constraints can also have a positive (or creative) effect. In fact, the example discussed in the previous section of the bacteria requiring a *potentiating* genetic background for the evolution of the capacity to use citrate as a source of energy in aerobic conditions could be seen in this more positive light (see also Blount, this volume, chap. 10). Evolutionary constraints, thus more neutrally and generally understood, are like attractors in the evolutionary state space that produce a shift in the probability of outcomes.

The work of Price and Carr (2000) and Price (2003) on phylogenetic constraints and oviposition in sawflies illustrates this very well. According to their account, phylogenetic constraints possess three essential properties. First, they should themselves be conserved evolutionarily, which means that they persist in lineages. The persistence of phylogenetic constraints may result from the fact that the character is "tightly integrated into developmental programs" (Price and Carr 2000, 646). Being tightly integrated results in some sort of phylogenetic inertia because changing them would most likely produce malfunctions. We will come back to this hypothesis later, which closely resembles the notion of "generative entrenchment" developed by Wimsatt (1986, 2001) and Schank and Wimsatt (1986). Second, phylogenetic constraints should set key aspects of the ecological interactions and the selective regime of a taxon. For example, they can be features that determine species' modes of alimentation, locomotion, or reproduction. Third and finally, by virtue of the first two properties, phylogenetic constraints limit the major adaptive options available to a lineage. With this view of phylogenetic constraints in hand, Price and Carr (2000) formulate the Phylogenetic Constraint Hypothesis:

> Macroevolutionary patterns provide the basis for understanding broad ecological patterns in nature involving the distribution, abundance and population dynamics of species. A phylogenetic constraint is a critical plesiomorphic character [a shared character, derived from a common ancestor], or set of characters, common to a major taxon. . . . Such characters limit the ecological and thus the major adaptive options in a lineage, but many minor adaptations are coordinated to maximize the ecological opportunities that can be exploited given the constraint. Such a set of adaptations is called the "adaptive syndrome." These characters in the adaptive syndrome, which evolve in response to the constraint, then result in inevitable ecological consequences, called "emergent properties." (Price and Carr 2000, 645)

They tested this hypothesis with sawflies, named after their sawlike, plant-piercing ovipositor. This organ meets the criteria of phylogenetic constraints: it is common to the entire family of sawflies and traces back to some of the earliest wingless insects, the silverfish, from Devonian time (more than 350 million years ago). The ovipositor's architecture is complex and delicate and limits its oviposition to the inside of soft plant tissues (endophytic oviposition), thus minimizing wear on the saw. So females tend to lay their eggs in the youngest shoots of their host plant. This further limits the evolution of alternative life cycles, because females must emerge when their host plant phenology is appropriate, i.e., when the host plant has reached a certain stage of development relative to climatic conditions. Bearing a sawlike ovipositor thus generates a certain type of adaptive syndrome because it defines the range of options available for ecological interactions and some major adaptations. A further interesting and fairly novel aspect of their work is the importance given to the impact of phylogenetic constraints on ecological phenomena—the so-called emergent properties that result from the adaptive syndrome evolved by the species. The close relationship between the endophytic oviposition and the host plant phenology seems to generate certain patterns in abundance, distribution, and population dynamics. The distribution of sawflies depends on the availability of young and vigorous shoots. Data show a strong correlation between perturbations (e.g., fire, flood, heavy mammals browsing) and population abundance. Disturbed sites tend to increase the availability of young plants and vigorous growth, which in turn increases the survival and reproduction rate of sawflies. Conversely, if host plant populations become more stable and older, sawflies become rare and may even go locally extinct. Thus, the main emergent ecological effects of having a sawlike ovipositor are: (1) very patchy distribution (on young, vigorous shoots), (2) generally low abundance at the landscape level, but rare dense populations in very favorable sites, and (3) relatively stable and predictable dynamics for many generations.

In order to show the relevance of path dependence, we have to consider alternative adaptive syndromes and emergent ecological properties. Fortunately for us, Price and Carr further support their hypothesis by doing just that, i.e., by showing that species with a different mode of oviposition do not display the same adaptive syndromes and consequently present different emergent properties. For instance, the spruce budworm—whose outbreaks in Canadian Boreal Forests are well known—lacks a plant-piercing ovipositor and instead lays eggs on the surface of mature foliage. The spruce budworm oviposition is also a phylogenetically primitive trait and is commonly found in the order Lepidoptera. But the constraint in this case has to be extended to the behavior instead of the organ. The posterior opening of the vagina of most lepidopterans is not an ovipositor per se, because it merely serves for discharging eggs. Price and Carr (2000) argue, however, that their *mode* of oviposition can nevertheless be a constraint in the sense that it limits the amount of information available to females. Because females simply lay their eggs on the surface, they receive less information from the host plant about its content in nutrients.

The authors infer that this results in nonspecific utilization of foliage at the oviposition site.[8] This, in turn, has important effects on the ecological emergent properties, which diametrically differ from the ones observed in sawflies. Resources are often abundant and not limiting for the spruce budworm, so populations can build to high densities and adopt nonpatchy spatial distribution. Moreover, because their abundance and density can reach very high proportions, the functional and numerical responses of natural enemies can be strong. That is, species feeding on the spruce budworms will be able to proliferate during periods of high abundance in budworms. This, and the fact that epidemic disease can be common when abundance is high, will cause eruptive, as opposed to stable, population dynamics.

This work on the Phylogenetic Constraint Hypothesis is interesting for at least two reasons. First, it uses constraints in a positive way. The different modes of oviposition create, more than limit, evolutionary paths. Second, it clearly shows how the occurrence of phylogenetic constraints entails historicity. Species of insects undergoing one evolutionary path (endophytic oviposition) tend to evolve a given set of adaptive syndromes and emergent ecological properties, but had they taken another evolutionary path (exophytic oviposition), they would have most likely acquired a different set of adaptive syndromes and emergent ecological properties. This is a clear case of path dependence.

The work done by Wimsatt (1986, 2001) and Schank and Wimsatt (1986) on generative entrenchment (GE) can be extremely useful in investigating more deeply how such evolutionary constraints, and consequently path dependence, arise. Recall how Price and Carr (2000) argue that features become phylogenetic constraints if they persist in lineages, which can happen if they become *tightly integrated into the developmental program*. One of the best conceptual resources for spelling out this phenomenon is GE.

Just like Jacob (1977) and Gould (1989), Wimsatt (2001) maintains that evolution is historical by virtue of its contingent nature. Here again, chance and historicity are closely related. In order to mark history, he says, "an event must cause cascades of dependent events that affect evolution" (2001, 227). This clearly refers to some species of causal dependence. But he also suggests that history matters to evolution because "minor unrelated 'accidents' or 'incidents' can massively change evolutionary history" (226). And both, chance and historicity, come together in the claim that the evolution of life forms is comparable to a "successive layered patchwork of contingencies." But Wimsatt's true contribution to this discussion is how he defends this inglorious conception of evolution by appealing to GE: "GE provides an explanation, perhaps the only possible explanation, for how and why this [successive layered patchwork of contingencies] is possible. In reproducing heritable systems, GE and selection may provide sufficient conditions for the incorporation and growing importance over time of contingency, and of history, in the explanation of form" (Wimsatt 2001, 226). I would not go as far as saying that GE is the only explanation possible here, but I think that Wimsatt touches on something important. As discussed in the context of the Phylogenetic Constraint Hypothesis, the evolutionary past can often become the framing principle for subsequently acquired adaptations. Let me now try to unpack this idea that GE makes it possible for some chance variations to accumulate, become integrated, and have long-lasting effects on the evolution of life forms. If we can establish this, then we can also substantiate the claim that GE can be an important source of phylogenetic constraints and path dependence at the macroevolutionary and macroecological levels.

The notion of GE combines two elements: "generativity" and "entrenchment." A *generative* structure typically possesses multiple elements that come together as a whole over time and implies that later elements presuppose the presence and proper assemblage of earlier ones. To give a very simple example, one can imagine how we build a tower with blocks. The building process extends in time; even the simplest tower will need at least two steps. In a metaphorical sense, we can say that our tower grows or develops each time a piece is added. Moreover, the position and stability of the pieces at a given instant depend on the position and stability of the pieces assembled earlier. For example, a wide and solid basis offers the potential for a higher and more stable tower, whereas a narrow and weak basis will offer less support and less potential for development. So building a tower is a generative process: it takes place during an extended time period, and later events causally depend on earlier ones.

Let us now look at the other component of GE, *entrenchment.* Something is entrenched when difficult to change. In the present discursive context, this stability can be interpreted at the developmental and evolutionary scale. Take the tower example again. The pieces in the tower that come earlier in the building process and constitute the foundations will generally have the highest degree of entrenchment. Anyone familiar with the game Jenga would know that removing or changing those pieces (even slightly) may be catastrophic for the whole tower, which would become highly unstable. The same is not true of the top and superficial pieces because they present a lesser degree of generative entrenchment. Thus the architect has more freedom with the top pieces, because the stability of the whole tower does not depend as much on them.

GE can also explain the stability of certain traits at the population level. Imagine a city in which a population of towers is generated by replication of successful designs. When an architect finds a viable design, she or he gets more contracts and her or his plans are more likely to become copied by other architects. Some modifications will be added here and there, but we should observe in the long run that most towers in that city have similar foundations. It will be easier to be creative with the coat than the core of towers, because modifying the core could result in unstable constructions. Thus, having a higher degree of entrenchment at the developmental level can have a stabilizing/constraining effect at the population level, creating a phenomenon of inertia for these structures.

A similar story can be told about biological organisms. Like building a tower, biological development is a generative process. The whole organism comes into being as a result of assembling various elements. As in the building of the tower, the proper development of an organism at a later stage causally depends upon earlier stages (this fundamental property of biological development is of course more prominent in multicellular organisms with differentiated parts). Each step of its development is causally connected to some antecedent events, and the form resulting from the whole process depends on the presence and proper assemblage of several developmental features. Evolution is essential to this story, too. It is the process by which the integration of features takes place. Schank and Wimsatt (1986) explain that certain features become increasingly integrated and entrenched in lineages by the mechanism of *accretion*, i.e., by the accumulation of new features at later developmental stages through evolution. The first life forms were relatively simple, but they evolved into more complex species by accumulating new features on top of the ones they had previously integrated. With enough time, this process of accretion results in complex developmental networks, with certain parts being more or less deeply integrated in a whole developmental process. Thus, evolution by accretion will most likely result in organisms with elements presenting different degrees of entrenchment, with older features being (on average) more entrenched than more recent ones.

Entrenchment comes in degrees. A feature will become more or less entrenched in a developmental process depending on the extent to which it is integrated into a complex, generative structure. Schank and Wimsatt (1986) explain that the degree of entrenchment of a given developmental feature depends on the magnitude of its "downstream effect." They "speak of features as being 'generatively entrenched' in proportion to the number of 'downstream' features which depend on them" (Schank and Wimsatt 1986, 38).[9] A feature with many important downstream effects possesses a high degree of GE, and vice versa. Rightly, they infer that these effects will most likely be detrimental (Schank and Wimsatt 1986, 37). Conversely, a feature with a low degree of GE does not tend to produce important effects (or malfunctions) in the developmental process if modified.

This phenomenon of differential entrenchment is at the very basis of the type of phylogenetic constraints discussed earlier. It explains why some features display more evolutionary inertia than others. In a nutshell, if we grant the idea that modifications of highly GE'd features most likely result in malfunctions, and therefore reduce fitness, then GE implies that evolution should tend to be more conservative with highly GE'd features. Schank and Wimsatt (1986) and Wimsatt (1986, 2001) emphasized the higher degree of generative entrenchment of early developmental features and of regulatory genes. Early developmental and regulatory features are like the foundations in a tower. Although they can undergo mutations, their important developmental downstream effects will confer upon them a higher evolutionary stability.

Let us now return to the idea that different phylogenetic constraints lead to different adaptive syndromes and emergent ecological properties and thus result in path dependent evolution. GE has a tendency to stabilize certain regions of the evolutionary state space. As such, GE renders evolutionary outcomes more or less likely, depending on the accretion path taken by a lineage. But these accretion paths are contingent, in the sense that they could have been otherwise. In other words, there is rarely only one possible adaptive syndrome for a given lineage. Yet as a certain adaptive syndrome makes its way, it becomes increasingly difficult to substantially change it. This means that GE, by creating attractors in the evolutionary space, renders evolution much less shifty. It also makes it less random. This brings us back to the point made in the previous sections about contingent evolution being not merely a chance process. GE, by making the occurrence of various evolutionary outcomes dependent on a certain accretion path, prevents evolution from becoming a mere random walk, where chance erases any traces of history.

CONCLUDING REMARKS

I opened this chapter with some passages from François Jacob (1977) that established a connection between Darwinian evolution on the one hand and a world full of contingencies and imperfections on the other. I did not address the claim of imperfection here (see Beatty and Desjardins 2009), but it should be clear by now that chance *and* history are essential to this view of evolution, and I think that the notion of path dependence is a fruitful way to bring these factors together. I would like to conclude by opening, albeit superficially, another connection between evolution and path dependence. If path dependence is a reason for interpreting evolution as an historical process, then it can also be involved in our understanding of biological entities as historical. Marc Ereshefsky (2014), looking at how species become distinct, makes an interesting argument to this effect. In a nutshell, he emphasizes that species are historical (path dependent) entities because speciation is a path dependent process. After Mayr (1970), Ereshefsky presents speciation as the result of isolating mechanisms that are by-products of new adaptations in new species. Two incipient species become distinct because they "have different mutations and mutation order (as well as differences in the effects of genetic drift) even if these two populations start with identical clones and identical environments. The upshot is that speciation is a path dependent process: vary the path and it is very, very unlikely the same species will be produced" (Ereshefsky 2014, 72). This view resonates with the interpretation of species entertained in the work of Brooks and McLennan (1991, 1993, 1994), two ecologists who argue that ecology needs to pay more attention to evolution in explaining ecological phenomena. At the basis of their approach lies a historical view of species that embraces many of the themes elaborated here. Species, they claim, are "vessels of future potential, living legacies of past modifications and stasis shaped by millennia of biotic and abiotic interactions. They are history embodied" (Brooks and McLennan 1993, 267).

NOTES

1. The occurrence of a state can be understood as some kind of holistic event that defines the overall conditions of the system at a given time.

2. A detail-oriented reader could object that the initial conditions and subsequent environmental conditions must have included some small differences—after all, no two clones are perfectly identical, and the conditions in the petri dish could vary slightly from time to time. Still, this experiment is perhaps the closest thing to Gould's Replaying Life's Tape thought experiment.

3. A suitably modified Replay experiment could create the conditions that would allow for the discrimination of the chance and history hypotheses as well. But as that experiment is usually described, one replay from a single point and a mere comparison of outcomes does not render this possible.

4. See also Blount (this volume, chap. 10). An accessible summary of some of the different divergences and convergences recorded in the LTEE can also be found here:

https://telliamedrevisited.wordpress.com/2013/12/12/what-weve-learned-about
-evolution-from-the-ltee-number-3/.

5. The history hypothesis has been tested at least twice during the course of the LTEE project. The first time was rather early, and the results were reported in Travisano et al. 1995. But as discussed in Desjardins 2011, the historical hypothesis tested in this experiment corresponds to historicity in the form of dependence on initial conditions. Although the conceptual work done in that paper is worthy of mention, my focus in this chapter is more on path dependence. So I will leave the readers to explore this experiment by themselves, and I will present a more recent experiment instead.

6. I wish to thank the anonymous reviewer who raised this third possibility.

7. In the 2012 paper, Blount et al. call the history hypothesis "epistasis," whereas the other alternative is called "physical promotion." To avoid confusion, I will maintain the terminology set in the earlier, 2008, paper.

8. There is no clear evidence demonstrating whether this is a result of natural selection or a mere collateral effect of the oviposition mode. Nevertheless, the data clearly show no preference in oviposition site.

9. This aspect of GE clearly presupposes some historical dependence. But it does not automatically entail path dependence. It is possible, although unlikely, that there is but one viable organization for a given system.

CHAPTER 10

History's Windings in a Flask: Microbial Experiments into Evolutionary Contingency

Zachary D. Blount

September 1862 was a dark time for the United States of America. The Civil War to bring the breakaway Confederacy back into the Union was well into its second bloody year, and Confederate general Robert E. Lee had invaded Maryland on September 4. The dispirited Union army could not manage to find Lee's army, and the United States seemed assured further humiliation at a time when elections were looming. Worse, the British and French governments were considering formally recognizing and aiding the Confederacy. Victory for the Confederacy looked to be in sight (McPherson 1988).

So things stood on the morning of September 13, 1862, as the US Army sluggishly marched into Frederick, Maryland. During a stop from marching, an Indiana corporal took the opportunity to rest under a tree at the edge of a field. As he lay down, the corporal happened to see an envelope in the tall grass nearby. In this envelope, he found a piece of paper entitled "Special Order 191" that detailed Lee's marching orders (Sears 1983; McPherson 1988).

The copy of Special Order 191 soon came to Major General George B. McClellan, commanding general of the Union Army. With Lee's plans in hand, McClellan was able to bring his opponent to pitched battle at Antietam on September 17. The bloody battle forced Lee to retreat into Virginia. Morale improved in the North, and supporters of the war maintained their power in Congress after the fall elections (McPherson 1988). Antietam also allowed President Lincoln to issue the Emancipation Proclamation, making the war a crusade to end slavery. The war's new moral character effectively ended the chance for foreign intervention (Sears 1983; McPherson 1988). The Confederacy was never again close to victory, and Antietam is now recognized as the key turning point of the war. Remarkably, this momentous event occurred because of the happenstances of a dropped envelope, where a tired man happened to rest, and where he happened to look while doing so.

Of all the ways the world could be, one of these the world is. The world is as it is because history occurred as it did, with all the many coincidences, happenstances, and freak events that played out in the tangled interplay of chance and necessity that history always involves. The range of possible futures that may result from a given present must always collapse down into a single actual outcome that in turn determines the range of possible later futures. The future is therefore dependent upon the particular causal chain of innumerable, interacting, and often small antecedent factors leading up to it (Beatty and Carrera 2011; Desjardins, this volume, chap. 9). This is to say that history is path dependent and subject to contingency, the property of historical sequences that makes history matter.

Biological evolution is also subject to a profound tension between chance and necessity. Evolutionary outcomes are determined by a complex interplay of stochastic and deterministic processes (Monod 1971; Mayr 1988). Natural selection systematically adapts organisms to the environmental conditions they encounter, but it must act on heritable variation introduced stochastically by gene flow, recombination, and, ultimately, mutation. Moreover, genetic drift can cause random loss of even the most beneficial variation that may arise. Finally, mutations with similar adaptive value in a given environment can differ greatly in their correlated effects on adaptation to other environments, as well as their effects on a variety of traits (pleiotropy) and their interactions with other mutations (epistasis), altering the genomic and organismal context in which future evolution takes place (Gould and Lewontin 1979; Jacob 1977). Consequently, what mutations occur and the order in which they occur can profoundly impact evolutionary trajectories, evolvability, and correlated fitness in environments that may be later encountered (Lenski et al. 1991; Mani and Clarke 1990; V. S. Cooper and Lenski 2000; Weinreich, Watson, and Chao 2005). Seemingly trivial differences between lineages can even determine survival and extinction. Moreover, chaotic interactions between geological, astronomical, and climatological processes and the biosphere can cause rapid and capricious environmental changes that trigger mass extinctions in which only those lineages fortuitously preadapted to the new conditions survive (Lewontin 1966; Jablonski 1986; Gould 2002). Evolution is clearly a process that takes place in lineages shaped by unique evolutionary histories billions of years long that have occurred within the Earth's singular history. It is therefore a fundamentally historical phenomenon that, just like human history, is subject to path dependence that arises from its core processes and the broader planetary context in which it occurs.

How important is the historical nature of evolution? Evolutionary historicity has been recognized since Darwin, but this question received little attention for much of evolutionary biology's history. However, Stephen Jay Gould began to highlight its importance in the 1980s.[1] Gould's answer to the question, most forcefully in his 1989 book *Wonderful Life*, was a resounding "Very!" Gould suggested that evolutionary outcomes are fundamentally subject to contingency. As in human history, Gould asserted, the quirks and happenstances of the complex causal chains of evolutionary history play a critical role in determining what evolutionary outcomes result, and so small changes along the way could lead to very different outcomes (Gould 1989, 2002). He famously suggested that replaying the "tape of life" from various points in the distant past would each time result in a very different biological world (1989). Gould argued that this contingency renders evolution inherently unpredictable, and therefore explicable and understandable only in retrospect using narrative, actual sequence explanations (Gould 1985b, 1989, 2002; Beatty 1993, 2006b; Blaser 1999; Sterelny and Griffiths 1999; Desjardins 2011).

Others have strongly disagreed with Gould. Many have suggested that widespread evolutionary convergence indicates that contingency's scope is highly limited (Conway Morris 2003, 2010; Parker et al. 2013). Simon Conway Morris has argued that organisms can occupy only a limited number of possible niches that present biological challenges to which biological and physical constraints provide a limited set of solutions. Natural selection then deterministically finds these solutions (Conway Morris 2003). Consequently, were one to replay the tape of life many times, very similar outcomes would be observed (Conway Morris 2003, 2010; Vermeij 2006). Therefore, while contingency might provide some indeterminacy, evolution is broadly repeatable, predictable, and regular enough to be potentially described by the sorts of history-insensitive robust process explanations found in physics and chemistry (Sterelny and Griffiths 1999).

The contingency debate was hampered in its early stages by a lack of focused empirical research, in large part because evolutionary contingency is a tricky phenomenon to study. However, evolutionary contingency should have empirically evaluable effects. First, contingency should reduce or preclude evolutionary repeatability (Gould 1989, 2002). This prediction is drawn from what Beatty has identified as Gould's "unpredictability" notion of contingency (Beatty 2006b; Beatty and Desjardins 2009; Beatty and Carrera 2011). Beatty also identified a second, "causal dependence," notion of contingency in Gould's writings that describes the historical path dependence of evolutionary outcomes (Beatty and Desjardins 2009; Beatty and Carrera 2011; Desjardins, this volume, chap. 9). This notion predicts that at least some evolutionary outcomes are highly sensitive to history (Gould 2002; Beatty 2006b). Finally, an outcome contingent upon a particular prior historical path should be delayed compared to an outcome driven principally by selection (Foote 1998; Dick et al. 2009).

In recent decades, researchers have begun to evaluate evolutionary contingency at a variety of levels. These studies have included the examination of the timing and phylogenetic distribution of evolutionary innovations (Vermeij 2006), the repeatability of *Anolis* lizard ecomorph evolution on Caribbean islands (Losos et al. 1998; Losos 2010), and the effects of history on the evolution of Southeast Asian fanged frogs (Emerson 2001) and snake diets (de Queiroz and Rodríguez-Robles 2006). Some of the most intriguing work on contingency has been done using microbial evolution experiments. In this chapter, I will focus on these microbial experiments and the implications their findings hold for the issue of evolutionary contingency.

EXPERIMENTAL EVOLUTION WITH MICROBES

Experimental evolution with microbes involves maintaining populations of microorganisms, typically bacteria, yeast, or viruses, under laboratory conditions to examine evolutionary processes as they occur (Bennett and Hughes 2009; Kawecki et al. 2012). In a typical serial transfer microbial evolution experiment, a population is founded from an ancestral clone and grown in a nutrient medium under controlled conditions (fig. 10.1). Following a defined incubation time, a fraction of the culture is transferred to fresh medium, and this pattern is then continued, potentially indefinitely. Experimental evolution goes back to Rev. Henry Dallinger, who conducted experiments into the evolution of microbial thermotolerance in the 1880s (Dallinger 1887). Despite this early start, experimental evolution with microorganisms did not become a significant approach to studying evolution until the 1980s. In the decades since, microbial evolution experiments have proved to be a powerful way to examine a variety of difficult evolutionary issues, including historical contingency (Elena and Lenski 2003; Kacar, this volume, chap. 11; Kawecki et al. 2012).

Much of the power of microbial evolution experiments comes from the simple fact that microbes have many characteristics that make them excellent organisms for evolution research. Many microbes have rapid generation

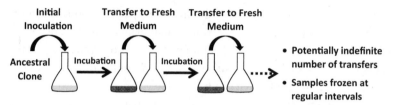

FIGURE 10.1. Basic serial transfer regime used in microbial evolution experiments.

times, some as short as 20 minutes, so hundreds to tens of thousands of generations of evolution can be studied in the course of experiments that take only weeks or years. Microbes typically remain viable after freezing, meaning that samples of evolving microbial populations may be frozen indefinitely in "fossil records" from which ancestral and evolved forms may be revived for later study. Microbial cultures can also reach extremely high population sizes despite small volumes, which ensures rich supplies of variation from mutation during experiments. Moreover, because microbes primarily reproduce by cloning, genetically identical replicate populations can be founded, which permits rigorous statistical analysis of variation that arises between populations exposed to the same conditions. Fitness assays also allow direct determination of changes in relative reproductive fitness by comparison of the growth rate of an evolved population or clone to that of its ancestor (Lenski et al. 1991). Remarkable levels of control can also be maintained, so that differences in population size, mutation supply, biotic and abiotic environmental factors, and even evolutionary history can be manipulated and examined (Levin and Lenski 1985; de Visser et al. 1999; Bohannan and Lenski 2000; Burch and Chao 1999, 2000; Elena et al. 2001; Fukami et al. 2007; Meyer and Kassen 2007; Kacar, this volume, chap. 11). Finally, modern advances in DNA sequencing and genetic engineering enable researchers to identify evolved genetic changes and to then manipulate and directly link them to changes in fitness and phenotype (Hegreness and Kishony 2007; Barrick et al. 2009; Barrick and Lenski 2009, 2013; Blount et al. 2012).

Evolution experiments with microorganisms can be used to evaluate evolutionary contingency in ways that are impossible with other approaches. After all, this method permits one to come as close to replaying the "tape of life" as is foreseeably possible, albeit on a much smaller scale than Gould envisioned. Microbial evolution studies into evolutionary contingency may be divided into two general categories, which may be called "Parallel Replay" experiments, or "PREs," and "Historical Difference" experiments, or "HDEs"

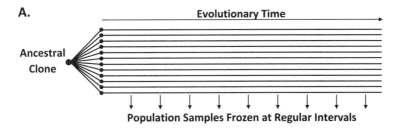

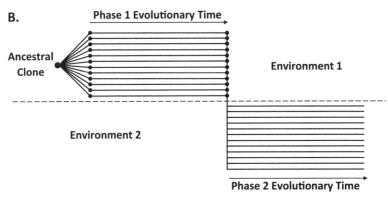

FIGURE 10.2. Contingency Experiment Structures. In a Parallel Replay Experiment (*A*), multiple, genetically identical replicate populations are founded from a single ancestral clone and then evolved under the same conditions. At regular intervals, samples of each population are frozen for later analysis and comparison to examine evolutionary parallelism and divergence. This design is open-ended and may be carried on indefinitely. In a typical two-step Historical Difference Experiment (*B*), phase one evolution consists of a PRE carried out in a given environmental condition to allow the parallel populations to accrue different evolutionary histories. Clones derived from the evolved populations are used to found new populations that are then evolved under a new environmental condition in phase 2. Typically, analysis and comparison in an HDE is done of clones used to found the second phase populations and then of populations or clones after the completion of phase 2.

(fig. 10.2). PREs involve founding identical replicate populations and evolving them under identical conditions. This setup permits examination of the range of parallelism and divergence in evolutionary trajectories and outcomes that emerge from the effects of contingency inherent to the core evolutionary processes, as it involves what are essentially simultaneous replays of evolution from a single evolutionary point. By contrast, HDEs involve examining the impact of different evolutionary histories on subsequent evolution. In both, the frozen fossil record and the analytic tools available often make it possible to identify the genetic and ecological interactions upon which later evolutionary outcomes may be contingent. In the following, I will examine experiments and findings in each category. Space limitations preclude a thorough review of all the relevant experiments, so I will instead focus on one major experiment from each category, explicate its findings, and compare them to those of other experiments.

PARALLEL REPLAY EXPERIMENTS

If evolution is highly contingent in the unpredictability sense, then one might expect evolutionary outcomes to be fundamentally unrepeatable even from the same starting point. Parallel replay microbial evolution experiments are the simplest way to test this prediction. In these experiments, replicate populations are founded from the same genotype and then are evolved under identical conditions (fig. 10.2A). PREs therefore involve replaying the same tape of life from the same evolutionary point multiple times simultaneously (Lenski et al. 1991; Desjardins, this volume, chap. 9). The only differences that should arise between the populations in these experiments should be those due to the stochastic influx of variation via mutation, and the action of drift and selection upon that variation.

The longest-running and best studied PRE is Richard Lenski's *E. coli* Long-Term Evolution Experiment (LTEE). The LTEE began on February 24, 1988, with the founding of twelve populations from a single clone of *E. coli* B (Lenski et al. 1991; Lenski 2004). Save for a neutral genetic marker that allowed six populations to grow on arabinose (Ara$^+$), all were initially identical. The founding strain is strictly asexual and lacks any intrinsic capacity for horizontal gene transfer (Jeong et al. 2009; Studier et al. 2009). The populations are grown at a stable 37°C with 120 rpm aeration in DM25, a carbon-limited minimal medium supplemented with 25 μg/mL glucose (Davis and Mingioli 1950). Each day, 1 percent of each population is transferred to fresh medium, after which each grows 100-fold, or ~6.7 generations (Lenski et al. 1991). This serial transfer regime produces a "seasonality" in which a short feast of abundant glucose after transfer is followed by famine until the next transfer (Vasi, Travisano, and Lenski 1994; Lenski 2004). Each population has evolved for more than 60,000 generations since the experiment began, and population samples have been frozen every 500 generations throughout, providing an extensive fossil record that represents a rich resource for research (Lenski 2004).

Remarkable evolutionary parallelism has been observed in the LTEE. The fitness of all twelve populations rose rapidly and then decelerated without plateauing (Lenski and Travisano 1994; Lenski et al. 1991; Lenski 2004; Wiser, Ribeck, and Lenski 2013). All twelve have evolved faster growth rates, shorter lag phases, increased average cell size, and lower population sizes (Lenski and Travisano 1994; Vasi, Travisano, and Lenski 1994; Lenski 2004; Philippe et al. 2009). All have lost some or all of their capacity to grow on a variety of substrates other than glucose (Travisano, Vasi, and Lenski 1995; V. S. Cooper and Lenski 2000; T. F. Cooper, Rozen, and Lenski 2003). The populations have also evolved parallel changes in gene expression and regulation (T. F. Cooper, Rozen, and Lenski 2003; Philippe et al. 2007; Crozat et al. 2010; Crozat et al. 2011), protein profiles (Pelosi et al. 2006), epistatic interactions with the CRP regulon (T. F. Cooper et al. 2008), and resistance to phages T6* and lambda (Meyer et al. 2010). Ten populations have evolved parallel changes in DNA supercoiling (Crozat et al. 2005; Crozat et al. 2010; Crozat et al. 2011). In numerous instances, the same genes have been mutated in multiple populations (V. S. Cooper et al. 2001; T. F. Cooper, Rozen, and Lenski 2003; Crozat et al. 2005; Crozat et al. 2010; Pelosi et al. 2006; Woods et al. 2006; Barrick et al. 2009; Philippe et al. 2009). Most remarkably, IS150-mediated deletions caused complete loss of capacity to grow on maltose across all populations (T. F. Cooper, Rozen, and Lenski 2003). Such parallelism is a hallmark of selection, and, indeed, a number of the parallel mutations have been demonstrated to confer increased fitness under the conditions of the experiment (V. S. Cooper et al. 2001; T. F. Cooper, Rozen, and Lenski 2003; Barrick et al. 2009; Philippe et al. 2009).

The populations have also diverged. Despite overall similarity, the populations' actual fitness trajectories show significant and persistent differences (Lenski et al. 1991; Lenski and Travisano 1994; Wiser, Ribeck, and Lenski 2013). Moreover, fitness trajectories in other environments have varied considerably (Travisano, Vasi, and Lenski 1995). Few beneficial mutations have fixed in all populations (Stanek, Cooper, and Lenski 2009; Blount et al. 2012). Indeed, each population has accumulated unique sets of mutations, including gross changes such as IS insertions, inversions, and deletions, as well as SNPs (Papadopoulos et al. 1999; Barrick et al. 2009). Six populations have evolved high mutation rates due to substitutions in their mutation repair pathways (Sniegowski et al. 2000; Barrick and Lenski 2009; Barrick et al. 2009; Blount et al. 2012). Even those genes mutated in parallel across the populations typically differ in the location and type of mutation involved (V. S. Cooper et al. 2001; T. F. Cooper, Rozen, and Lenski 2003; Crozat et al. 2005; Pelosi et al. 2006; Woods et al. 2006). Moreover, while all populations have experienced the decay of some metabolic capacities, the capacities impacted, as well the extent of the decay, and their genetic bases have varied (V. S. Cooper and Lenski 2000; Ostrowski, Woods, and Lenski 2008).

The pattern of extensive parallelism with some divergence observed during the LTEE is typical of other PREs. Convergence to similar levels of fitness in the experimental environment (Bull et al. 1997; Fong, Joyce, and Palsson 2005; Le Gac et al. 2013; Riley et al. 2001; Treves, Manning, and Adams 1998), though not in other environments (Melnyk and Kassen 2011), is remarkably common across experiments (Korona et al. 1994; Dettman et al. 2012; Kawecki et al. 2012). As in the LTEE, variation in other environments among evolved populations is common (Melnyk and Kassen 2011), suggesting that parallel adaptation can be accomplished via different genetic routes. Indeed, while parallelism is seen at the level of adaptive genetic changes, it is not generally pervasive, though it is sometimes seen even at the level of specific nucleotide changes in organisms with very simple genomes (C. J. Brown, Todd, and Rosenzweig 1998; Herring et al. 2006; Bollback and Huelsenbeck 2009; Betancourt 2009; Wichman and Brown 2010). This preponderance of gross convergence in PREs strongly suggests that there are often multiple available genetic paths to similar phenotypic and adaptive states. Such multiple genetic paths mean that phenotypic similarities can mask significant genetic differences that may carry significant consequences for subsequent evolution (Bedhomme, Lafforgue, and Elena 2013). Two LTEE populations, designated Ara-2 and Ara-3, have diverged profoundly from the other ten, demonstrating that seemingly incidental genetic differences can significantly impact evolutionary outcomes.

The Ara-2 population evolved a unique ecology that has supported a long-term polymorphism. Two monophyletic cell lineages that arose before generation 6,000 have coexisted for more than 50,000 generations. The coexistence between these two lineages, referred to as "S" and "L" for their respective small and large colony morphologies, has been maintained by negative frequency-dependent selection based on their ecological differences (Rozen and Lenski 2000). L cells are more fit than S cells in the abundant glucose environment encountered immediately after transfer. However, L cells experience higher mortality after glucose exhaustion, when S cells are more fit due to cross-feeding on substances released by lysed L cells (Rozen, Schneider, and Lenski 2005; Rozen et al. 2009). New evolution occasionally allows L to encroach on the S niche, after which S has invariably evolved to counter the encroachment. The result is a dynamic, fluctuating relationship driven by recurrent rounds of evolution that maintains diversity and is perhaps leading to incipient speciation (Rozen, Schneider, and Lenski 2005; Cohan and Perry 2007; Rozen et al. 2009; Le Gac et al. 2012). The case of Ara-2 suggests that, just as potentially important genetic differences can be masked by the superficially convergent adaptive states of evolving populations, so too can differences in population structure and ecology that should be taken into account when evaluating convergence and divergence in PREs.

The divergence of Ara-3 is even more striking. A cell lineage in Ara-3 evolved the capacity to exploit an open ecological opportunity provided by the large amount of citrate added to DM25 as a chelator to facilitate the bacteria's acquisition of iron in the medium (Cox et al. 1970; Frost and Rosenberg 1973; Hussein, Hantke, and Braun 1981). The concentration of citrate is far higher than is necessary for this role due to DM having been developed for a particular set of experiments conducted before citrate's biological role in *E. coli* medium was known (Davis 1949; Davis and Mingioli 1950; Blount, n.d.). The citrate is a potential carbon and energy source, but *E. coli* is unable to transport citrate into the cell during aerobic growth, preventing it from being used as a food source despite having a complete TCA cycle and the ability to ferment citrate anaerobically (Lara and Stokes 1952; Lütgens and Gottschalk 1980; Pos, Dimroth, and Bott 1998). This Cit$^-$ phenotype is a very stable diagnostic characteristic of *E. coli* as a species, and spontaneous aerobic citrate using (Cit$^+$) mutants are extraordinarily rare (B. G. Hall 1982; Scheutz and Strockbine 2005).

After 33,000 generations, the Ara-3 population became several-fold larger, as Cit$^+$ variants rose to dominance in the population. These variants had evolved approximately 2,000 generations earlier due to a 2933 bp duplication that contains part of the *cit* operon regulating citrate fermentation (Pos, Dimroth, and Bott 1998; Blount et al. 2012). The duplication placed the anaerobically expressed *citT* gene, which encodes a citrate-succinate antiporter, under the control of a promoter that normally regulates an aerobically expressed gene, *rnk*. The new *rnk-citT* regulatory module supports weak aerobic expression of the CitT transporter, which provides marginal access to the citrate resource (Blount et al. 2012). The Cit$^+$ variants remained at low frequency until refining mutations arose, including further amplification of the duplicated segment that increased the dosage of the *rnk-citT* module, changes in expression of the DctA succinate transporter, and changes in carbon flow through central metabolism, all of which improved growth on citrate, producing a stronger citrate-using phenotype referred to as Cit^{++} (Blount et al. 2012; Quandt et al. 2014; Quandt et al., forthcoming). The Cit$^+$ variants did not sweep to fixation when they rose to numerical dominance. Instead, a small

Cit⁻ subpopulation persisted through at least generation 40,000 by evolving to cross-feed on succinate and other substances released into the medium by the Cit⁺ cells (Blount, Borland, and Lenski 2008; Blount et al. 2012; C. B. Turner et al., forthcoming). This long-term coexistence, along with the fact that Cit⁺ exceeds the accepted range of variation for *E. coli*, suggests that the Cit⁺ lineage may be an incipient species.

The Cit⁺ trait has been experimentally demonstrated to be historically contingent. Contingent traits require multiple, nonuniquely beneficial mutations before manifestation (Blount, Borland, and Lenski 2008). Cumulative selection cannot directly facilitate the accumulation of these necessary mutations, which must instead occur as a chance product of a population's evolutionary history. Contingent traits should be rare, as the necessary antecedent history is unlikely, and delayed with respect to the presentation of the ecological opportunity or environmental challenge to which they provide access or adaptation (Foote 1998). The evolution of Cit⁺ was therefore hypothesized to have been multistep and contingent upon the prior accumulation of one or more mutations that produced a "potentiating" genetic background in which the rate of mutation to Cit⁺ was much higher than in the ancestor. Consistent with this hypothesis, a series of experiments in which evolution was "replayed" from clonal genotypes isolated from various time points in Ara-3's fossil record showed that reevolution of the Cit⁺ trait was much more likely to occur in replays started from later generation clones (Blount, Borland, and Lenski 2008). Fluctuation tests later showed that the ancestor's Cit⁺ mutation rate is immeasurably small, with an *upper bound* of ~3.6×10^{-13} per cell per generation, while later clones have a measurable rate with a point estimate of ~6.6×10^{-13}. The ancestral and potentiated clones have the same background mutation rate, so potentiation is not attributable to general hypermutability. Although the potentiated rate is still orders of magnitude lower than a typical mutation rate, the increase was sufficient to make the Cit⁺ function mutationally reachable (Blount, Borland, and Lenski 2008).

The genetic basis of potentiation has not yet been determined, but suggestive details about potentiation have been uncovered. A population phylogeny based on fossil genome sequences shows that Ara-3 was diverse over much of its pre-Cit⁺ history. Three clades, C1, C2, and C3, arose between 10,000 and 20,000 generations and then coexisted until some point after Cit⁺ became dominant. Ecological divergence between the clades may explain this coexistence, though pre-Cit⁺ ecology has not yet been investigated. C1 diverged before 15,000 generations, and then C2 and C3 diverged from each other before 20,000 generations. The Cit^+ lineage later arose from C3. Clones from all three clades yielded Cit^+ mutants during the replay experiments, but C3 is significantly overrepresented among them. These findings suggest that potentiation involved at least two mutations, the first of which occurred prior to C1's divergence, and a second that occurred in C3 (Blount et al. 2012).

In principle, Cit^+ evolution might have been potentiated either by physical promotion of the *cit* duplication or by functional epistatic interactions that made the *rnk-citT* module effective when it occurred, likely by improving citrate metabolism (Blount, Borland, and Lenski 2008; Blount et al. 2012).[2] Cit^- clones transformed with a high copy number plasmid containing a complete *rnk-citT* module display a Cit^+ phenotype. However, C3 transformants show much stronger and consistent Cit^+ phenotypes than do those from the other two clades. The second potentiating mutation therefore appears to have worked by functional epistasis. Moreover, with one exception, mutations responsible for the Cit^+ phenotypes of the Cit^+ mutants isolated during the replay experiments are all different, though all involve *citT*. Many of the mutations involved the capture of other promoters, which is difficult to explain by physical promotion (Blount et al. 2012). It thus appears that the first potentiating mutation was also functionally epistatic, and the variety of promoters that can be co-opted for *CitT* expression argues that potentiation was at the level of improved citrate metabolism. These findings pose the interesting possibility that the potentiating mutations were originally adaptive to the pre-Cit^+ ecological conditions of Ara-3 (Quandt et al. 2014). A similar interplay between ecology, coevolution, and epistatic genetic changes has been implicated in the contingent evolution of the capacity of phage lambda to infect through an alternate host cell surface receptor (Meyer et al. 2012).

HISTORICAL DIFFERENCE EXPERIMENTS

The causal dependence aspect of contingency holds that a lineage's prior evolutionary, ecological, and environmental history should leave an indelible mark in its genome that can alter and constrain its evolutionary potential (Gould 2002; Beatty 2006b). Historical difference experiments examine this aspect of contingency (Beatty 2006b). While PREs examine populations evolving from the same genotypic and historical point, HDEs examine populations evolving from different points. HDEs typically involve variations on a two-step design (fig. 10.2B). Initially identical populations are first founded and evolved for some length of time under a given condition or conditions

(Travisano et al. 1995; Collins, Sültemeyer, and Bell 2006). In the second step, new sets of populations are founded from those evolved in the first step, are evolved under one or more new conditions, and then various traits are compared. The first step therefore serves to generate different histories for the test organisms, while the second tests the evolutionary impacts of those historical differences. In some experiments, different strains or organisms with long histories outside the lab are used, which circumvents the need for the first step (F. B.-G. Moore and Woods 2006). In another experiment populations founded with clones isolated from the same population that had evolved different competing alleles were used to examine the evolutionary consequences of these alleles (Woods et al. 2011). The experiment described by Kacar (this volume, chap. 11) can also be seen as a form of HDE, albeit with the differences in history between phase 2 genotypes being limited to a single locus. In all cases, the central prediction from contingency is that differences accrued during prior histories will have detectable consequences during the second phase of evolution (Travisano et al. 1995).

The HDE design was originally developed by Travisano et al. (1995) to evaluate the roles of adaptation, chance, and history in evolution. In one experiment, a single clone was isolated from each of the LTEE populations discussed above after 2,000 generations of evolution on glucose. Each derived clone was used to found new populations that were then evolved for 1,000 generations in maltose-limited medium. The founding clones had similar fitness on glucose but varied significantly in both cell size and fitness on maltose (Travisano et al. 1995; Travisano, Vasi, and Lenski 1995). These two traits were again assessed after the maltose evolution phase. Adaptation was expected to cause the populations to evolve by approximately the same magnitude in the same direction, while the persistence of significant trait differences between the clones, noted before maltose evolution, would suggest the lingering effects of history, and chance would be indicated by variation in mean trait value for populations founded from the same founding clones. A nested ANOVA was used to determine the relative contributions of the three factors to the observed trait evolution. All populations evolved similar fitness on maltose after 1,000 generations, which was overwhelmingly attributable to adaptation. While prior history did contribute significantly to the final fitness, its effect was largely swamped by adaptation, and chance showed no significant effect. By contrast, adaptation, chance, and history were all found to have significantly contributed to the final cell size. These findings led to the conclusion that adaptation can largely overcome history's effects for traits that are subject to strong selection, but that history's mark persists far more in those traits not under selection.

A number of subsequent HDEs have found similar patterns (Pérez-Zaballos et al. 2005; Collins, Sültemeyer, and Bell 2006; Bennett and Lenski 2007; Saxer, Doebeli, and Travisano 2010; Bedhomme, Lafforgue, and Elena 2013). Deviations from this pattern have also been observed (Bollback and Huelsenbeck 2009; Flores-Moya et al. 2012). Perhaps most significantly, populations that were founded from natural isolates of *E. coli* and evolved for 2,000 generations in a novel environment reached significantly different final fitness values, and they did so at significantly different tempos (F. B.-G. Moore and Woods 2006). The isolates had diverged far more during prior history than those used in other studies, which suggests that, as Travisano et al. noted, longer prior histories might produce deeper effects than they had observed.

The above suggestion points to certain difficulties inherent in the HDE design. The HDE approach is designed around the idea that the prior history provided by the first phase of evolution may alter evolutionary potential that may be detected in the second phase. This design is predicated on the assumptions that the first phase of evolution will provide sufficient history to allow sufficient resolution of history's effect, that significant divergence between populations observed during the second phase may be attributed to the different histories that arise from the first phase, and that the second phase will be long enough to adequately examine the effects of history. While performing both phases of HDE evolution in the lab provides control, it may preclude accruing sufficient history in the first phase to produce detectable effects. It is possible that the findings suggesting that history is generally swamped by adaptation are artifacts of this trade-off. Using natural isolates can provide sufficient history to yield better resolution, but that history is unknown. Of course, as ongoing PREs like the LTEE that may be used as the first evolution phase in HDEs accrue more history, this trade-off may be ameliorated in the course of time.

Another problem is that the second phase itself constitutes a sort of PRE, so it is always possible that divergent evolution during it may reflect different histories of the populations within the second phase of evolution rather than any differences stemming from the first phase. (It is conceivable, for instance, that adaptation to the first phase condition would not involve any mutations that would impact final fitness or subsequent fitness in the second phase condition.) It is therefore problematic that few HDEs in the literature include control populations that are evolved only under the second phase condition, which would permit disentanglement of differences arising from the two sources of history. Finally, consideration should be given to whether or not the second evolution phase in an HDE is long enough to detect the impact of divergent first phase history. For example, while a prior history of evolution at high CO_2 levels was not found to significantly impact fitness evolution in *Chlamydomonas* populations, these populations did not return to normal CO_2 uptake characteristics following back adaptation to ambient CO_2 levels (Collins, Sültemeyer, and Bell 2006). Similarly, while prior evolution of the tobacco etch potyvirus to different hosts did not prevent later and equivalent phenotypic adaptation to a common host, substantial historical effects on genotype were detected (Bedhomme, Lafforgue, and Elena 2013). Both of these findings suggest that prior history can have subtle effects that may impact evolution over longer time spans. These methodological shortcomings of currently documented HDEs should be considered in the design and performance of future experiments.

CONCLUSIONS AND IMPLICATIONS

The question of the scope of historical contingency's impact on evolution is one with major implications for how evolution should be approached and explained (Beatty 1993; Sterelny and Griffiths 1999). A growing body of microbial evolution experiments explicitly designed to examine historical contingency have begun to shed much needed empirical light on this question. However, the empirical examination of evolutionary contingency is still at a relatively early stage, and it would be premature to declare that current findings have actually resolved the question. Indeed, it remains to be seen how the findings from microbial evolution experiments into contingency may apply to the broader biological world and its greater complexity and opportunities for iteration and the interaction between lineages. Nonetheless, these experimental studies have improved our understanding of contingency's role in evolution, and they point to considerations for future research. The following are the five points that I think are the most important going forward.

1. **The role of contingency in evolution is constrained.** PREs have shown that initially identical populations maintained in the same environment evolve remarkably in parallel along highly similar fitness trajectories (Lenski 2004; Kawecki et al. 2012; Stern 2013). This parallelism partially extends to the genetic level, where similar adaptive mutations often accrue across multiple populations. Similarly, HDEs have generally shown that short periods of lab evolution in one environment do not impede adaptation to another environ-

ment (Travisano et al. 1995). Natural selection therefore often seems to be capable of deterministically driving broadly similar evolutionary outcomes in spite of the historicity imparted by the core processes of evolution or the effects of prior evolution.

2. **Parallelism at one level can mask divergence at another.** Parallel evolution to highly similar or identical fitness is commonly seen in evolution experiments (Lenski 2004; Kawecki et al. 2012). However, similarities in fitness can conceal important differences between evolving populations. Parallelism at one level of assessment does not necessarily imply parallelism at other levels. Variation in both genotype and unused functions among populations with similar fitness is commonly observed in both PREs and HDEs. More dramatically, LTEE population Ara-2 evolved a complex ecology that maintained a balanced polymorphism. At the level of fitness, Ara-2 was evolving in parallel with nonpolymorphic populations, and yet it did so by following a different adaptive path (Wiser, Ribeck, and Lenski 2013). These observations highlight the difficulties inherent in defining evolutionary convergence and parallelism when there are multiple levels of analytic granularity (Currie 2012). Experiments that involve assessment of parallelism and divergence would benefit from better and more extensive means of doing so. Future research will be strengthened by the development of standard measures for quantifying the parallelism and divergence identifiable in PREs and HDEs at multiple levels, including fitness, phenotype, genotype, and evolved ecology.

3. **The role of historical contingency intrinsic to the evolutionary process can be understood in terms of evolvability.** Evolvability has come to be a topic of great interest in evolutionary biology in recent years. In general, evolvability refers to the capacity of a genotype to produce new heritable variation by mutation, though some more restrictive definitions focus on the capacity to produce adaptive variation (Pigliucci 2008b). Though it has received little notice, there is clearly a deep connection between evolvability and historical contingency arising from core evolutionary processes, as both deal with evolutionary potential.

In the absence of gene flow, natural selection can act only on the variation that arises by mutation of the genomes extant within a population. However, not all variation is equally reachable from all genotypes. From any given genotype, there is a range of possible variants that can arise at frequencies determined by the number of mutations necessary to reach them and the rates at which those mutations occur. Prior history constructs the genotype and impacts future evolution by determining that from which variation arises, and thereby what variation can reasonably arise. Prior history can also alter background mutation rates, which is most obvious in the case of mutator geno-types. In so doing, history can have two possible effects. The first is potentia-tion, as was demonstrated in the case of Cit^+ evolution in the LTEE. In the case of a variant that requires multiple mutations, history can potentiate that variant's evolution by coincidental accumulation of one or more of the neces-sary mutations, thereby increasing the frequency at which the variant arises. Such a variant may be of no consequence evolutionarily, but it might confer a rare refinement of an existing trait. It might also confer a novel trait that can, like Cit^+, cause lineages to chance upon previously inaccessible novel func-tions that can grant access to wildly different evolutionary paths. The second effect is depotentiation, in which history reduces the likelihood of evolving a given variant. The historical accumulation of mutations that must either be re-verted or compensated for in order for a variant to occur is one mechanism by which depotentiation may occur. Depotentiation is therefore related to func-tional loss due to mutation accumulation (V. S. Cooper and Lenski 2000). Once a mutation disables a gene required for a given function, for instance, the reevolution of that function would be depotentiated by the accumulation of further disruptive mutations. (Deletion, of course, may completely elimi-nate the possibility of reevolution, unless there is another gene or pathway that might be co-opted for restored function.) Epistatic interactions have also been implicated in depotentiation. Woods et al. (2011) showed that one of two competing alleles in an LTEE population went extinct despite a significantly higher fitness benefit because its epistatic effects eliminated the fitness benefit and depotentiated the selection of a potential later mutation. Potentiation and depotentiation therefore contribute to the changing capacity of a population to generate variation that can alter the potential for adaptation in a range of environments. This is to say that potentiation and depotentiation alter evolv-ability, and evolvability may therefore be a valuable way to understand and approach evolutionary contingency (Lenski, Barrick, and Ofria 2006; Cole-grave and Collins 2008; Pigliucci 2008b). A synthesis between the concepts of evolvability and contingency would no doubt be helpful to future empirical and theoretical work on evolutionary contingency.

4. **History is a progressive factor in evolution.** Building on the above, as history accumulates, the effects of potentiation and depotentiation should in-crease and vary, altering the evolutionary potential of the lineage to follow various trajectories. History's impacts on evolution should therefore be pro-gressive, leaving deeper marks over time. As Desjardins (this volume, chap. 9) has observed, historicity comes in degrees. Moreover, history's impacts may be subject to threshold effects, in which avenues for future evolution may be

qualitatively closed or opened by the accumulation of interacting mutations in a manner unpredictable from their subsets. In other words, history may not matter until it does matter. This is all the more true when it is considered that contingency can have impacts on multiple levels of the evolutionary process (Erwin, this volume, chap. 12). It is therefore difficult to come to valid conclusions about contingency's potential impact on evolution from short-term experiments, and this issue should be kept in mind in future experiments.

5. **Evolutionary contingency researchers need to collaborate more broadly.** Empirical research into evolutionary contingency would benefit greatly by improved interdisciplinarity. For instance, greater collaboration between researchers using the microbial experimental evolution approach I have discussed here, paleontologists, and those who study contingency in macroorganisms is certainly called for. Systems biologists could help to develop a better and more theoretically grounded understanding of how the complex interactions within evolving organisms impact and are impacted by contingency arising from the core evolutionary processes. Collaboration with molecular biologists and geneticists would similarly lead to better understanding of how particular molecular events and mutational processes factor into evolutionary contingency. Similarly, these researchers could help greatly in better defining convergence and examining how divergence at the smallest level can impact later evolution.

This interdisciplinary approach to understanding contingency should not be limited to scientific fields. Historical contingency is a complex, multifaceted concept that is difficult to fully grasp and define. Gould's writings on evolutionary contingency display some of the confusion this complexity engenders. Gould never offered a single, technical definition of evolutionary contingency, and he, apparently unknowingly, articulated at least two, very different notions of contingency (Gould 1989, 2002; Beatty 2006b; Beatty and Carrera 2011; D. D. Turner 2011). Most experimental work on contingency is based on individual readings of Gould's writings, leading to a situation in which different researchers have designed experiments based on different notions of contingency without necessarily making this clear. This semantic discord has produced understandable confusion that has made it difficult to meaningfully synthesize the various interesting and illuminating experimental findings about evolutionary contingency. I propose that these conceptual difficulties could best be overcome and the study of contingency advanced by collaboration between evolutionary biologists and philosophers of science. The complexity and difficulty inherent to historical contingency makes it the sort of conceptual tangle philosophers excel at analyzing and parsing

(Pigliucci 2008a). Indeed, historical contingency is a philosophically rich area that has been the subject of much recent work (Beatty 1993, 2006b; Beatty and Desjardins 2009; Beatty and Carrera 2011; Currie 2012; Desjardins 2011; Desjardins, this volume, chap. 9; Sterelny and Griffiths 1999; D. D. Turner 2011). These philosophers have thought deeply about the issues contingency researchers seek to address, and they might help to unpack ideas of evolutionary contingency, delineate subsidiary issues within the concept, develop rigorous definitions, and trace interesting conceptual implications while also perhaps helping to guide and structure productive interdisciplinary collaborations. This conceptual work could then be used to develop more precise empirical questions and design better experiments, the results of which could then be integrated into a more coherent and unified understanding of evolutionary contingency. Indeed, a full understanding of evolutionary contingency may be contingent on such a collaboration.

ACKNOWLEDGMENTS

I would like to thank my mentor Richard Lenski and lab manager Neerja Hajela for years of support and guidance, and Justin Meyer, Caroline Turner, Erik Quandt, John Beatty, Robert Pennock, David Bryson, Sabrina Mueller-Spitz, and Betul Kacar for discussions that were helpful in the development of the ideas behind this chapter. I thank Richard Lenski and two anonymous reviewers for their suggestions on this chapter. I acknowledge funding from a John Templeton Foundation Foundational Questions in Evolutionary Biology grant (FQEB #RFP-12-13), the BEACON Center for the Study of Evolution in Action (NSF Cooperative Agreement DBI-0939454), and the US National Science Foundation, which has funded the LTEE (NSF; DEB-1019989).

NOTES

1. Gould was actually part of a broader trend of increased interest in contingency and historicity that developed during the second half of the twentieth century, the causes of which have yet to be fully explored. This interest can be seen in popular culture, where contingency and counterfactual history came to be a mainstay of TV and film, and the science fiction subgenre of alternate history experienced strong growth. In academia, counterfactual analysis came to be an accepted methodological tool in a variety of fields, including philosophy (Lange 2005), sociology (Harding 2003; S. L. Morgan and Winship 2007), and economics (Cowan and Foray 2002; Cartwright 2007). Increased appreciation for historical contingency and counterfactual

methods also developed in both professional and popular historiography (McPherson 1988; Ferguson 1997; Bulhof 1999).

2. Contra Desjardins (this volume, chap. 9), both hypotheses implicate history, and they differ only in whether the importance of history was in rendering possible the final necessary mutation itself (physical promotion) or the final mutation's effect (functional epistasis). Either way, historical sequences that included the prior occurrence of the potentiating mutation or mutations and the events and processes that maintained them in the population were necessary for the Cit^+ function to evolve.

CHAPTER 11

Rolling the Dice Twice: Evolving Reconstructed Ancient Proteins in Extant Organisms

Betul Kacar

Living organisms are historical systems, and their evolutionary history is one of the vital determinants of their capacity to respond to their environment. Historical contingency is a property of living systems that assigns past events as important factors that shape a system's current state. In his book *Wonderful Life*, Stephen Jay Gould famously posed a thought experiment to address whether other possible evolutionary trajectories could have taken place, meaning that current biota are the contingent product of random luck, or whether the observed evolutionary trajectory was inevitable. Assigning historical contingency as a fundamental influence in shaping evolutionary outcomes, Gould posited that if life's tape were rewound and replayed from various points in the distant past, the resulting living world would be very different than it is now:

> I call this experiment "replaying life's tape." You press the rewind button and, making sure you thoroughly erase everything that actually happens, go back to any time and place in the past—say, to the seas of the Burgess Shale. Then let the tape run again and see if the repetition looks at all like the original. If each replay strongly resembles life's actual pathway, then we must conclude that what really happened pretty much had to occur. But suppose that the experimental versions all yield sensible results strikingly different from the actual history of life? What could we then say about the predictability of self-conscious intelligence? or of mammals? or of vertebrates? or of life on land? or simply of multicellular persistence for 600 million years? (Gould 1989, 48, 50)

Understanding to what degree historical contingency shapes evolutionary trajectories would be possible by performing Gould's thought experiment. We cannot, of course, carry out this experiment at the global scale Gould envisioned; however, methods of experimental evolution allow researchers to tackle some aspects of rewinding and replaying evolution in controlled environments (Blount, Borland, and Lenski 2008; Fortuna et al. 2013; Losos 1994; Travisano et al. 1995; Desjardins, this volume, chap. 9). A recent advance in experimental biology also offers an approach that permits the tape of life to be rewound for individual genes and proteins. Commonly referred as "ancestral sequence reconstruction," or paleogenetics, this method integrates molecular phylogeny with experimental biology and in vitro resurrection of inferred ancestral proteins (fig. 11.1; see also Dean and Thornton 2007). Here I present a novel approach that merges ancestral sequence reconstruction with experimental evolution. In this method, my colleagues and I have engineered a reconstructed ancestral gene directly into a modern microbial genome with the intention of observing the immediate effects of this ancient protein in a modern bacterial context (Kacar and Gaucher 2012). This approach therefore initially follows a backward-from-present-day strategy in which we essentially recover a previous point on the tape of life, reconstruct it (rewind), and then observe the interaction of this ancient protein with the present (replay). Before presenting the details of our system, and where we would like to go with the in vivo resurrection of ancient genes, I will first detail how ancestral sequence reconstruction provides insights into the past.

ANCESTRAL SEQUENCE RECONSTRUCTION: USING PHYLOGENY TO UNRAVEL EVOLUTIONARY HISTORY

More than a half century ago, Pauling and Zuckerkandl recognized the potential of phylogenetics to provide information about the molecular history of life. In a pioneering paper titled "Chemical Paleogenetics: Molecular 'Restoration Studies' of Extinct Forms of Life," they suggested a relatively straightforward methodology in which gene or protein sequences obtained from existing organisms would be determined, aligned, and then used to construct phylogenetic trees from which the ancestral states of modern genes and protein sequences could be inferred (Pauling and Zuckerkandl 1963). In 1990, Benner et al. realized the vision of Pauling and Zuckerkandl by reconstructing, synthesizing, and characterizing an ancestral ribonuclease from an extinct bovid ruminant in the laboratory (Stackhouse et al. 1990).

Ancestral sequence reconstruction has since been used to examine the

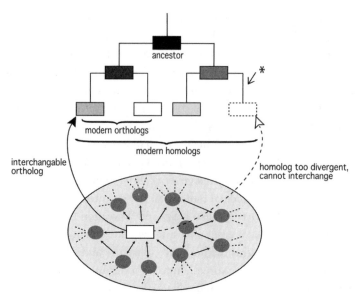

FIGURE 11.1. *At top:* A simplified phylogenetic tree is shown. Phylogenetic trees help us to visualize the evolutionary relationships among organisms. Modern gene and/or protein components are at the tip of the tree. Genes and/or proteins that are connected by a branching point are more closely related to each other and may exhibit more closely related functional properties. In ancestral sequence reconstruction, homologous characters of modern gene or protein components are used to infer the character state of the common ancestor (located at the root of the tree, in black).

At bottom: A central component of a cell (represented by the white box) will exhibit various interactions with other ancillary cellular partners (represented by gray circles) that also exhibit interactions with many other cellular components (represented by dashed lines). To understand the evolutionary mechanisms that underlie protein function, we need to acknowledge that intrinsic and extrinsic properties of the protein within its interaction network, not just the properties of an individual protein, contribute to and perhaps define how a protein performs its function in a given environment. Replacing a modern gene with its modern homologous or orthologous counterpart from another organism would allow us to identify the functional constraints that shape the two proteins. However, such an approach cannot directly speak to how the replaced protein evolved because its replacement is not on the same line of descent. By replacing a fine-tuned member of a networked system with another component that has coadapted to another networked system (shown in dashed box), the interactions between the replaced protein and other native proteins may be damaged to the point of producing a nonfunctional system due to evolved incompatibilities in the homolog (represented by an asterisk (*)).

molecular history of many proteins, allowing biologists to test various hypotheses from molecular evolutionary theory by tracing the evolutionary paths proteins have taken to acquire their particular function (Benner, Sassi, and Gaucher 2007). This backward-from-today strategy has also provided helpful insights about various environmental conditions of the ancient Earth (Galtier, Tourasse, and Gouy 1999; Gaucher 2007; Ogawa and Shirai 2013) and has paved the way for emerging paradigms such as functional synthesis (Dean and Thornton 2007), evolutionary synthetic biology (Cole and Gaucher 2011), and evolutionary biochemistry (Harms and Thornton 2013). Moreover, it has led to the development of models of molecular evolution that have been used to guide applications of evolutionary theory in medical and industrial settings (Kratzer et al. 2014; Risso et al. 2013).

Our work has focused on reconstructing the molecular past of Elongation Factor-Tu (EF-Tu) proteins. EF-Tu proteins function to deliver aminoacylated-tRNA molecules into the A-site of the ribosome and are thus essential components of the cell (Czworkowski and Moore 1996). Moreover, EF-Tu proteins also interact with a variety of other proteins functioning outside of translation machinery and thus serve in activities outside of their role in the ribosome (Defeu Soufo et al. 2010; Kacar and Gaucher 2013; Pieper et al. 2011).

Various ancestral sequences of bacterial EF-Tu, ranging from approximately 500 million to 3.6 billion years old, have been previously inferred and reconstructed (Gaucher, Govindarajan, and Ganesh 2008). Useful to these studies was the information that EF-Tu proteins exhibit a high correlation with their organisms' environmental temperature (Gromiha, Oobatake, and Sarai 1999). For example, EF-Tus from thermophilic organisms (organisms living in high temperature environments) exhibit high temperature tolerance and thermostability, while EF-Tus obtained from organisms that live in moderate conditions (i.e., mesophiles) exhibit much lower temperature tolerance and stability. EF-Tus therefore appear to be subject to strong selection to be optimized to their host's thermal environment.

The combination of EF-Tu's essential role in the cell, the strong selection acting on EF-Tu phenotype, and the availability of various phenotypically and genotypically altered ancient EF-Tu proteins created an ideal situation for us to rewind the molecular tape of life for one gene. A rather obvious question is whether a living organism could function if its native EF-Tu was replaced with an ancient form. How and to what degree would the reconstructed ancestral EF-Tu proteins interact with the components of the modern cellular machinery? Moreover, can we provide insights into the factors that determine whether a reconstructed ancient EF-Tu, with very distinct genotypic and phenotypic properties, can interact with modern components?

We therefore set out to first identify whether a recombinant bacteria whose endogenous EF-Tu is replaced with an ancestral EF-Tu at the precise genomic location would result in a viable bacteria. First to assess was whether ancient EF-Tu proteins would interact with the members of the translational machinery and thus perform their primary function. For this, we measured the activity of various ancient EF-Tu proteins in an in vitro translation system. This system is composed of translation machinery components recombinantly purified from modern *E. coli* bacteria and individually presented in a test tube, providing us control over which molecules are presented or omitted (Shimizu et al. 2001). We removed the modern *E. coli* EF-Tu from this cell-free system, individually inserted ancient EF-Tu proteins one by one, and measured the activity of the hybrid system by following fluorescent protein production by the translation machinery.

Our in vitro findings show that foreign EF-Tus (both ancient counterparts and modern homologs of *E. coli* EF-Tu) exhibit function, albeit not equal to the *E. coli* EF-Tu, in a modern translation system where all the other components of the translation system are obtained from *E. coli* (Zhou et al. 2012). Therefore, while replacing the native, modern EF-Tu with an ancient EF-Tu seems likely to put stress on *E. coli*, the recombinant organism is also likely to be viable. Although this is an exciting observation, replacing the EF-Tu of a natural system with another EF-Tu protein that has coadapted to a foreign cellular system (be it a modern homolog or an ancient one) will always carry the risk that the interactions between the foreign protein and its ancillary cellular partners may be damaged to the point of complete dysfunction (fig. 11.1). Such a possibility restricts our ability to move genes between organisms; we are fundamentally limited by the historical adaptive mutations and the epistatic relationships that define any particular organism. Before I discuss our success at creating such a recombinant, I will address a question that has no doubt occurred to the reader: Why even bother performing such an experiment with a reconstructed ancestral EF-Tu variant instead of a modern homolog from a different living organism?

WHY ENGINEER A MODERN BACTERIAL GENOME WITH A RECONSTRUCTED ANCIENT GENE?

One common way to assess how a protein may function in a foreign host is to do so directly by removing that protein from its host and inserting it into the foreign organism. One might ask why our new approach does not involve replacing a modern gene with a homolog or ortholog from another modern organism. Ultimately, swapping a protein with its homolog has been providing and will continue to provide valuable insights into functional constraints that shape the two proteins, regardless of whether they share functional identity (Applebee et al. 2011; Couñago, Chen, and Shamoo 2006). However, this approach has the drawback that swapped proteins may not share a direct line of descent that connects them over evolutionary time. This lack of parity presents the possibility of "functional nonequivalence," meaning that the two proteins may have traversed two separate and possibly functionally divergent adaptive paths that now prevent the two homologs from functioning properly after being swapped between organisms (fig. 11.1). Moreover, it is important to note that not all of the mutations of the modern homologs are adaptive; random mutations that are results of stochastic events, for instance, can lead to the accrual of mutations in a modern homolog that could prevent the functionality of the modern homologs in another organism once swapped, even though they were neutral under the conditions in which they were accumulated (Camps et al. 2007; Romero and Arnold 2009).

Despite the problems involved, studies in which modern genes are replaced by resurrected ancestral genes indicate that the results are worth the effort. In their article on the mechanistic approaches to study molecular evolution, Dean and Thornton remark that "the functional synthesis should move beyond studies of single genes to analyze the evolution of pathways and networks that are made up of multiple genes. By studying the mechanistic history of the members of an interacting gene set, it should be possible to reconstruct how metabolic and regulatory gene networks emerged and functionally diversified over time" (Dean and Thornton 2007, 686). It is not yet feasible to rewire a whole network, but replacing a key component of a complex network with its ancestral counterpart complements these efforts. It is expected that the ability of the interaction network to function with the ancestral component will be limited by incompatibilities that have accrued by the rest of the network over time since the point from which the ancestor was resurrected. A remaining question is how the modern components of a cellular system would cofunction with their ancestral interaction partners, if given a chance.

AN ANCIENT-MODERN HYBRID TEST ORGANISM

Following up on our in vitro work, we were set to generate a strain of modern *E. coli* harboring an approximately 500-million-year-old EF-Tu protein at the precise chromosomal location of the modern EF-Tu. The ancestral protein is inferred to have been functioning within the common ancestor of γ-proteobacteria, has 21 (of 394) amino acid differences from *E. coli* EF-Tu, and its melting temperature is comparable to that of *E. coli* EF-Tu (39.5°C versus 37°C, respectively).

Marking the first genomic resurrection of a reconstructed ancestral gene in place of its modern counterpart within an extant organism, we were able to obtain a viable *E. coli* strain that contains an ancient EF-Tu as the sole genomic copy (Kacar and Gaucher 2012). We next measured the doubling time of the recombinant organism hosting the ancestral EF-Tu. Consistent with our expectations, the hybrid organism exhibits lower fitness, as demonstrated by a doubling time twice as long as its wild type parent strain (fig. 11.2). The hybrid is viable, showing that the ancient EF-Tu can complement essential functions of a descendant in a modern organism. Moreover, the mutations that have accumulated in the components of the modern bacteria over time do not prohibit the modern components that are essential for viability from interacting with the ancient EF-Tu protein. This is an intriguing result, especially considering the vast number of proteins EF-Tu interacts with beyond its primary interaction partners in the translation machinery. The hybrid is also less fit, demonstrating that the ancestral component triggered a stress on

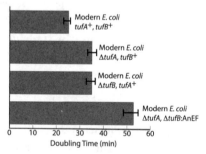

FIGURE 11.2. Replacement of a modern essential gene with its ancestral counterpart reduces host bacterium fitness. Precise replacement of a modern bacterial EF-Tu gene with its ~500-million-year-old ancestor doubles the bacterial strain REL606's doubling time. Two genes, tufA and tufB (varying by just one amino acid) code for EF-Tu proteins in modern *E. coli*. We deleted the tufA gene from the chromosome and replaced the tufB copy with the ancient EF. Deletions of tufA or tufB in the *E. coli* REL606 strain have similar effects (~ 34 minutes) when deleted individually. Measurements are performed in rich growth media at 37°C in triplicates. Figure adapted from Kacar and Gaucher 2012.

the modern organism, creating an ideal case to monitor the coevolution of the ancient component and the recombinant bacteria.

One interesting aspect of this modern-ancient recombinant approach is that it presents a new means of understanding how protein function evolves as a consequence of changes in protein sequence, which in turn may allow us to elucidate how changes in these functions correlate to the overall cellular context within living organisms. Although this is not the primary motivation for setting up these experiments, the approach uniquely complements other studies focused on understanding in vivo evolution of proteins. Here I would like to briefly discuss why this peripheral aspect carries biological significance.

Modern cells are the products of immensely long evolutionary histories and are thus the heirs of a biological heritage that was shaped by the genetic and developmental characteristics of their ancestors, which were themselves shaped by the environments in which they lived. Within this heritage are highly conserved proteins, the functions of which are so crucial that the cellular machinery will not tolerate much change, and which therefore serve as functional fossils. Other proteins are either highly resilient, of less crucial function, or else encode some other flexible potential, and so can readily change in response to exigencies. These proteins show higher rates of evolution and, by definition, lower levels of conservation over time and across taxa. The two groups may be readily distinguished by comparison of proteins across taxa (Baker and Šali 2001; Kominek et al. 2013; Martí-Renom et al. 2000; Papp, Notebaart, and Pál 2011; Wellner, Raitses Gurevich, and Tawfik 2013).

In order to assess how a protein performs its function, a basic approach is to couple molecular biology and biochemistry by removing the protein from the cellular context and measuring the protein's activity through its interactions with a defined reactant in a test tube. To experimentally identify what role the a priori predicted functionally important sites play, the predicted sites are altered via tools such as site-directed mutagenesis and the mutant protein's function is measured in vitro. Two central points come from this; first, properly predicting the functionally important sites, and second, properly analyzing the protein's function so that studies reflect the protein's in vivo, and thus biologically relevant, function in the cell.

To address the first point, one common way to predict what sites or domains of a protein carry functional importance is to analyze the evolutionary conservancy of the protein sequence. Briefly, this method follows the alignment of the amino acid sequences of multiple protein homologs, determining the conserved, different, and similar sites through comparison of the aligned taxa. As expected, functionally important sites of a protein will be the most highly conserved across homologs due to their crucial role in protein function (Benner 1989). However, this straightforward methodology does not fully confront the fact that a protein site that is not necessarily conserved across homologous organisms can also carry a functionally crucial role that goes beyond the immediate primary sequence of the protein itself, and it may represent an adaptation to a host's specific intracellular and extracellular environments (Fraser 2005; R. A. Jensen 1976; Khersonsky, Roodveldt, and Tawfik 2006). Recent approaches that include trajectory-scanning mutagenesis and identify regions in which coevolving proteins interact with each other provide valuable measures addressing this challenge (Ashenberg and Laub 2013; Capra et al. 2010). Indeed, a single amino acid substitution far away from the primary functional site can greatly impact that site's function (Copley 2003), indicative of various other factors shaping the evolution of protein function.

This observation leads to the second point that needs to be carefully addressed when studying how proteins evolve. To reveal how a protein performs its function by biologically realistic means, protein genotype and organismal phenotype need to be directly connected. Engineering strains lacking a particular protein of interest and then reintroducing a homolog or synthetically reconstructed variants of the protein in these mutant strains holds considerable value. Going one step further and observing the coadaptation between reconstructed proteins and microbial organisms through laboratory evolution would allow us to examine the protein's function in vivo. Finally, this approach would allow us to identify functionally important sites that are not necessarily conserved across taxa but are specific to a host's intercellular environment, thus allowing us to consider the context-dependent protein adaptation within a specific lineage.

EXPERIMENTAL EVOLUTION OF BACTERIA
CONTAINING AN ANCIENT GENE COMPONENT

The above discussion provides a context for understanding how functionality is connected to protein divergence. Only after site-specific evolutionary constraints are defined can we truly begin to understand how a protein "functions." One alternative way to determine protein functionality is to let an organism mutate its proteome in response to some intra- or intercellular environmental pressure. Such pressures can arise from the availability of novel energy sources or even from the manipulation of internal cellular components. Toward this end, experimental evolution with microbes would provide information on the organismal level by monitoring the real-time evolution of microbial populations to some unique pressure. In these evolution experiments, microbial populations are evolved in the lab through either continuous culture or serial transfer under controlled conditions, permitting evolution to be studied in unprecedented detail in near real time (Elena and Lenski 2003).

The experimental evolution approach is particularly powerful because of the high level of control it permits. Samples of evolving populations may be frozen at regular intervals while still remaining viable, providing "frozen fossil records" that can be analyzed to examine a variety of important questions (Elena and Lenski 2003). Experimental evolution may be used for highly detailed study of a variety of questions, such as whether evolution follows contingent or deterministic paths, whether mutations accumulate neutrally or adaptively, and how epistatic interactions impact evolution. Indeed, laboratory evolution experiments using microbes have provided deep insights into the interplay between contingency and deterministic processes in evolution (Blount, Borland, and Lenski 2008; Blount, this volume, chap. 10; Losos 2011; Travisano et al. 1995; Vermeij 2006).

The experimental evolution approach is key in our next step with the ancient-modern hybrid organism we have constructed to replay the tape. We are currently evolving the reconstructed organism in the laboratory to study how compensatory evolution alters the ancient gene and members of its epistatic network. Of particular interest is how much and how quickly the construct's fitness and phenotype will change, and how much of these changes are mediated by the ancient EF-Tu's direct accumulation of beneficial mutations versus their accumulation in other, associated genes. Moreover, we wish to assess the extent to which the mutational trajectory is guided by random and unpredictable events (i.e., attributable to chance) and how much it is determined directly by the genotype and the phenotype of the ancient EF-Tu engineered inside the bacteria (i.e., attributable to contingency).

Evolving a bacterium containing a single ancient gene in its genome is analogous to replaying a particular track on the tape of life within the context of the modern organism. To play with the metaphor a bit, we made a mix-tape by splicing a very old track from the tape of life and are now playing it on a modern tape deck (the modern organism). Both the tape and the tape deck can evolve in this case, and we will be able to examine how they evolve and adapt to each other, potentially recapitulating some of the evolution that occurred in the first place. While this sort of replay experiment is not exactly what Gould had in mind, how the organism adapts to the ancestral protein and vice versa promise to shed light on the molecular evolution that took place during EF-Tu's evolution.

This approach does have limitations as well. It is important to recognize that the ancient component resurrected in a modern organism's adaptation will be shaped by its interactions with modern components adapted to both its modern counterpart and a cellular environment that is millions of years ahead of the conditions of the ancient component. It is also important to note that the environmental conditions in the laboratory are unlikely to be comparable to the conditions and circumstances under which the ancestral protein evolved into its modern descendant.

Despite its limitations, reevolving ancient genes in modern organisms permits us to investigate historical aspects of repeatability and parallelism in evolution that replacing a gene with a modern homolog does not. Multiple aspects of the evolutionary processes, including mutation and genetic drift, are inherently stochastic, which makes it challenging to predict how a system hosting a reconstructed gene (as well as a modern homolog) would coevolve. By setting up laboratory evolution experiments of microbial organisms that carry an ancient component in replicate populations, we are able to observe whether or not multiple populations carrying the ancient component evolve the same way, or in parallel ways, under identical environmental conditions. Engineering constructs with genes encoding different reconstructed ancestral states of the same protein (e.g., 500-million-year-old, 1-billion-year-old, etc.) would permit investigation of whether or not evolutionary trajectories are strictly dependent on evolutionary starting points (fig. 11.3; see also Travisano et al. 1995). For instance, certain adaptive zones may be more readily accessible as we go further back in time; that is, the older the component we resurrect, the greater the likelihood of finding a mutational point that is capable of functional innovation by resetting the epistatic ratchet (Bridgham, Ortlund, and Thornton 2009).

CONCLUDING REMARKS

To directly examine the relationship between historical constraints and evolutionary trajectories, and to assess the role historical contingency played in shaping protein evolution within a biologically realistic framework, it is important to determine the historical paths of proteins, both genotypic and phenotypic levels, and then study how these paths affected the organism. Many properties of protein-protein interaction networks, not just those of individual proteins, contribute to and perhaps define how a protein performs its

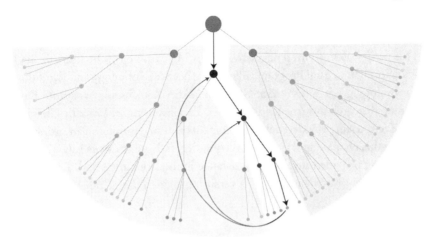

FIGURE 11.3. A hypothetical scheme of evolutionary patterns: Interaction between an organism containing ancestral variants of a protein (represented by different colored circles) and its environment will produce a series of adaptive zones. Replacing a modern protein with an ancestral counterpart may potentially lead the organism to explore a wider potential adaptive landscape. Replaying evolution starting from successively older time-points (shown by successively longer arrows) would then allow access to points corresponding in time to historic evolutionary states that preceded the accumulation of mutational steps, thus allowing the engineered organism to operate in different adaptive zones.

function in a given environment. Consequently, the biochemistry of proteins may be intimately linked to their host organisms' behavior.

The only lesson history teaches is that no one learns from history (Hegel 1953). This playful quote is not the case for understanding the role history plays in shaping evolution. Integrating molecular-level theory into studies of evolutionary history and merging ancestral sequence reconstruction studies with experimental evolution will help to provide mechanistic explanations for several long-standing questions in evolutionary biology: How does an organism's history shape its future trajectories, and are there deterministic paths along these trajectories? Does evolution inevitably lead to the same endpoints? How much does the past restrict the future? Moreover, how do answers to these questions depend on the genetic and the environmental conditions of the system examined? Synthesis of the fields of synthetic biology, biochemistry, and experimental evolution holds great promise for generating a new understanding of functional, structural, and historical constraints that shape biological evolution. Indeed, this synthesis holds great promise for learning much of and from life's history.

ACKNOWLEDGMENTS

This work was supported partly by the NASA Astrobiology Institute through a NASA Postdoctoral Program Fellowship, NASA Exobiology and Evolutionary Biology grant NNX13AI08G, and a NASA Astrobiology Institute Early Career Research Award. Eric Gaucher, Zachary Blount, W. Seth Childers, and Zach Adam provided helpful comments and suggestions on an earlier version of this chapter, and Georgia Tech undergraduate students Lily Tran and Jen Zhang provided invaluable assistance in the laboratory. Rich Lenski and Neerja Heala (Michigan State University) provided the *E. coli* REL606 and REL607 strains.

CHAPTER 12

Wonderful Life Revisited: Chance and Contingency in the Ediacaran-Cambrian Radiation

Douglas H. Erwin

Although the importance of determinism versus chance was not one of the "eternal metaphors" in Stephen Jay Gould's essay of the same name (Gould 1977), it might well have been, for the debate over how much of the history of life is deterministic, the result of physical processes, scaling laws, or various biological constraints, versus how much of it reflects either chance or just one of many possible evolutionary histories (contingency) dates at least to Darwin (Beatty 2008; Depew, this volume, chap. 1). Although the topic became less prominent with the modern synthesis, chance and necessity served as the title of Monod's eponymous book (Monod 1971), arose as "frozen accidents" in Crick's early discussion of the structure of the genetic code (Crick 1968), and played a significant role in Kauffman's arguments that many aspects of evolution may be "frozen accidents," preserved not for their adaptive value but rather because they generate complexity that is subsequently preserved (Kauffman 1993). The tension between determinism and chance continues today (S. B. Carroll 2001). It appears in discussions over the ubiquity of allometric scaling relationships (J. H. Brown et al. 2004) and in structuralist arguments for the importance of physical forces (Salazar-Ciudad, Jernvall, and Newman 2003). In astrobiology a fundamental assumption is that sufficient regularities exist that studies on earth will have implications for life elsewhere in the universe. The ubiquity of convergence of morphological form has often been used as evidence against contingency (Conway Morris 2003; McGhee 2011), generating a suite of issues. But chance and contingency are different issues. Since 1989, much of the focus of this tension over chance, contingency, convergence, and determinism has revolved around Gould's book *Wonderful*

Life (Gould 1989), his recounting of the history of research into the extraordinary fossils of the Burgess Shale and their implications for the history of life.

As I discuss further below, one of the meanings of contingency involves the unpredictability of evolutionary history. Let me state at the outset that there seems little point in disputing this facet of life. My favorite example comes not from the Burgess Shale or the Cambrian, but from the great end-Permian mass extinction 252 million years ago (Ma) when over half of all families and perhaps 90 percent of marine species disappeared (Erwin 2006b). Echinoids (sea urchins, sand dollars, and their allies) were a relatively insignificant component of Paleozoic marine ecosystems, with perhaps eight genera in the Permian. Only one or two species of echinoid appear to have survived the mass extinction, but they gave rise to the incredible diversity and morphological range of post-Paleozoic echinoids with almost one thousand living species. But the survival of these one or two species was almost certainly fortuitous and illustrates the role contingency can play in the history of life. More interesting perhaps, is how contingency plays out at different levels of evolution and the trade-offs with deterministic processes.

Gould published *Wonderful Life* in the early years of a tremendous increase of research interest in the Cambrian explosion of animal life, which, coupled with the spread of new, more rigorous methods of phylogenetic reconstruction through cladistic methods, considerably altered our understanding of many of the empirical issues raised by Gould. To take a few examples of changes that occurred within a few years of the publication of *Wonderful Life*, our understanding of the phylogenetic relationships among animal clades was fundamentally altered by new techniques in phylogenetic analysis (Jefferies 1979), particularly the introduction of molecular data (Field et al. 1988), and the application of molecular clocks (Runnegar 1982; Wray, Levinton, and Shapiro 1996). New radiometric dates began to emerge in the early 1990s, changing our understanding of the tempo of the Cambrian explosion and the association between the earlier Ediacaran macrofossils and the Cambrian (Bowring et al. 1993). The Chengjiang Fauna had been discovered by Hou Xianguang in 1984, providing a slightly older, early Cambrian Chinese analogue to the Burgess Shale (Hou, Ramsköld, and Bergström 1991), but most publications of the new taxa came during the 1990s and continue today. Many other new fossil discoveries have occurred as well. New techniques were developed to deal with many issues of macroevolutionary dynamics, such as quantitative approaches to the study of morphological disparity (Briggs, Fortey, and Wills 1992; Foote 1992), in direct response to the claims made by Gould. Finally, the growth of comparative evolutionary developmental biology ("evo-devo") has shown that many of the developmental genes responsible for body plan formation in animals are broadly shared across the major clades and thus must have been present early in animal history. In short, since 1989 almost every aspect of our understanding of the early history of animals and the Cambrian explosion has been fundamentally altered, and recent descriptions of the Cambrian explosion are much different from those of the late 1980s or early 1990s (Erwin et al. 2011; Erwin and Valentine 2013). These developments have altered, often significantly, the evidentiary basis for his arguments.

The theoretical framework of Gould's argument has changed as well. The arguments for contingency that Gould advanced in *Wonderful Life* were only a part of a larger landscape of his scholarship on the role of history in evolution. Indeed, there is a curious and I think largely unresolved tension in Gould's work between his earlier interest in the role of lawlike processes (Gould 1970; Raup et al. 1973; Raup and Gould 1974; Gould et al. 1977) and later papers addressing how historical constraints may limit the power of adaptive evolution (Gould and Lewontin 1979). A central theme in this later phase is limits to the role of natural selection in structuring the waxing and waning of diversity in different clades of organisms. These threads were the focus of a 1980 *Paleobiology* article in which Gould explicitly addressed the tension between idiographic studies (descriptions of historically unique events) and nomothetic research (the search for general principles or laws) (Gould 1980b). Gould later returned to the theme, arguing for the importance of mass extinctions in limiting the power of adaptive evolutionary trends (Gould 1985b). (See also discussions in Baron 2009; Beatty and Desjardins 2009; Sepkoski 2012; Sepkoski and Ruse 2009; the philosopher Chris Haufe is currently studying this issue.) There can be, of course, no certain resolution to idiographic versus nomothetic approaches, or to the tension between historicity and a search for general laws. As Gould realized, paleontology is a combination of both idiographic and nomothetic approaches. Each approach plays an important role in understanding aspects of the history of life, and both have been addressed in many discussions of macroevolution since 1989, as well as in papers by a number of philosophers and historians.

My objective here is to reconsider the arguments Gould makes in *Wonderful Life* in light of the discoveries and conceptual advances since 1989, both on the Burgess Shale fossils specifically and on the Cambrian explosion more generally. I begin with a summary of the principal arguments that Gould developed in *Wonderful Life*, focusing on those associated with the role of contingency. I turn next to a necessarily abbreviated discussion of the scientific developments on the events of the Ediacaran and Cambrian since 1989 before assessing the status of Gould's arguments in *Wonderful Life* (see Erwin 2015b for a related discussion). The issues of contingency raised by Gould have received extensive discussion by philosophers and historians of science. Lacking a union card for either discipline, I will confine my comments to those issues where knowledge of the fossil record or evolution is relevant.

MAJOR ARGUMENTS OF *WONDERFUL LIFE*

In *Wonderful Life*, Gould employed the fossils of the Burgess Shale and the history of research on the fossils to further his arguments for the importance of historical contingency relative to selection and adaptation in the history of life. Gould made three primary, albeit overlapping, arguments: First, that rather than the "cone of increasing diversity," which he describes as the primary iconography of evolution, the fauna of the Burgess Shale supports a model of rapid increase in morphological diversity or disparity, followed by elimination of many lineages and diversification of the successful lineages. His critical point here is that maximal morphological disparity was a primary feature of the Cambrian metazoan radiation, not a consequence of subsequent evolutionary history. Thus evolutionary innovation was primarily focused in the events of the Cambrian, with later history largely, in Gould's view, "generating endless variants upon a few surviving models" (1989, 47). He describes this as a model of decimation and diversification. The book as a whole celebrates the incredible morphological disparity of the Burgess Shale fossils, suggesting at one point that in addition to the thirty or so animal phyla recognized in 1989, "the Burgess Shale, one small quarry in British Columbia, contains the remains of some fifteen to twenty organisms so different one from the other, and so unlike anything now living, that each ought to rank as a separate phylum. . . . The fifteen to twenty unique Burgess designs are phyla by virtue of anatomical uniqueness" (99–100). This statement accurately captures the zeitgeist of the times. As a personal example, Jim Valentine and I made approximately the same estimate of phyla and new classes (Valentine and Erwin 1987), and Conway Morris and others made similar arguments.

The model of maximal increase in morphological diversity (now described as disparity) early in the history of Metazoa as a whole, and of specific clades, leads naturally to the second major argument of the book: Gould's highly contentious thought experiment of "playing the tape of life again." (The metaphor of "playing the tape of life again" as applied to the Cambrian actually originated with Conway Morris [1985].) If the rate of evolutionary experimentation documented by the fossils of the Burgess Shale was as high as claimed by Gould, and if the success of different clades was based on factors other than adaptive value, indeed if success had been largely a contingent phenomenon, then an Ordovician ocean could have been composed of very different clades. But as Gould recognized, this argument depends upon the survivors' having greater morphological complexity and competitive ability. Gould's argument represents the acme of his fight against adaptive storytelling and inferred evolutionary progress. In a critical passage in *Wonderful Life* Gould argues for an alternative view:

> Any replay of the tape would lead evolution down a pathway radically different from the road actually taken. But the consequent differences in outcome do not imply that evolution is senseless, and without meaningful pattern; the divergent route of the replay would be just as interpretable, just as explainable *after* the fact, as the actual road. But the diversity of possible itineraries does demonstrate that eventual results cannot be predicted at the outset. Each step proceeds for cause, but no finale can be specified at the start and none would even occur a second time in the same way, because any pathway proceeds through thousands of improbable stages. Alter any early event, ever so slightly and without apparent importance at the time, and evolution cascades into a radically different channel. (Gould 1989, 51)

For Gould, this is the essence of contingency, in which the outcome of any historical process is dependent upon a sequence of largely unpredictable prior events. (Gould's discussion of post hoc evolutionary explanations is described by Taleb 2007 as the "narrative fallacy"; see also Kahneman 2011.) Gould contrasted the fate of priapulids, a group of carnivorous worms that persist today but at very low diversity, with the polychaete annelids, the clade that includes earthworms and is today a major component of animal biodiversity. Although priapulids are today ecologically insignificant, they were far more abundant than annelids among the fauna of the Burgess Shale specimens, and there are five species of each clade (Briggs, Erwin, and Collier 1994).

Near the end of *Wonderful Life*, Gould considers other examples that illustrate the power of contingency, events where the history of life might have turned out differently: the diversification of large flightless birds versus placental mammals after the end-Cretaceous mass extinction, survival during mass extinctions, the evolution of the eukaryotic cell, the failure of the

Ediacara fauna of soft-bodied animals that serves as the prelude to the Cambrian explosion, the small shelly fossils of the earliest Cambrian, terrestrial vertebrates, mammals, and finally humans.

Finally, Gould's third argument in *Wonderful Life* builds from the pattern of disparity and the ubiquity of contingency to conclude that selection and adaptation play a much less significant role, at least over the great spans of deep time, than acknowledged by many evolutionary biologists. This theme was first enunciated in the "Spandrels of San Marco" paper (Gould and Lewontin 1979) and developed through much of the middle part of Gould's career. Gould's views on adaptation figure in evaluating just what he meant by contingency.

THE EDIACARAN-CAMBRIAN RADIATION
SINCE *WONDERFUL LIFE*

Although *Wonderful Life* was intended as much for a general as a professional audience, publication led to considerable controversy among paleontologists over Gould's interpretations of the fossil data, among evolutionary biologists and philosophers over his claims for the role of contingency, and among many scholars about the nature of contingency itself. In this section my interest is in how subsequent research on the Burgess Shale and, more broadly, on the Cambrian explosion have affected Gould's conclusions.

Several decades of work on fossil and geological data through the Ediacaran (635–541 Ma) and Cambrian (541–489 Ma) have established that the Ediacaran–Cambrian Radiation (ECR) began with the appearance of the first soft-bodied Ediacaran macrofossils after 579 Ma. The suite of Ediacara fossils includes a variety of different clades, likely of metazoan affinities but lacking obvious guts, appendages, or other features characteristic of post-Cambrian metazoan clades (Erwin et al. 2011; Xiao and Laflamme 2009). The Cambrian explosion itself begins after 542 Ma with the appearance in the fossil record of a diverse array of bilaterian groups, including arthropods, molluscs, and most other major clades of marine animals. This evolutionary diversification clearly encompasses more than just bilaterian animals, however, as it is associated with the appearance of a diversity of sponges (Botting and Butterfield 2005), organic walled microfossils (Butterfield 2001), and a diverse array of burrows and other trace fossils (S. Jensen, Droser, and Gehling 2005). With additional study of the fossil record and new radiometric dating, the abruptness of the fossil appearances has only increased (Erwin et al. 2011). Our understanding of the morphological breadth of the Cambrian explosion has been greatly enhanced by numerous occurrences of extraordinary preservation where soft-bodied organisms have been exquisitely preserved. The Burgess Shale was just the first of these to be discovered. It is now dated to Cambrian Stage 5, about 511 Ma. In particular, the Chengjiang fauna contains many fossils representing taxa very similar to the Burgess Shale, as well as some new forms (Hou et al. 2004).

The chordate *Pikaia* is the final taxon discussed in *Wonderful Life*, with Gould claiming that he "saved the best for last" (Gould 1989, 321). Recognizing the apparent segmentation, Walcott had placed it among the polychaete annelids, but as a graduate student Conway Morris had recognized it as the earliest-known chordate. Following a rather prolonged gestation, the paper redescribing *Pikaia* recently appeared (Conway Morris and Caron 2012). Gould closes the book with the following rumination, worth quoting at some length:

> I do not, of course, claim that *Pikaia* itself is the actual ancestor of vertebrates, nor would I be foolish enough to state that all opportunity for a chordate future resided with *Pikaia* in the Middle Cambrian; other chordates, as yet undiscovered, must have inhabited Cambrian seas. But I suspect from the rarity of *Pikaia* in the Burgess and the absence of chordates in other Lower Paleozoic *Lagerstätten*, that our phylum did not rank among the great Cambrian success stories, and that chordates faced a tenuous future in Burgess times.
>
> *Pikaia* is the missing and final link in our story of contingency—the direct connection between Burgess decimation and eventual human evolution. . . . Wind the tape of life back to Burgess times, and let it play again. If *Pikaia* does not survive in the replay we are wiped out of future history—all of us, from shark to robin to orangutan. And I don't think that any handicapper, given the Burgess evidence as known today, would have granted very favorable odds for the persistence of *Pikaia*. (Gould 1989, 332–33)

I wish I had taken the bet, for Gould's suspicion that other chordates inhabited Cambrian seas has been abundantly confirmed by the Chengjiang biota (J.-Y. Chen 2008; Shu et al. 2010). Representatives of each of the extant chordate subphyla have been described from the Chengjiang, including vertebrates. *Cathymyrus* is, like *Pikaia*, a cephalochordate (Shu, Conway Morris, and Zhang 1996); several craniates have been described as well as *Shankouclava*, a urochordate. Other clades of crown or stem-group chordates are also known, including vetulicolids and the more problematic yunnanozoans. And several of these, such as *Haikouichthys*, a jawless fish, are far from rare. As a good Bayesian, this forces me to recalculate the odds on the persistence, if not of *Pikaia*, then of chordates as a clade, and they certainly seem a great bit higher.

The evidence for relatively abrupt diversification of Cambrian fauna contrasts sharply with evidence from molecular clock studies, which compare DNA sequences of living taxa, calibrated against divergence times estimated from the fossil record, to estimate older divergence estimates. The rigor of such analyses has improved greatly in the past decade, and in a recent study we estimated the origin of animals at about 780 Ma and of bilaterians at about 660 Ma (Erwin et al. 2011). These results indicate a long, largely hidden history of early metazoan radiation, but the molecular clock study also confirms a burst of diversification of crown group of bilaterians during the late Ediacaran and Cambrian, consistent with fossil evidence. Thus we appear to have evidence for both a 200 million year early history of animals and a burst of diversification associated with the ECR. The resolution of the apparent conflict between the fossil and molecular clock models can be found in a model of gradual divergence of metazoan lineages up to the Ediacaran, followed by the rapid establishment of a variety of larger, macroscopic stem- and crown-group bilaterian clades in the latest Ediacaran and early Cambrian (Erwin and Valentine 2013; Erwin 2015a). Thus the Cambrian explosion is a real and significant macroevolutionary event, but it is not the same thing as the origin and early diversification of Metazoa, a process that played out over some 150 million years during the Cryogenian and early Ediacaran.

Molecular methods have also revolutionized our understanding of the overall topology of metazoan relationships, such that current views of a tripartite division of bilaterians among deuterostomes, lophotrochozoans, and ecdysozoans (Aguinaldo et al. 1997; Halanych 2004) bears little relationship to the views of the late 1980s (e.g., Erwin and Valentine 2013). For example Walcott, Whittington, and Gould each accepted a fairly close relationship between arthropods and annelids, yet molecular data has since shown that the two clades are distantly related, belonging to the Ecdysozoa and Lophotrochozoa, respectively.

One of the most profound changes impacting Gould's arguments was the introduction of cladistics or phylogenetic methods for rigorously assessing the evolutionary relationships among taxa. The introduction of cladistics or phylogenetic analysis was in its infancy in the 1980s, and in fact neither word appears in the index to Gould's book. Today, most researchers would rightly criticize a revision of Burgess Shale taxa, or a description of new species, that did not include at least an initial phylogenetic analysis. Early in *Wonderful Life*, Gould claimed that the intellectual straightjacket of the cone of increasing diversity necessarily forced Walcott to classify the fossils of the Burgess Shale "either as primitive forms within modern groups, or as ancestral animals that might, with increased complexity, progress to some familiar form of the modern seas" (1989, 46). Much of the middle part of the book describes (as an "intellectual revolution") the changes brought about by Whittington, Briggs, and Conway Morris and the recognition that many of the Burgess animals represented "weird wonders" unlike any modern phyla. But in a turnaround that many of the critics of *Wonderful Life* must relish, the introduction of phylogenetic methods has altered the perspective of paleontologists yet again, with many seemingly distinct phyla now recognized as stem lineages of extant clades. Clades are defined as monophyletic groups of taxa, incorporating an ancestor and all of its descendants. This is a very different (and far more useful) approach than reliance upon some poorly characterized inference of "morphologic distinctiveness" as a basis for distinguishing taxa. As a consequence of this emphasis on "tree-thinking," the 1980s boom in extinct phyla and classes was replaced by a focus on phylogeny and the identification of stem and crown groups. The morphologies of these groups are no less distinctive, but the evolutionary relationships of most of the Burgess Shale and Chengjiang fauna are now well established.

Stem groups are extinct representatives from early in the history of a clade, branching before the crown group. The crown group, in contrast, represents the last common ancestor of the *living* members of a clade and all of its descendants both extinct and living (Jefferies 1979). The total group comprises the crown group plus various stem groups back to the last common ancestor of all members of the clade. Derek Briggs (one of the chief protagonists of *Wonderful Life*) was the principal advocate for the introduction of phylogenetic methods to the study of the Burgess Shale (Briggs and Fortey 1989). Brysse (2008) presents a discussion of the impact of this development on interpretations of the Burgess Shale. As she emphasizes, many of the Burgess arthropods, while clearly arthropods, possessed combinations of morphological characters that were otherwise unknown and thus had proved impossible to classify. The introduction of phylogenetic analysis and the recognition of stem and crown groups focused attention on shared morphological characters and led to the recognition that most of the canonical "weird wonders" of the Burgess Shale represent stem groups of well-established clades. Thus *Opabinia*, the various anomalocarids, and some arthropods such as *Marrella*,

Yohoia, and *Naraoia* represent various basal stem groups of the Panarthropoda (Daley et al. 2009; Edgecombe 2010). With *Hallucigenia* joining *Ayshaeia* as a lobopodian, relatives of the extant Onycophora, and the discovery of many lobopodians among the Chengjiang biota (Liu et al. 2008), it became apparent that the diversity of Cambrian lobopodians was much greater than previously realized. Subsequent studies have suggested that the lobopods form a paraphyletic grade leading to the base of the Panarthropoda (Budd 1996, 1998; Edgecombe 2010).

This cladistic approach to phylogeny has by now decisively replaced the earlier approach, known as evolutionary systematics, in which morphological distinctiveness often played a significant role in taxonomic assignments. As Brysse (2008) observes, Gould might well have rejected a cladistic approach because it focused on shared derived characters and rejected the use of unique morphological features (apomorphies). This was a point of serious contention through the 1980s and into the 1990s but eventually faded with the introduction of quantitative techniques to assess morphological diversity, better known now as disparity.

Through the 1980s, paleontologists and many evolutionary biologists used taxonomic ranks as a rough proxy for morphological disparity (e.g., Erwin, Valentine, and Sepkoski 1987). Paleontologists were not unaware of the limitations of such an approach, but the development of quantitative morphometrics during the 1980s (Benson, Chapman, and Siegel 1982; Bookstein et al. 1985) and the controversies engendered by the publication of *Wonderful Life* led directly to quantitative methods to assess morphological disparity (Briggs and Fortey 2005; Briggs, Fortey, and Wills 1992; Foote 1992, 1993; Fortey, Briggs, and Wills 1996; McShea 1993; Wills 1998; reviewed by Erwin 2007; Foote 1997). When combined with phylogenetic analyses, such approaches constitute a powerful suite of methods with which to analyze the occupation of evolutionary space. Although the early studies by Briggs, Fortey, and colleagues were intended to refute Gould's claims about disparity, the general conclusion from a variety of studies (reviewed by Erwin 2007; Foote 1997) has been that many large clades, particularly those involved in the Cambrian explosion, rapidly defined the morphospace occupied by the clade (early maximal disparity), followed by subsequent filling out of the space through taxonomic diversification. This pattern is not universal, however. Post-Cambrian priapulids, for example, appear to occupy a morphospace adjacent to the Cambrian representatives (Wills 1998; Wills et al. 2012). Thus, Gould's claims that Cambrian morphospace was larger than later in the Phanerozoic have not been supported by subsequent studies, but these studies have not supported the "cone of expanding diversity" model that Gould questioned. Gould deserves credit (along with Valentine and others who made similar observations [Valentine 1980]) for driving this conceptual shift in macroevolution.

The discovery of highly conserved developmental regulatory genes has also dramatically altered our understanding of the ECR, with implications for some of the arguments about contingency in *Wonderful Life*. Briefly, analysis of the genomic basis of development in many living animals has revealed that they share a common "developmental toolkit" of genes responsible for patterning the developing embryo. These shared genes are generally transcription factors or elements of signaling pathways and include the HOX genes as well as others involved in anterior-posterior and dorsal-ventral patterning and the formation of eyes, brains, heart, appendages, and the gut (S. B. Carroll 2008; S. B. Carroll, Grenier, and Weatherbee 2001; Davidson and Erwin 2009; Erwin 2006a; Erwin and Davidson 2002; Knoll and Carroll 1999; Tweedt and Erwin 2015). In contrast to views of development through the 1980s, these discoveries reveal a common developmental underpinning shared among all bilaterian animals and extending into basal clades such as cnidarians and even sponges. Consequently, similarities in developmental patterning (for example, limb development in lobopodians and arthropods) often reflect shared developmental mechanisms. The critical issue here for discussions of contingency is that although the morphological expression of eyes, appendages, etc. may have arisen independently in different clades (after all, the eyes of a cat and of a fly share few similarities), the underlying developmental foundation is shared. This raises some tricky issues for assessing claims of historical uniqueness versus convergence, for even distantly related groups may share common developmental pathways.

As research on the ECR has progressed, there are two components, discussed in the final chapter of *Wonderful Life*, that might seem even better exemplars of the role of contingency than the animals of the Burgess Shale: the soft-bodied forms of the Ediacaran fauna (579–542 Ma) and the small shelly fossils that dominated the earliest Cambrian (542–520 Ma). The Ediacara macrofossils encompass a diverse array of centimeter- to meter-sized fronds, disks, and more complex architectures, some vaguely resembling bilaterians (Fedonkin et al. 2008; Xiao and Laflamme 2009). Although many exhibit bilateral symmetry, none show evidence of appendages, eyes, a gut, or other bilaterian features, which has led to a long history of controversy over their phylogenetic affinities (discussed by Erwin and Valentine 2013). Gould relied on Seilacher's reinterpretation of the Ediacara fossils as a separate and independent multicellular clade (Seilacher 1984). Seilacher's novel perspective, while insightful, has relatively few adherents today (see discussion in Erwin and Valentine 2013). Laflamme's recent analysis suggests there are perhaps six different clades of Ediacarans distributed across the Metazoan tree, as well as a number of as yet unresolved groups (Laflamme, n.d.; Erwin et al. 2011). One of the clades includes *Kimberella* and likely represents a molluscan stem group. The causes of the disappearance of the Ediacara fossils remains unclear (Laflamme et al. 2013), but Gould suggests that in an alternative world we might wind up with nothing but Ediacarans. I think we can now reject this possibility. Molecular clock evidence indicates that many metazoans, including bilaterian clades, were present during the Ediacaran, although most are not represented in the Ediacara macrofauna. But once the physical environment and ecological interactions triggered the Cambrian explosion, the majority of the Ediacara clades were done for—predation alone would have seen to their disappearance. So if one is resetting the clock to the Ediacaran, evolution might have played out much the way it did, at least at the macroscale.

The small shelly fossils (SSFs) are assemblages of minute (<2mm) plates, spines, tubes, and other fossils. Some of these skeletal elements are the shells of whole animals, but many represent parts (sclerites) of larger animals; when the sclerites are preserved as part of the whole animal rather than disarticulated, they are described as the scleritome. The spines of *Hallucigenia* from the Burgess Shale, if disarticulated, would be SSF elements, for example. A great diversity of SSFs have been described, appearing near the base of the Cambrian and increasing steadily in diversity through stages 1 and 2 (Bengtson 2005; Kouchinsky et al. 2012; Li et al. 2007; Maloof et al. 2010). The SSFs represent many different clades of lophotrochozoans and some ecdysozoans; deuterostomes seem to be poorly represented. But the numerous clades of SSFs provide us many phylogenetically independent tests of Gould's claims for contingency—and relatively few passed. Having a multiplate scleritome just doesn't seem to have been a viable strategy in the face of increased predation, particularly when most of these clades were capable of forming a single larger, and more protective, shell.

Thus the Ediacara macrofossils and the SSF share many similarities with Burgess Shale–type faunas: Both exhibit high degrees of disparity with relatively low diversity (although assessing taxonomic diversity in the SSFs is hampered by disarticulation); their preservation has been strongly influenced by rather narrow taphonomic windows, from microbial mats for Ediacara fossils (Gehling 1999; Laflamme et al. 2011) and an abundance of phosphate deposits for the SSFs (S. M. Porter 2004). The "disappearance" of both groups from the fossil record at least partly reflects these preservational issues, clouding their true evolutionary duration. The recent report of an anomalocarid from the Lower Ordovician of Morocco (Van Roy and Briggs 2011) revealed that some of the Burgess Shale lineages may have had quite respectable durations but are simply not well represented in the fossil record. Nonetheless, for both the Ediacara macrofossils and the small shelly fossils at the macroscale, a pure contingency argument seems difficult to sustain.

In summary, research into the ECR since the publication of *Wonderful Life* has altered the empirical foundation for some but not all of Gould's arguments. Phylogenetic methods have made most if not all of the "weird wonders" less weird as they have found homes as stem groups of larger clades. This even extends to such peculiar forms as *Opabinia*. Quantitative assessments of morphological disparity, however, have largely confirmed much of Gould's argument about maximal early disparity and the limitations of "the cone of expanding diversity" as an appropriate metaphor for the history of life. As Conway Morris, Valentine, I, and many others argued during the 1980s, the ECR was an interval of extraordinary morphological innovation. Indeed, evidence from the body and trace fossil record and the molecular clock estimates of divergences of crown-group bilaterians independently support the conclusion that the Cambrian explosion was a significant and rapid episode of evolutionary innovation. The completely unexpected discovery of deep conservation of developmental pathways across all major clades of animals has recast arguments about morphological patterning. The early origin of much of the developmental toolkit required to construct bilaterians (probably by about 680 Ma: Erwin et al. 2011; Tweedt and Erwin 2015) strongly suggests that the Cambrian explosion of bilaterian forms in the fossil record reflects ecological feedbacks and possibly environmental changes (Erwin and Valentine 2013).

CONTINGENCY

In Chapter 3 of *Wonderful Life*, Gould focuses his argument for the contingent nature of life on the more iconic specimens, particularly the arthropods *Marrella*, *Yohoia*, *Naraoia*, *Sanctacaris*, and some of the bivalve arthropods, as well as *Pikaia*, *Opabinia*, *Aysheaia*, *Anomalocaris*, and *Wiwaxia*. This argument is central to his thesis and has generated considerable controversy and discussion among paleontologists, evolutionary biologists, and philosophers. Here I want to sharpen my evaluation of Gould's claims in light of the discussions over the nature of contingency.

Five different senses of contingency have been identified:

1. *Drift or sampling error*, which Gould explicitly rejected as a form of contingency, although not always to the satisfaction of some (Beatty 2006b; Travisano et al. 1995).
2. *Unpredictability*, in which the outcome of a process cannot be determined from a prior state (Beatty 2006b).
3. *Causal dependence*, or sensitivity to initial conditions, in which a prior state is necessary to reach a particular outcome (Beatty 2006b; Ben-Menahem 1997, 2009). Derek Turner (2011) suggests that contingency as causal insufficiency might be a better term. Beatty (2006b) argued that although unpredictability and causal dependence can be complementary, Gould failed to distinguish between them and often conflated the two (see Desjardins, this volume, chap. 9, for a perceptive discussion of this issue).
4. *Sensitivity to external disturbance*, which is related to the resilience of a historical process and is distinct from sensitivity to initial conditions (Inkpen and Turner 2012).
5. *Macroevolutionary stochasticity*, which centers on unbiased species sorting. Derek Turner (2011), in a discussion of Beatty (2006b) suggested that Gould's argument in *Wonderful Life* is largely focused on macroevolutionary sorting among species and that he viewed contingency as an issue of unbiased species sorting, rather than the sampling error problem in point 1 above. Turner claimed: "Evolutionary contingency is the random or unbiased sorting of entire lineages. It just *is* the macroevolutionary analogue of random drift" (2011, 69, emphasis in original).

Finally, Inkpen and Turner (2012) offered a preliminary discussion of how the "topography" of historical contingency may change over time. This seems quite probable, and they discuss how history might involve conditional inevitability (a concept originally introduced by Sterelny 2005). This important issue is one that has received too little attention (Erwin 2011).

Although Gould viewed testing his contingency hypothesis as difficult, several different groups have attempted just that, using a variety of theoretical, field-based, and experimental approaches. For example, Fontana and Buss (1994) applied λ-calculus, a form of abstract chemistry, and concluded that general patterns of self-organization would emerge even in the absence of selection. Fontana and Buss suggest that these results are applicable to early stages in the history of life, perhaps up to the origin of eukaryotes. The structure of repeated adaptive radiation on islands provides an interesting "natural" experiment testing Gould's contingency hypothesis, for ecologists expect similar ecological communities to arise in similar environments. Losos and colleagues used their studies of repeated production of *Anolis* lizards in the Caribbean to evaluate the predictability or determinism of adaptive radiations (Losos et al. 1998). The similarity of ecomorphs on different islands suggested that contingency was less significant than ecologically generated constraints, although within a single island the order of evolution of ecomorphs did constrain the available evolutionary options for other species. A recent paper evaluated replicate radiations of *Anolis* across Caribbean islands, including unique species, to evaluate the overall convergence using a model of the adaptive landscape (Mahler et al. 2013). The results showed that convergence of morphologies onto shared adaptive peaks was the most favored model. This suggests that the adaptive landscape has persisted for perhaps 30–40 million years, more than long enough to influence macroevolutionary patterns. A similar pattern of ecomorphological constraint has been found among the various ecomorphs of the spider *Tetragnatha* in Hawaii (R. Gillespie 2004), and Losos (2010) evaluates other cases. The implications of these highly structured radiations for Gould's argument is mixed, for while they strongly support a deterministic view, the radiations are also confined to closely related species that share common developmental and genetic systems (Losos 2010), a situation most unlike the Cambrian explosion. Beatty (2006b) reports that Losos heard that Gould was unimpressed by their initial results, suggesting that such recent divergences had little relevance for the Cambrian explosion. If true (and I suspect it was), this may provide further insight into Gould's perspective and support Turner's argument for the relevance of macroevolutionary sorting to the contingency argument in *Wonderful Life*.

Lenski's long-term experimental evolution project with *E. coli* was in part intended to test Gould's assertions. In Lenski and Travisano (1994) and Travisano et al. (1995), they conducted several experiments to test the role of contingency and showed that different evolutionary outcomes could result from the effects of random mutation and the order of mutation (see discussion by Blount, this volume, chap. 10; Beatty and Desjardins 2009 discuss these issues as well). Thus, in contrast to the ecological studies reported by Losos, in Lenski's experimental work where the *E. coli* populations were in identical environments, different clones exhibited very different evolutionary trajectories. These results seem to support Gould's arguments for contingency, but as Derek Turner (2011) points out, as beautiful as the work of Lenski and Travisano has been, it really does not address the issue of macroevolutionary sorting that is at the heart of Gould's argument (Desjardins, this volume, chap. 9, reaches the same conclusion).

A more macroevolutionary critique of Gould's contingency argument by Vermeij (2006) utilized a compilation of both unique or singular and repeated evolutionary innovations, ranging from the origins of the genetic code and arthropod wings to vertebrate teeth and plant alkaloids. Vermeij acknowledges that contingency (in Beatty's sense of unpredictability) is "an essentially universal property of dynamic systems" (2006, 1804) but suggests that an examination of unique evolutionary innovations provides an opportunity to examine the ubiquity of contingency. According to Vermeij, unique innovations support claims of contingency, but if innovations occurred multiple times, this supports a more deterministic system in which the innovation would occur eventually, even if in a different clade. Vermeij shows that the "purportedly unique" innovations are significantly older than those that occurred multiple times. However, he concludes that this pattern is more likely to reflect the loss of information from other clades; with a more complete record, many of these unique innovations would be revealed to have occurred several times as well. Vermeij's list of both singular and repeated innovations seems highly idiosyncratic, however, reflecting the absence of a metric for evolutionary innovations and difficulty in granularity. Thus, bilaterian pattern formation is counted as a singular innovation but eyes as a repeated innovation since the phenotypic expression has arisen multiple times. More problematic is the inclusion of many items among the repeated innovations (e.g., a gastropod labral tooth) that seem incommensurate with the origin of eukaryotes, or of sex. Vermeij concludes that most innovations arise multiple times in many clades because their adaptive value is sufficiently high that selection will favor their development even if the pathways to produce the innovations differ in their particulars.

Perhaps the most sustained critique of Gould's contingency argument has come from Simon Conway Morris and his proposals that the ubiquity of convergence suggests that despite apparent contingency, long-term determinism regulates patterns of evolutionary change (Conway Morris 1998, 2003, 2009). Conway Morris's argument rests on the ubiquity of convergence—the appearance of similar patterns in different groups. He has exhaustively compiled examples of convergence and employed them to argue that evolution is far more deterministic than admitted by Gould. Curiously, some of the best evidence for the extent of convergence comes from phylogenetic analyses, yet Conway Morris has consistently, and somewhat perversely, refused to use such examples. As several authors have noted, it is often difficult to discriminate between convergence between unrelated taxa and parallelism (Pearce 2012; Powell 2012). The tension between chance and determinacy is plagued by issues of granularity as well. At a coarse-grained level, two features may seem convergent, while they look very different in a more fine-grained analysis, as both Sterelny (2005) and Inkpen and Turner (2012) suggest for Conway Morris's (2003) discussion of agriculture in humans and leafcutter ants. In the absence of a concrete metric for the evaluation of apparent convergence, in terms of phylogenetic distance, morphological similarity (or similarity of other features), and developmental similarity, it seems difficult to adjudicate competing claims.

I think Conway Morris too readily dismissed the issues raised by the discovery of deep homology underlying developmental mechanisms in distantly related animal groups. Work over the past two decades has challenged distinctions between convergence and parallelism and has led to both scientific and philosophical discussions about the nature of homology (Abouheif 1999; Arendt and Reznick 2008; B. K. Hall 2003; Losos 2011; Pearce 2012; Wake, Wake, and Specht 2011). Reliably discriminating between convergence and parallelism is critical to evaluating claims of contingency, since convergence focuses attention on the power of external selection, while parallelism emphasizes the power of internal developmental mechanisms and constraints on the generation of variation. In a sense, developmental patterning may impose a structure on available evolutionary variation not unlike the ecological landscape of Losos's *Anolis* lizards and similarly support deterministic processes. Unfortunately, clearly identifying developmental homologies is difficult, particularly with increasing focus on developmental gene regulatory networks (Arendt and Reznick 2008). Pearce (2012) proposes a "neo-Gouldian" approach, emphasizing morphology in distinguishing between homology and homoplasy, and development in separating convergence and parallelism. He suggests: "Convergent traits are realized by non-homologous underlying generators, whereas parallel traits are realized by homologous underlying generators" (Pearce 2012, 445). Understanding where, within a hierarchical GRN, developmental changes lie may provide a mechanistic basis for Pearce's generators (Davidson and Erwin 2009, 2010; Erwin and Davidson 2009). There may be no final resolution to the debate over deep homology and how it applies to convergence versus parallelism, in part because researchers' use of the terms may often reflect differing rhetorical strategies.

With these discussions of contingency in hand, we are now in a position to evaluate Gould's discussion of the Burgess Shale and the Cambrian explosion in light of the five different definitions identified at the beginning of this section (see also Erwin 2015b).

Gould dismissed the issue of sampling error as contingency with some justification, because sampling error, drift, and similar problems did not really encompass the issues that he wanted to address. But I want to raise the sampling issue because it may be relevant for some of Gould's claims. Raup distinguished between three different modes of extinction selectivity: a "field of bullets" scenario in which extinction is random without regard to differences in fitness; the "fair game" scenario in which extinction selectivity is Darwinian, based on differential fitness or adaptation, and promoting long-term increases in same; and "wanton extinction," where extinction is selective but not based on factors other than those that promote survival in a species' normal environment (Raup 1991). For example, during many mass extinctions (but not during background intervals), geographic range at the generic level increases survival probability (Jablonski 1986).

Gould's discussion of the "decimation" of the Burgess Shale fauna in *Wonderful Life* (1989, 233–39) contrasts assumptions that survival was based on "superior competitive ability" with his suggestion that widespread evolutionary experimentation led to contingent survival (see also Derek Turner's [2011] discussion of the differential sorting of the Burgess Shale fossils as a lottery). Gould argues: "But if we face the Burgess fauna honestly, we must admit that we have no evidence whatsoever—not a shred—that losers in the great decimation were systematically inferior in adaptive design to those that survived" (1989, 236; I'll address the issue of competitive ability again below). But in Raup's field of bullets scenario, extinction selectivity is a function of the species diversity of a group, or abundance of a species, depending on the level of analysis, and many of the "weird wonders" of the Burgess were very rare. Very few specimens of *Aysheaia*, *Hallucigenia*, or *Opabinia* are found in the collections at the Smithsonian, while more than 9,000 individuals of *Marrella* (closely related to trilobites) were recovered by Walcott. Thus, even if *Hallucigenia* and *Marella*, or one of the abundant euarthropods, had equivalent competitive abilities, and if the fossil abundances are a reasonable proxy for abundances of living species (correcting for arthropod molts and time-averaging of sediment, for example), we would expect the persistence of the more abundant group on purely statistical grounds. And the more abundant or diverse a group is, and the longer it persists, the more likely it is to leave descendants rather than to go extinct. Gould, I think, rightly questioned the assumption of superior competitive ability as an explanation for selective extinction (Raup's "fair game" scenario), but he does not seem to have considered whether sampling effects might be a sufficient explanation for the pattern observed (see also the discussion of historicity versus random chance in Desjardins, this volume, chap. 9). I do not want to argue that sampling effects are sufficient to explain the disappearance of the Burgess fauna as a whole. Because the fossil record preserves very few Burgess Shale–type exceptionally preserved faunas, we have little information on when the more extraordinary organisms of the Burgess died out. Thus, the question of extinction is essentially unresolvable with current data, but it is likely more complex than Gould described.

I accept Beatty's argument that Gould used contingency in the two senses of unpredictability and of causal dependence on initial conditions, although I think he largely intended it in the former sense. Inkpen and Turner (2012) suggest that contingency could also apply to sensitive dependence to external disturbance (developing the perceptive argument of Ben-Menahem 1997). While they are correct that this is one compelling use of the term *contingency*, it is not obvious that this applies to Gould's claims for the Cambrian (and since we have no data on the extinction of these groups, we can't apply it in any case).

Derek Turner's (2011) suggestion that Gould's discussion of contingency needs to be seen as a macroevolutionary phenomenon and allied to issues of species selection and clade sorting is an interesting one. Turner relies overly much on a suite of papers written by Raup, Gould, Schopf, and Simberloff during the 1970s for support of his thesis, evidently unaware that since Stanley et al. (1981), paleontologists have largely discounted the significance of these models because of the scaling problems (although, see Slowinski and Guyer 1989 for an improved approach). Nonetheless, Turner is at least partly correct that Gould's discussion of contingency focuses on the differential success of clades (priapulids versus annelids, or eurathropods versus onycophorans), but focusing entirely on the macroevolutionary aspect of contingency would ignore Gould's long-standing critique of adaptationism, which is better encompassed by uses 2 and 3. But I do not view Turner's macroevolutionary stochasticity as entirely separate from Beatty's unpredictability and causal dependence. Rather, in a hierarchical view of evolution with sorting and (less frequently) selection at many levels from genes through populations to species and clades (Jablonski 2007), contingency may arise at different levels. The contingency of Travisano and Lenski applies to genes, Losos's studies to species, and Vermeij's at the level of macroevolutionary innovations among clades. Thus, the question of "contingency versus determinism" is poorly posed. More accurately, the issue is the relative importance of contingency and determinism at a particular focal level. In any case, as Turner points out, Gould provides no operational means to evaluate unbiased versus biased sorting among the Burgess Shale taxa, and this remains a compelling research question.

In the case of the Ediacara macrofossils and the small shelly fossils of the early Cambrian, I have argued that there are fairly good reasons for thinking that both groups would have disappeared in any alternative worlds in which active, particularly predatory, animals appeared. Thus, Gould's claims of contingency would be limited to some suite of alternative worlds in which metazoan evolution for some reason stops with either of these assemblages (a fairly unlikely scenario).

Evaluating the claims of contingency for the Burgess Shale is more difficult. While I think it is plausible that contingency in the sense of both unpredictability and causal dependence are involved in aspects of the Cambrian explosion, the appropriate focal level for reconsidering *Wonderful Life* is the differential success of various clades of early metazoans. Here I think Gould's claims run into three critical difficulties: First, as described above, the apparently low diversity and low abundance of some of the more unusual forms of the Burgess Shale renders their long-term success implausible simply on statistical grounds, with no need to consider other factors. Second, the adoption of a cladistic approach to systematics, and particularly the concept of stem clades, has generated phylogenetic homes for most, if not all, of the "weird wonders" of the late 1980s. This reduces the number of independent phylum-grade lineages and considerably alters the evaluation of claims of contingency. With the identification of many lobopod lineages and their close association with *Opabinia*, *Anomalocaris*, and related forms and with basal arthropods, the evaluation of their differential success becomes more interesting. Third, key to Gould's claims for contingency is his claim of "apparently equivalent anatomical promise—over twenty arthropod designs later decimated to four survivors, perhaps fifteen or more unique anatomies available for recruitment as major branches, or phyla, of life's tree" (1989, 288). The recognition of stem clades has already made a hash of the claims of "fifteen or more unique anatomies," but what of the claims for "apparently equivalent anatomical promise"? In my discussion of the Ediacra macrofossils and the small shelly fossils, I suggested that it was difficult to argue that either clade was adaptively superior to later evolving bilaterians, particularly in the face of increased predation. Gould's argument is also difficult to sustain for some of the other Burgess organisms. Rigorously formulating such an argument would require functional studies, but euarthropods do share a number of important adaptations (jointed appendages, for example) that might have enhanced their "anatomical promise."

DISCUSSION

The debate over contingency versus determinism touches another fundamental issue in evolution—why is there such a phenomenal underdispersion of genetic, developmental, and morphological possibilities? Although Gould barely touched on the issue, Sean Carroll (2001) posed the question of why morphological complexity is so much less than what seems possible based on the number of genes and their possible developmental deployment. Following the logic of Kauffman (1995), a genome of 25,000 genes, with two inputs per gene, has $2^{25,000}$ possible states. Although accounting for the hierarchical structure of developmental regulatory interactions would reduce this state space, the fact remains that evolutionary complexity at all levels is so much less than the possible. Lewontin (2003) viewed the underdispersion of morphologies as one of the great unresolved problems in evolution. How much of this reflects determinism, and how much chance or contingency? Koonin usefully rephrased the question as one about "the fraction of all possible trajectories in the genotype space that are open for exploration by the evolutionary process" (2012, 405). In many molecular settings, the range of possible pathways may be very limited, yielding what appear to be fairly deterministic pathways. More importantly, Koonin argues that the balance between contingency and determinism is critically dependent upon the selection pressure and thus population size. With sufficiently large populations, evolution may be largely deterministic, while contingency will become increasingly important with smaller population sizes.

It seems unlikely that there will be any general answer to questions over the role of contingency versus necessity in evolution, either with the Burgess Shale fossils and the Cambrian radiation specifically, or more generally in the history of life, in large part because at such a coarse level the question is wrongly specified. The relative importance of contingency depends upon whether one is interested in molecular processes, development, phenotype, or macroevolutionary patterns, and the answer may differ at these different levels, even for the same event.

The role of contingency versus historical laws in evolutionary processes is not only of great interest in understanding the trajectories of the history of life on the Earth, but critical for the nascent field of astrobiology (S. B. Carroll 2001). Essential to astrobiology as a field of inquiry is the assumption that there are sufficient underlying generalities to the history of life, a "Book of Rules," if you will, to guide an exploration strategy. Among these assumptions are that life beyond Earth will be carbon-based, will rely on aqueous chemistry, and will employ other metabolic, ecological, and evolutionary principles that evolutionary biologists have established here, such as natural selection. Any of these assumptions might be false: life could be silicon-based, exist in gaseous clouds in some noncellular form, and employ a truly Lamarckian evolutionary scheme. For a variety of chemical and biological reasons, the likelihood of any of these alternatives is low relative to the probability of carbon- and water-based life. Thus, the astrobiology community has focused its attention and exploration strategy on the more reasonable assumptions. Much astrobiological research, either explicitly or implicitly, assumes that the nature of chance and contingency facing life on other planets is largely confined to particular domains.

In trying to understand the roles of chance and of contingency in evolution more generally (either on Earth or elsewhere), we can distinguish between a set of questions implying that laws are likely to play a major role in controlling general patterns and a domain in which chance and contingency are likely to play an important role; there is also a set of questions for which we currently lack sufficient information to allot them to these two different categories.

ACKNOWLEDGMENTS

I appreciate comments from Charles Pence and two anonymous reviewers of a previous draft of the manuscript. This research was supported by the NASA National Astrobiology Institute.

Abouheif, Ehab. 1999. "Establishing Homology Criteria for Regulatory Gene Networks: Prospects and Challenges." In *Homology (Novartis Foundation Symposium 222)*, 207–25. Chichester: Wiley.

Abrams, Marshall. 2009. "What Determines Biological Fitness? The Problem of the Reference Environment." *Synthese* 166 (1): 21–40. doi:10.1007/s11229-007-9255-9.

———. 2012. "Mechanistic Probability." *Synthese* 187 (2): 343–75. doi:10.1007/s11229-010-9830-3.

Achaz, Guillaume, Alejandra Rodriguez-Verdugo, Brandon S. Gaut, and Olivier Tenaillon. 2014. "The Reproducibility of Adaptation in the Light of Experimental Evolution with Whole Genome Sequencing." In *Ecological Genomics*, edited by Christian R. Landry and Nadia Aubin-Horth, 211–31. Advances in Experimental Medicine and Biology 781. Dordrecht: Springer Netherlands.

Agrawal, Aneil F., Edmund D. Brodie III, and Loren H. Rieseberg. 2001. "Possible Consequences of Genes of Major Effect: Transient Changes in the G-Matrix." *Genetica* 112–13 (1): 33–43. doi:10.1023/A:1013370423638.

Aguinaldo, Anna Marie A., James M. Turbeville, Lawrence S. Linford, Maria C. Rivera, James R. Garey, Rudolf A. Raff, and James A. Lake. 1997. "Evidence for a Clade of Nematodes, Arthropods and Other Moulting Animals." *Nature* 387 (6632): 489–93. doi:10.1038/387489a0.

Alfaro, Michael E., Daniel I. Bolnick, and Peter C. Wainwright. 2005. "Evolutionary Consequences of Many-to-One Mapping of Jaw Morphology to Mechanics in Labrid Fishes." *American Naturalist* 165 (6): E140–E154. doi:10.1086/an.2005.165.issue-6.

Alphey, Luke. 1997. *DNA Sequencing: From Experimental Methods to Bioinformatics*. Oxford: Bios Scientific.

Alter, Robert. 1981. *The Art of Biblical Narrative*. New York: Basic Books.

Amundson, Ron. 1994. "Two Concepts of Constraint: Adaptationism and the Challenge from Developmental Biology." *Philosophy of Science* 61 (4): 556–78.

———. 2005. *The Changing Role of the Embryo in Evolutionary Thought: Roots of Evo-Devo*. Cambridge: Cambridge University Press.

Applebee, M. Kenyon, Andrew R. Joyce, Tom M. Conrad, Donald W. Pettigrew, and Bernhard Ø. Palsson. 2011. "Functional and Metabolic Effects of Adaptive Glycerol Kinase (GLPK) Mutants in *Escherichia coli*." *Journal of Biological Chemistry* 286 (26): 23150–59. doi:10.1074/jbc.M110.195305.

Aquinas, Saint Thomas. 1947. *Summa Theologica*. New York: Benziger Brothers.

Arendt, Jeff, and David Reznick. 2008. "Convergence and Parallelism Reconsidered: What Have We Learned about the Genetics of Adaptation?" *Trends in Ecology & Evolution* 23 (1): 26–32. doi:10.1016/j.tree.2007.09.011.

Ariew, André. 2007. "Under the Influence of Malthus's Law of Population Growth: Darwin Eschews the Statistical Techniques of Aldolphe Quetelet." *Studies in History and Philosophy of Biological and Biomedical Sciences* 38 (1): 1–19. doi:10.1016/j.shpsc.2006.12.002.

———. 2008. "Population Thinking." In Ruse 2008, 64–86.

Ariew, André, Collin Rice, and Yasha Rohwer. 2014. "Autonomous-Statistical Explanations and Natural Selection." *British Journal for the Philosophy of Science*. doi:10.1093/bjps/axt054.

Aristotle. 1984. *The Complete Works of Aristotle*. Edited by Jonathan Barnes. 2 vols. Princeton, NJ: Princeton University Press.

Arthur, Richard T. W. 2006. "Animal Generation and Substance in Sennert and Leibniz." In Smith 2006, 147–74.

Ashenberg, Orr, and Michael T. Laub. 2013. "Using Analyses of Amino Acid Coevolution to Understand Protein Structure and Function." In *Methods in Enzymology*, edited by Amy E. Keating, 523:191–212. Methods in Protein Design. London: Academic Press.

Augustine. 1992. *Confessions*. Edited by Henry Chadwick. Oxford: Oxford University Press.

Baker, David, and Andrej Šali. 2001. "Protein Structure Prediction and Structural Genomics." *Science* 294 (5540): 93–96. doi:10.1126/science.1065659.

Bakker, Robert T. 1983. "The Deer Flees, the Wolf Pursues: Incongruencies in Predator-Prey Coevolution." In *Coevolution*, edited by Douglas J. Futuyma and Montgomery Slatkin, 350–82. Sunderland, MA: Sinauer Associates.

Baron, Christian. 2009. "Epistemic Values in the Burgess Shale Debate." *Studies in History and Philosophy of Biological and Biomedical Sciences* 40 (4): 286–95. doi:10.1016/j.shpsc.2009.09.008.

Barrett, Rowan D. H., and Dolph Schluter. 2008. "Adaptation from Standing Genetic Variation." *Trends in Ecology & Evolution* 23 (1): 38–44. doi:10.1016/j.tree.2007.09.008.

Barrick, Jeffrey E., and Richard E. Lenski. 2009. "Genome-Wide Mutational Diversity in an Evolving Population of *Escherichia coli*." *Cold Spring Harbor Symposia on Quantitative Biology* 74:119–29. doi:10.1101/sqb.2009.74.018.

———. 2013. "Genome Dynamics during Experimental Evolution." *Nature Reviews Genetics* 14 (12): 827–39. doi:10.1038/nrg3564.

Barrick, Jeffrey E., Dong Su Yu, Sung Ho Yoon, Haeyoung Jeong, Tae Kwang Oh, Dominique Schneider, Richard E. Lenski, and Jihyun F. Kim. 2009. "Genome Evolution and Adaptation in a Long-Term Experiment with *Escherichia coli*." *Nature* 461 (7268): 1243–47. doi:10.1038/nature08480.

Beatty, John H. 1984. "Chance and Natural Selection." *Philosophy of Science* 51:183–211. doi:10.1086/289159.

———. 1987. "Dobzhansky and Drift: Facts, Values and Chance in Evolutionary Biology." In Krüger, Gigerenzer, and Morgan 1987, 271–311.

———. 1993. "The Evolutionary Contingency Thesis." In *Concepts, Theories, and Rationality in the Biological Sciences: The Second Pittsburgh-Konstanz Colloquium in the Philosophy of Science*, 45–81. Pittsburgh, PA: University of Pittsburgh Press.

———. 2006a. "Chance Variation: Darwin on Orchids." *Philosophy of Science* 73 (5): 629–41. doi:10.1086/518332.

———. 2006b. "Replaying Life's Tape." *Journal of Philosophy* 103 (7): 336.

———. 2008. "Chance Variation and Evolutionary Contingency: Darwin, Simpson (The Simpsons) and Gould." In Ruse 2008, 189–210.

———. 2010. "Reconsidering the Importance of Chance Variation." In *Evolution—The Extended Synthesis*, edited by Massimo Pigliucci and Gerd B. Müller, 21–44. Cambridge, MA: MIT Press.

———. 2013. "Chance and Design." In *The Cambridge Encyclopedia of Darwin and Evolutionary Thought*, edited by Michael Ruse, 146–51. Cambridge: Cambridge University Press.

———. 2014. "Darwin's Cyclopean Architect." In *Evolutionary Biology: Conceptual, Ethical, and Religious Issues*, edited by R. Paul Thompson and Denis M. Walsh, 175–92. Cambridge: Cambridge University Press.

Beatty, John H., and Isabel Carrera. 2011. "When What Had to Happen Was Not Bound to Happen: History, Chance, Narrative, Evolution." *Journal of the Philosophy of History* 5 (3): 471–95. doi:10.1163/187226311X599916.

Beatty, John H., and Eric Cyr Desjardins. 2009. "Natural Selection and History." *Biology & Philosophy* 24 (2): 231–46. doi:10.1007/s10539-008-9149-3.

Bedhomme, Stéphanie, Guillaume Lafforgue, and Santiago F. Elena. 2013. "Genotypic but Not Phenotypic Historical Contingency Revealed by Viral Experimental Evolution." *BMC Evolutionary Biology* 13 (1): 46. doi:10.1186/1471-2148-13-46.

Beisel, Craig J., Darin R. Rokyta, Holly A. Wichman, and Paul Joyce. 2007. "Testing the Extreme Value Domain of Attraction for Distributions of

Beneficial Fitness Effects." *Genetics* 176 (4): 2441–49. doi: 10.1534/genetics.106.068585.

Bengtson, Stefan. 2005. "Mineralized Skeletons and Early Animal Evolution." In *Evolving Form and Function: Fossils and Development*, edited by Derek E. G. Briggs, 101–24. New Haven, CT: Peabody Museum of Natural History, Yale University.

Ben-Menahem, Yemima. 1997. "Historical Contingency." *Ratio* 10 (2): 99–107. doi:10.1111/1467-9329.00032.

———. 2009. "Historical Necessity and Contingency." In *A Companion to the Philosophy of History and Historiography*, edited by Aviezer Tucker, 120–30. Chichester: Blackwell.

Benner, Steven A. 1989. "Patterns of Divergence in Homologous Proteins as Indicators of Tertiary and Quaternary Structure." *Advances in Enzyme Regulation* 28:219–36. doi:10.1016/0065-2571(89)90073-3.

Benner, Steven A., Slim O. Sassi, and Eric A. Gaucher. 2007. "Molecular Paleoscience: Systems Biology from the Past." In *Advances in Enzymology*, edited by Eric J. Toone, 1–132. Hoboken, NJ: Wiley.

Bennett, Albert F., and Bradley S. Hughes. 2009. "Microbial Experimental Evolution." *American Journal of Physiology—Regulatory, Integrative and Comparative Physiology* 297 (1): R17–R25. doi:10.1152/ajpregu.90562.2008.

Bennett, Albert F., and Richard E. Lenski. 2007. "An Experimental Test of Evolutionary Trade-Offs during Temperature Adaptation." *Proceedings of the National Academy of Sciences* 104 (suppl. 1): 8649–54. doi:10.1073/pnas.0702117104.

Benson, Richard H., Ralph E. Chapman, and Andrew F. Siegel. 1982. "On the Measurement of Morphology and Its Change." *Paleobiology* 8 (4): 328–39.

Benzer, Seymour. 1959. "On the Topology of the Genetic Fine Structure." *Proceedings of the National Academy of Sciences* 45 (11): 1607–20.

———. 1961. "On the Topography of the Genetic Fine Structure." *Proceedings of the National Academy of Sciences* 47 (3): 403–15.

———. 1962. "The Fine Structure of the Gene." *Scientific American* 206 (1): 70–84.

Benzer, Seymour, and Ernst Freese. 1958. "Induction of Specific Mutations with 5-Bromouracil." *Proceedings of the National Academy of Sciences* 44 (2): 112–19.

Bergson, Henri. 1907. *L'évolution créatrice*. Paris: Felix Alcan.

Betancourt, Andrea J. 2009. "Genomewide Patterns of Substitution in Adaptively Evolving Populations of the RNA Bacteriophage MS2." *Genetics* 181 (4): 1535–44. doi:10.1534/genetics.107.085837.

Bierne, Nicolas, Pierre-Alexandre Gagnaire, and Patrice David. 2013. "The Geography of Introgression in a Patchy Environment and the Thorn in the Side of Ecological Speciation." *Current Zoology* 59 (1): 72–86.

Bigelow, John, Brian Ellis, and Robert Pargetter. 1988. "Forces." *Philosophy of Science* 55 (4): 614–30. doi:10.1086/289464.

Blaser, Kent. 1999. "The History of Nature and the Nature of History: Stephen Jay

Gould on Science, Philosophy, and History." *History Teacher* 32 (3): 411–30. doi:10.2307/494379.

Blount, Zachary D. n.d. "A Case Study in Evolutionary Contingency." Unpublished manuscript accepted for publication in *Studies in History and Philosophy of Biological and Biomedical Sciences*. In Zachary D. Blount's collection. doi:10.1016/j.shpsc.2015.12.007.

Blount, Zachary D., Jeffrey E. Barrick, Carla J. Davidson, and Richard E. Lenski. 2012. "Genomic Analysis of a Key Innovation in an Experimental *Escherichia coli* Population." *Nature* 488 (7417): 513–18. doi:10.1038/nature11514.

Blount, Zachary D., Christina Z. Borland, and Richard E. Lenski. 2008. "Historical Contingency and the Evolution of a Key Innovation in an Experimental Population of *Escherichia coli*." *Proceedings of the National Academy of Sciences* 105 (23): 7899–7906. doi:10.1073/pnas.0803151105.

Bohannan, Brendan J. M., and Richard E. Lenski. 2000. "Linking Genetic Change to Community Evolution: Insights from Studies of Bacteria and Bacteriophage." *Ecology Letters* 3 (4): 362–77. doi:10.1046/j.1461-0248.2000.00161.x.

Bollback, Jonathan P., and John P. Huelsenbeck. 2009. "Parallel Genetic Evolution within and between Bacteriophage Species of Varying Degrees of Divergence." *Genetics* 181 (1): 225–34. doi:10.1534/genetics.107.085225.

Bonnell, Michael L., and Robert K. Selander. 1974. "Elephant Seals: Genetic Variation and near Extinction." *Science* 184 (4139): 908–9. doi:10.1126/science.184.4139.908.

Bookstein, Fred L., Barry Chernoff, Ruth L. Elder, Julian M. Humphries Jr., Gerald R. Smith, and Richard F. Strauss. 1985. *Morphometrics in Evolutionary Biology: The Geometry of Size and Shape Change, with Examples from Fishes.* Philadelphia: Academy of Natural Sciences of Philadelphia.

Botting, Joseph P., and Nicholas J. Butterfield. 2005. "Reconstructing Early Sponge Relationships by Using the Burgess Shale Fossil *Eiffelia globosa*, Walcott." *Proceedings of the National Academy of Sciences* 102 (5): 1554–59. doi:10.1073/pnas.0405867102.

Bowler, Peter J. 1988. *The Non-Darwinian Revolution: Reinterpreting a Historical Myth*. Baltimore: Johns Hopkins University Press.

Bowring, S. A., J. P. Grotzinger, C. E. Isachsen, Andrew H. Knoll, S. M. Pelechaty, and P. Kolosov. 1993. "Calibrating Rates of Early Cambrian Evolution." *Science* 261 (5126): 1293–98. doi:10.1126/science.11539488.

Brakefield, Paul M. 2006. "Evo-Devo and Constraints on Selection." *Trends in Ecology & Evolution* 21 (7): 362–68. doi:10.1016/j.tree.2006.05.001.

Brandon, Robert N. 1981. "Biological Teleology: Questions and Explanations." *Studies in History and Philosophy of Science* 12 (2): 91–105. doi:10.1016/0039-3681(81)90015-7.

———. 2005. "The Difference between Selection and Drift: A Reply to Millstein." *Biology & Philosophy* 20 (1): 153–70. doi:10.1007/s10539-004-1070-9.

———. 2006. "The Principle of Drift: Biology's First Law." *Journal of Philosophy* 103 (7): 319–35.

Brandon, Robert N., and Scott Carson. 1996. "The Indeterministic Character of Evolutionary Theory: No 'No Hidden Variables Proof' but No Room for Determinism Either." *Philosophy of Science* 63 (3): 315–37. doi:10.1086/289915.

Brandon, Robert N., and Leonore Fleming. 2014. "Drift Sometimes Dominates Selection, and Vice Versa: A Reply to Clatterbuck, Sober and Lewontin." *Biology & Philosophy* 29 (4): 577–85. doi:10.1007/s10539-014-9437-z.

Brandon, Robert N., and Grant Ramsey. 2007. "What's Wrong with the Emergentist Statistical Interpretation of Natural Selection and Random Drift?" In *The Cambridge Companion to the Philosophy of Biology*, edited by David L. Hull and Michael Ruse, 66–84. Cambridge: Cambridge University Press.

Brandon, Robert N., and Alex Rosenberg. 2000. "Philosophy of Biology." In *Philosophy of Science Today*, edited by Peter Clark and Katherine Hawley, 147–80. Oxford: Oxford University Press.

Bridgham, Jamie T., Eric A. Ortlund, and Joseph W. Thornton. 2009. "An Epistatic Ratchet Constrains the Direction of Glucocorticoid Receptor Evolution." *Nature* 461 (7263): 515–19. doi:10.1038/nature08249.

Briggs, Derek E. G., Douglas H. Erwin, and Frederick J. Collier. 1994. *The Fossils of the Burgess Shale*. Washington, DC: Smithsonian Institution Press.

Briggs, Derek E. G., and Richard A. Fortey. 1989. "The Early Radiation and Relationships of the Major Arthropod Groups." *Science* 246 (4927): 241–43. doi:10.1126/science.246.4927.241.

———. 2005. "Wonderful Strife: Systematics, Stem Groups, and the Phylogenetic Signal of the Cambrian Radiation." *Paleobiology* 31 (suppl. 2): 94–112. doi:10.1666/0094-8373(2005)031[0094:WSSSGA]2.0.CO;2.

Briggs, Derek E. G., Richard A. Fortey, and Matthew A. Wills. 1992. "Morphological Disparity in the Cambrian." *Science* 256 (5064): 1670–73. doi:10.1126/science.256.5064.1670.

Brooke, John Hedley, and Geoffrey Cantor. 1998. *Reconstructing Nature: The Engagement of Science and Religion*. Oxford: Oxford University Press.

Brooks, Daniel R., and Deborah A. McLennan. 1991. *Phylogeny, Ecology, and Behavior: A Research Program for Comparative Biology*. Chicago: University of Chicago Press.

———. 1993. "Historical Ecology: Examining Phylogenetic Components of Community Evolution." In *Species Diversity in Ecological Communities: Historical and Geographical Perspectives*, edited by Robert E. Ricklefs and Dolph Schluter, 267–80. Chicago: University of Chicago Press.

———. 1994. "Historical Ecology as a Research Programme: Scope, Limitations and the Future." In *Phylogenetics and Ecology*, edited by Paul Eggleton and Richard Irwin Vane-Wright, 1–27. London: Academic Press.

Brown, Celeste J., Kristy M. Todd, and R. Frank Rosenzweig. 1998. "Multiple

Duplications of Yeast Hexose Transport Genes in Response to Selection in a Glucose-Limited Environment." *Molecular Biology and Evolution* 15 (8): 931–42.

Brown, James H., James F. Gillooly, Andrew P. Allen, Van M. Savage, and Geoffrey B. West. 2004. "Toward a Metabolic Theory of Ecology." *Ecology* 85 (7): 1771–89. doi:10.1890/03-9000.

Brysse, Keynyn. 2008. "From Weird Wonders to Stem Lineages: The Second Reclassification of the Burgess Shale Fauna." *Studies in History and Philosophy of Biological and Biomedical Sciences* 39 (3): 298–313. doi:10.1016/j.shpsc.2008.06.004.

Budd, Graham E. 1996. "The Morphology of *Opabinia regalis* and the Reconstruction of the Arthropod Stem-Group." *Lethaia* 29 (1): 1–14. doi:10.1111/j.1502-3931.1996.tb01831.x.

———. 1998. "The Morphology and Phylogenetic Significance of *Kerygmachela kierkegaardi* Budd (Buen Formation, Lower Cambrian, N Greenland)." *Earth and Environmental Science Transactions of the Royal Society of Edinburgh* 89 (4): 249–90. doi:10.1017/S0263593300002418.

Buettner, Victoria L., Kathleen A. Hill, William A. Scaringe, and Steve S. Sommer. 2000. "Evidence That Proximal Multiple Mutations in Big Blue® Transgenic Mice Are Dependent Events." *Mutation Research* 452 (2): 219–29. doi:10.1016/S0027-5107(00)00090-7.

Buffon, George-Louis Leclerc de. 1749. *L'histoire naturelle, générale et particulière, avec la description du Cabinet du Roi*. 15 vols. Paris: Imprimerie Royale.

Bulhof, Johannes. 1999. "What If? Modality and History." *History and Theory* 38 (2): 145–68.

Bull, J. J., M. R. Badgett, Holly A. Wichman, John P. Huelsenbeck, David M. Hillis, A. Gulati, C. Ho, and I. J. Molineux. 1997. "Exceptional Convergent Evolution in a Virus." *Genetics* 147 (4): 1497–1507.

Burch, Christina L., and Lin Chao. 1999. "Evolution by Small Steps and Rugged Landscapes in the RNA Virus φ6." *Genetics* 151 (3): 921–27.

———. 2000. "Evolvability of an RNA Virus Is Determined by Its Mutational Neighbourhood." *Nature* 406 (6796): 625–28. doi:10.1038/35020564.

Burian, Richard M. 1988. "Challenges to the Evolutionary Synthesis." *Evolutionary Biology* 23:247–69.

Bury, J. B. 1920. *The Idea of Progress: An Inquiry into Its Origin and Growth*. London: Macmillan.

Butterfield, Nicholas J. 2001. "Ecology and Evolution of Cambrian Plankton." In *The Ecology of the Cambrian Radiation*, edited by Andrey Yu. Zhuravlev and Robert Riding, 200–216. New York: Columbia University Press.

Cain, Joseph. 1994. "Ernst Mayr as Community Architect: Launching the Society for the Study of Evolution and the Journal *Evolution*." *Biology & Philosophy* 9 (3): 387–427. doi:10.1007/BF00857945.

———. 2009. "Rethinking the Synthesis Period in Evolutionary Studies." *Journal of the History of Biology* 42 (4): 621–48. doi:10.1007/s10739-009-9206-z.

Cain, Joseph, and Michael Ruse, eds. 2009. *Descended from Darwin: Insights into the History of Evolutionary Studies, 1900–1970*. Philadelphia: American Philosophical Society.

Camps, Manel, Asael Herman, Ern Loh, and Lawrence A. Loeb. 2007. "Genetic Constraints on Protein Evolution." *Critical Reviews in Biochemistry and Molecular Biology* 42 (5): 313–26. doi:10.1080/10409230701597642.

Capra, Emily J., Barrett S. Perchuk, Emma A. Lubin, Orr Ashenberg, Jeffrey M. Skerker, and Michael T. Laub. 2010. "Systematic Dissection and Trajectory-Scanning Mutagenesis of the Molecular Interface That Ensures Specificity of Two-Component Signaling Pathways." *PLoS Genetics* 6 (11): e1001220. doi:10.1371/journal.pgen.1001220.

Carroll, Sean B. 2001. "Chance and Necessity: The Evolution of Morphological Complexity and Diversity." *Nature* 409 (6823): 1102–9. doi:10.1038/35059227.

———. 2008. "Evo-Devo and an Expanding Evolutionary Synthesis: A Genetic Theory of Morphological Evolution." *Cell* 134 (1): 25–36. doi:10.1016/j.cell.2008.06.030.

Carroll, Sean B., Jennifer Grenier, and Scott Weatherbee. 2001. *From DNA to Diversity: Molecular Genetics and the Evolution of Animal Design*. Malden, MA: Blackwell Science.

Carroll, William E. 2000. "Creation, Evolution, and Thomas Aquinas." *Revue des Questions Scientifiques* 171 (4): 319–47.

Cartwright, Nancy. 2007. "Counterfactuals in Economics: A Commentary." In *Causation and Explanation*, edited by Joseph Keim Campbell, Michael O'Rourke, and Harry Silverstein, 191–216. Cambridge, MA: MIT Press.

Chambers, Robert. 1846. *Vestiges of the Natural History of Creation*. 5th ed. London: John Churchill.

Charlesworth, Brian. 2013. "Stabilizing Selection, Purifying Selection, and Mutational Bias in Finite Populations." *Genetics* 194 (4): 955–71. doi:10.1534/genetics.113.151555.

Chen, Jun-Yuan. 2008. "Early Crest Animals and the Insight They Provide into the Evolutionary Origin of Craniates." *Genesis* 46 (11): 623–39. doi:10.1002/dvg.20445.

Chen, Liangbiao, Arthur L. DeVries, and Chi-Hing C. Cheng. 1997. "Convergent Evolution of Antifreeze Glycoproteins in Antarctic Notothenioid Fish and Arctic Cod." *Proceedings of the National Academy of Sciences* 94 (8): 3817–22.

Chevin, Luis-Miguel, Guillaume Martin, and Thomas Lenormand. 2010. "Fisher's Model and the Genomics of Adaptation: Restricted Pleiotropy, Heterogenous Mutation, and Parallel Evolution." *Evolution* 64 (11): 3213–31. doi:10.1111/j.1558-5646.2010.01058.x.

Christin, Pascal-Antoine, Daniel M. Weinreich, and Guillaume Besnard. 2010.

"Causes and Evolutionary Significance of Genetic Convergence." *Trends in Genetics* 26 (9): 400–405. doi:10.1016/j.tig.2010.06.005.

Cohan, Frederick M., Elaine C. King, and Piotr Zawadzki. 1994. "Amelioration of the Deleterious Pleiotropic Effects of an Adaptive Mutation in *Bacillus subtilis*." *Evolution* 48 (1): 81–95. doi:10.2307/2410005.

Cohan, Frederick M., and Elizabeth B. Perry. 2007. "A Systematics for Discovering the Fundamental Units of Bacterial Diversity." *Current Biology* 17 (10): R373–86. doi:10.1016/j.cub.2007.03.032.

Cole, Megan F., and Eric A. Gaucher. 2011. "Exploiting Models of Molecular Evolution to Efficiently Direct Protein Engineering." *Journal of Molecular Evolution* 72 (2): 193–203. doi:10.1007/s00239-010-9415-2.

Colegrave, N., and Sinéd Collins. 2008. "Experimental Evolution: Experimental Evolution and Evolvability." *Heredity* 100 (5): 464–70. doi:10.1038/sj.hdy .6801095.

Collins, Sinéd, Dieter Sültemeyer, and Graham Bell. 2006. "Rewinding the Tape: Selection of Algae Adapted to High CO_2 at Current and Pleistocene Levels of CO_2." *Evolution* 60 (7): 1392–1401. doi:10.1111/j.0014-3820.2006 .tb01218.x.

Colosimo, Pamela F., Kim E. Hosemann, Sarita Balabhadra, Guadalupe Villarreal, Mark Dickson, Jane Grimwood, Jeremy Schmutz, Richard M. Myers, Dolph Schluter, and David M. Kingsley. 2005. "Widespread Parallel Evolution in Sticklebacks by Repeated Fixation of Ectodysplasin Alleles." *Science* 307 (5717): 1928–33. doi:10.1126/science.1107239.

Conte, Gina L., Matthew E. Arnegard, Catherine L. Peichel, and Dolph Schluter. 2012. "The Probability of Genetic Parallelism and Convergence in Natural Populations." *Proceedings of the Royal Society of London B* 279 (1749): 5039–47. doi:10.1098/rspb.2012.2146.

Conway Morris, Simon. 1985. "The Middle Cambrian Metazoan *Wiwaxia corrugata* (Matthew) from the Burgess Shale and Ogygopsis Shale, British Columbia, Canada." *Philosophical Transactions of the Royal Society of London B* 307 (1134): 507–82.

———. 1998. *The Crucible of Creation: The Burgess Shale and the Rise of Animals.* Oxford: Oxford University Press.

———. 2003. *Life's Solution: Inevitable Humans in a Lonely Universe.* Cambridge: Cambridge University Press.

———. 2009. "The Predictability of Evolution: Glimpses into a Post-Darwinian World." *Die Naturwissenschaften* 96 (11): 1313–37. doi:10.1007/ s00114-009-0607-9.

———. 2010. "Evolution: Like Any Other Science It Is Predictable." *Philosophical Transactions of the Royal Society of London B* 365 (1537): 133–45. doi:10.1098/ rstb.2009.0154.

Conway Morris, Simon, and Jean-Bernard Caron. 2012. "*Pikaia gracilens* Walcott, a Stem-Group Chordate from the Middle Cambrian of British Columbia." *Biological Reviews* 87 (2): 480–512. doi:10.1111/j.1469-185X.2012.00220.x.

Cooper, David N., Albino Bacolla, Claude Férec, Karen M. Vasquez, Hildegard Kehrer-Sawatzki, and Jian-Min Chen. 2011. "On the Sequence-Directed Nature of Human Gene Mutation: The Role of Genomic Architecture and the Local DNA Sequence Environment in Mediating Gene Mutations Underlying Human Inherited Disease." *Human Mutation* 32 (10): 1075–99. doi:10.1002/humu.21557.

Cooper, Tim F., Susanna K. Remold, Richard E. Lenski, and Dominique Schneider. 2008. "Expression Profiles Reveal Parallel Evolution of Epistatic Interactions Involving the CRP Regulon in *Escherichia coli*." *PLoS Genetics* 4 (2): e35. doi:10.1371/journal.pgen.0040035.

Cooper, Tim F., Daniel E. Rozen, and Richard E. Lenski. 2003. "Parallel Changes in Gene Expression after 20,000 Generations of Evolution in *Escherichia coli*." *Proceedings of the National Academy of Sciences* 100 (3): 1072–77. doi:10.1073/pnas.0334340100.

Cooper, Vaughn S., and Richard E. Lenski. 2000. "The Population Genetics of Ecological Specialization in Evolving *Escherichia coli* Populations." *Nature* 407 (6805): 736–39. doi:10.1038/35037572.

Cooper, Vaughn S., Dominique Schneider, Michel Blot, and Richard E. Lenski. 2001. "Mechanisms Causing Rapid and Parallel Losses of Ribose Catabolism in Evolving Populations of *Escherichia coli* B." *Journal of Bacteriology* 183 (9): 2834–41. doi:10.1128/JB.183.9.2834-2841.2001.

Copley, Shelley D. 2003. "Enzymes with Extra Talents: Moonlighting Functions and Catalytic Promiscuity." *Current Opinion in Chemical Biology* 7 (2): 265–72. doi:10.1016/S1367-5931(03)00032-2.

Couñago, Rafael, Stephen Chen, and Yousif Shamoo. 2006. "In Vivo Molecular Evolution Reveals Biophysical Origins of Organismal Fitness." *Molecular Cell* 22 (4): 441–49. doi:10.1016/j.molcel.2006.04.012.

Cowan, Robin, and Dominique Foray. 2002. "Evolutionary Economics and the Counterfactual Threat: On the Nature and Role of Counterfactual History as an Empirical Tool in Economics." *Journal of Evolutionary Economics* 12 (5): 539–62. doi:10.1007/s00191-002-0134-8.

Cox, G. B., F. Gibson, R. K. J. Luke, N. A. Newton, I. G. O'Brien, and H. Rosenberg. 1970. "Mutations Affecting Iron Transport in *Escherichia coli*." *Journal of Bacteriology* 104 (1): 219–26.

Coyne, Jerry A. 2009. *Why Evolution Is True*. New York: Viking.

Crick, Francis H. C. 1968. "The Origin of the Genetic Code." *Journal of Molecular Biology* 38 (3): 367–79. doi:10.1016/0022-2836(68)90392-6.

Crozat, Estelle, Thomas Hindré, Lauriane Kühn, Jérome Garin, Richard E. Lenski, and Dominique Schneider. 2011. "Altered Regulation of the OmpF Porin by Fis in *Escherichia coli* during an Evolution Experiment and between B and K-12 Strains." *Journal of Bacteriology* 193 (2): 429–40. doi:10.1128/JB.01341-10.

Crozat, Estelle, Nadège Philippe, Richard E. Lenski, Johannes Geiselmann, and Dominique Schneider. 2005. "Long-Term Experimental Evolution in *Escherichia coli*. XII. DNA Topology as a Key Target of Selection." *Genetics* 169 (2): 523–32. doi:10.1534/genetics.104.035717.

Crozat, Estelle, Cynthia Winkworth, Joël Gaffé, Peter F. Hallin, Margaret A. Riley, Richard E. Lenski, and Dominique Schneider. 2010. "Parallel Genetic and Phenotypic Evolution of DNA Superhelicity in Experimental Populations of *Escherichia coli*." *Molecular Biology and Evolution* 27 (9): 2113–28. doi:10.1093/molbev/msq099.

Currie, Adrian Mitchell. 2012. "Convergence, Contingency & Morphospace." *Biology & Philosophy* 27 (4): 583–93. doi:10.1007/s10539-012-9319-1.

Czworkowski, John, and Peter B. Moore. 1996. "The Elongation Phase of Protein Synthesis." *Progress in Nucleic Acid Research and Molecular Biology* 54:293–332.

Daley, Allison C., Graham E. Budd, Jean-Bernard Caron, Gregory D. Edgecombe, and Desmond Collins. 2009. "The Burgess Shale Anomalocaridid *Hurdia* and Its Significance for Early Euarthropod Evolution." *Science* 323 (5921): 1597–1600. doi:10.1126/science.1169514.

Dallinger, W. H. 1887. "The President's Address." *Journal of the Royal Microscopical Society* 7 (2): 185–99. doi:10.1111/j.1365-2818.1887.tb01566.x.

Darwin, Charles. 1837. *Notebook B: [Transmutation of Species (1837–1838)]. CUL-DAR121.* Edited by Kees Rookmaker. Darwin Online, http://darwin-online.org.uk/.

———. 1838a. *Notebook C: [Transmutation of Species (1838.02–1838.07)]. CUL-DAR122.* Edited by Kees Rookmaker. Darwin Online, http://darwin-online.org.uk/.

———. 1838b. *Notebook D: [Transmutation of Species (7–10.1838)]. CUL-DAR123.* Edited by Kees Rookmaker. Darwin Online, http://darwin-online.org.uk/.

———. 1838c. *Notebook E: [Transmutation of Species (10.1838–7.1839)]. CUL-DAR124.* Edited by Kees Rookmaker. Darwin Online, http://darwin-online.org.uk/.

———. 1838d. *Notebook M: [Metaphysics on Morals and Speculations on Expression (1838)]. CUL-DAR125.* Edited by Kees Rookmaker. Darwin Online, http://darwin-online.org.uk/.

———. 1838e. *Notebook N: [Metaphysics and Expression (1838–1839)]. CUL-DAR126.* Edited by Kees Rookmaker. Darwin Online, http://darwin-online.org.uk/.

———. 1845. *Journal of Researches into the Natural History and Geology of the Countries Visited during the Voyage of H. M. S. Beagle around the World.* 2nd ed. London: John Murray.

———. 1859. *On the Origin of Species.* 1st ed. London: John Murray.

———. 1860a. "Letter 2814—Darwin, C. R. to Gray, A., 22 May [1860]." https://www.darwinproject.ac.uk/entry-2814.

———. 1860b. "Letter 2998—Darwin, C. R. to Gray, A., 26 Nov. [1860]." https://www.darwinproject.ac.uk/entry-2998.

———. 1861. *On the Origin of Species.* 3rd ed. London: John Murray.

———. 1871. *The Descent of Man and Selection in Relation to Sex.* 2 vols. London: John Murray.

———. 1874. "Letter 9483—Darwin, C. R. to Gray, A., 5 Jun. [1874]." https://www.darwinproject.ac.uk/entry-9483.

———. 1875. *Insectivorous Plants.* London: John Murray.

Darwin, Erasmus. 1794. *Zoonomia; or, The Laws of Organic Life.* 2 vols. London: J. Johnson.

———. 1803. *The Temple of Nature; or, The Origin of Society: A Poem, with Philosophical Notes.* London: J. Johnson.

Davidson, Eric H., and Douglas H. Erwin. 2009. "An Integrated View of Precambrian Eumetazoan Evolution." *Cold Spring Harbor Symposia on Quantitative Biology* 74:65–80. doi:10.1101/sqb.2009.74.042.

———. 2010. "Evolutionary Innovation and Stability in Animal Gene Networks." *Journal of Experimental Zoology* 314B (3): 182–86. doi:10.1002/jez.b.21329.

Davis, Bernard D. 1949. "The Isolation of Biochemically Deficient Mutants of Bacteria by Means of Penicillin." *Proceedings of the National Academy of Sciences* 35 (1): 1–10.

Davis, Bernard D., and Elizabeth S. Mingioli. 1950. "Mutants of *Escherichia coli* Requiring Methionine or Vitamin B12." *Journal of Bacteriology* 60 (1): 17–28.

Dawkins, Richard. 1982. *The Extended Phenotype: The Long Reach of the Gene.* Oxford: Oxford University Press.

———. 1986. *The Blind Watchmaker.* New York: Norton.

———. 1996. *Climbing Mount Improbable.* New York: Norton.

Dawkins, Richard, and John R. Krebs. 1979. "Arms Races between and within Species." *Proceedings of the Royal Society of London B* 205 (1161): 489–511. doi:10.1098/rspb.1979.0081.

Dean, Antony M., and Joseph W. Thornton. 2007. "Mechanistic Approaches to the Study of Evolution: The Functional Synthesis." *Nature Reviews Genetics* 8 (9): 675–88. doi:10.1038/nrg2160.

de Duve, Christian. 1995. *Vital Dust: The Origin and Evolution of Life on Earth.* New York: Basic Books.

Defeu Soufo, Hervé Joël, Christian Reimold, Uwe Linne, Tobias Knust, Johannes Gescher, and Peter L. Graumann. 2010. "Bacterial Translation Elongation Factor EF-Tu Interacts and Colocalizes with Actin-like MreB Protein." *Proceedings of the National Academy of Sciences* 107 (7): 3163–68. doi:10.1073/pnas.0911979107.

Delahaye, Jean-Paul. 1999. *Information, complexité et hasard*. 2nd ed. Paris: Hermès.

Delbrück, Max. 1971. "Aristotle-Totle-Totle." In *Of Microbes and Life*, edited by Ernest Borek and Jacques Monod, 50–55. New York: Columbia University Press.

Delisle, Richard G. 2009. "The Uncertain Foundation of Neo-Darwinism: Metaphysical and Epistemological Pluralism in the Evolutionary Synthesis." *Studies in History and Philosophy of Biological and Biomedical Sciences* 40 (2): 119–32. doi:10.1016/j.shpsc.2009.03.004.

———. 2011. "What Was Really Synthesized during the Evolutionary Synthesis? A Historiographic Proposal." *Studies in History and Philosophy of Biological and Biomedical Sciences* 42 (1): 50–59. doi:10.1016/j.shpsc.2010.11.005.

Depew, David J. 2008. "Consequence Etiology and Biological Teleology in Aristotle and Darwin." *Studies in History and Philosophy of Biological and Biomedical Sciences* 39 (4): 379–90. doi:10.1016/j.shpsc.2008.09.001.

———. 2009. "The Rhetoric of Darwin's *Origin of Species*." In Ruse and Richards 2009, 237–55.

———. 2010. "Incidentally Final Causation and Spontaneous Generation in Aristotle's Physics II and Other Texts." In *Was ist Leben? Aristoteles' Anschauungen zur Entstehung und Funktionsweise von Leben*, edited by Sabine Föllinger, 285–97. Stuttgart: Franz Steiner Verlag.

———. 2015. "Accident, Adaptation, and Teleology in Aristotle and Darwinism." In *Darwin in the Twenty-First Century: Nature, Humanity, God*, edited by Phillip R. Sloan, Gerald McKenny, and Kathleen Eggleston, 116–43. Notre Dame, IN: Notre Dame University Press.

Depew, David J., and Bruce H. Weber. 1995. *Darwinism Evolving: Systems Dynamics and the Genealogy of Natural Selection*. Cambridge, MA: Bradford Books.

de Queiroz, Alan, and Javier A. Rodríguez-Robles. 2006. "Historical Contingency and Animal Diets: The Origins of Egg Eating in Snakes." *American Naturalist* 167 (5): 684–94. doi:10.1086/503118.

Desjardins, Eric Cyr. 2011. "Historicity and Experimental Evolution." *Biology & Philosophy* 26 (3): 339–64. doi:10.1007/s10539-011-9256-4.

Desmond, Adrian. 1989. *The Politics of Evolution: Morphology, Medicine, and Reform in Radical London*. Chicago: University of Chicago Press.

Desmond, Adrian, and James Moore. 2009. *Darwin's Sacred Cause: How a Hatred of Slavery Shaped Darwin's Views on Human Evolution*. New York: Houghton Mifflin Harcourt.

Dettman, Jeremy R., Nicolas Rodrigue, Anita H. Melnyk, Alex Wong, Susan F. Bailey, and Rees Kassen. 2012. "Evolutionary Insight from Whole-Genome Sequencing of Experimentally Evolved Microbes." *Molecular Ecology* 21 (9): 2058–77. doi:10.1111/j.1365-294X.2012.05484.x.

de Visser, J. Arjan G. M., Clifford W. Zeyl, Philip J. Gerrish, Jeffrey L. Blanchard, and Richard E. Lenski. 1999. "Diminishing Returns from Mutation Supply Rate in Asexual Populations." *Science* 283 (5400): 404–6. doi:10.1126/science.283.5400.404.

Dewey, John. 1910. *The Influence of Darwin on Philosophy and Other Essays*. New York: Henry Holt.

Dick, Matthew H., Scott Lidgard, Dennis P. Gordon, and Shunsuke F. Mawatari. 2009. "The Origin of Ascophoran Bryozoans Was Historically Contingent but Likely." *Proceedings of the Royal Society of London B* 276 (1670): 3141–48. doi:10.1098/rspb.2009.0704.

Dietrich, Michael R. 2003. "Richard Goldschmidt: Hopeful Monsters and Other 'Heresies.'" *Nature Reviews Genetics* 4 (1): 68–74. doi:10.1038/nrg979.

Dobzhansky, Theodosius. 1937. *Genetics and the Origin of Species*. 1st ed. New York: Columbia University Press.

———. 1951. *Genetics and the Origin of Species*. 3rd ed. New York: Columbia University Press.

———. 1956. *The Biological Basis of Human Freedom*. New York: Columbia University Press.

———. 1962. *Mankind Evolving: The Evolution of the Human Species*. New Haven, CT: Yale University Press.

———. 1967. *The Biology of Ultimate Concern*. New York: New American Library.

———. 1970. *Genetics of the Evolutionary Process*. New York: Columbia University Press.

———. 1973. "Nothing in Biology Makes Sense Except in the Light of Evolution." *American Biology Teacher* 35:125–29.

Dobzhansky, Theodosius, and Ernest Boesiger. 1987. *Human Culture: A Moment in Evolution*. New York: Columbia University Press.

Drake, John W. 2007a. "Mutations in Clusters and Showers." *Proceedings of the National Academy of Sciences* 104 (20): 8203–4. doi:10.1073/pnas.0703089104.

———. 2007b. "Too Many Mutants with Multiple Mutations." *Critical Reviews in Biochemistry and Molecular Biology* 42 (4): 247–58. doi:10.1080/10409230701495631.

Drake, John W., Anna Bebenek, Grace E. Kissling, and Shyamal Peddada. 2005. "Clusters of Mutations from Transient Hypermutability." *Proceedings of the National Academy of Sciences* 102 (36): 12849–54. doi:10.1073/pnas.0503009102.

Dwyer, Peter D. 1984. "Functionalism and Structuralism: Two Programs for Evolutionary Biologists." *American Naturalist* 124 (5): 745–50.

Earman, John. 1986. *A Primer on Determinism*. Dordrecht: D. Reidel.

Edgecombe, Gregory D. 2010. "Arthropod Phylogeny: An Overview from the Perspectives of Morphology, Molecular Data and the Fossil Record." *Arthropod Structure & Development* 39 (2–3): 74–87. doi:10.1016/j.asd.2009.10.002.

Edwards, Denis. 2010. *How God Acts: Creation, Redemption and Special Divine Action*. Minneapolis: Fortress.

Eimer, Theodor. 1898. *On Orthogenesis: And the Impotence of Natural Selection in Species-Formation*. Translated by Thomas J. McCormack. Chicago: Open Court.

Elena, Santiago F., and Richard E. Lenski. 2003. "Evolution Experiments with Microorganisms: The Dynamics and Genetic Bases of Adaptation." *Nature Reviews Genetics* 4 (6): 457–69. doi:10.1038/nrg1088.

Elena, Santiago F., Rafael Sanjuán, Antonio V. Bordería, and Paul E. Turner. 2001. "Transmission Bottlenecks and the Evolution of Fitness in Rapidly Evolving RNA Viruses." *Infection, Genetics and Evolution* 1 (1): 41–48. doi:10.1016/S1567-1348(01)00006-5.

Elez, Marina, Andrew W. Murray, Li-Jun Bi, Xian-En Zhang, Ivan Matic, and Miroslav Radman. 2010. "Seeing Mutations in Living Cells." *Current Biology* 20 (16): 1432–37. doi:10.1016/j.cub.2010.06.071.

Elmer, Kathryn R., and Axel Meyer. 2011. "Adaptation in the Age of Ecological Genomics: Insights from Parallelism and Convergence." *Trends in Ecology & Evolution* 26 (6): 298–306. doi:10.1016/j.tree.2011.02.008.

Elton, C. S. 1924. "Periodic Fluctuations in the Numbers of Animals: Their Causes and Effects." *Journal of Experimental Biology* 2 (1): 119–63.

Emerson, Sharon B. 2001. "A Macroevolutionary Study of Historical Contingency in the Fanged Frogs of Southeast Asia." *Biological Journal of the Linnean Society* 73 (1): 139–51. doi:10.1006/bijl.2001.0532.

Ereshefsky, Marc. 2014. "Consilience, Historicity, and the Species Problem." In *Evolutionary Biology: Conceptual, Ethical, and Religious Issues*, edited by R. Paul Thompson and Denis M. Walsh, 65–86. Cambridge: Cambridge University Press.

Erwin, Douglas H. 2006a. "The Developmental Origins of Animal Body Plans." In *Neoproerozoic Geobiology and Paleobiology*, edited by Shuhai Xiao and Alan J. Kaufman, 157–97. Dordrecht: Kluwer.

———. 2006b. *Extinction: How Life on Earth Nearly Ended 250 Million Years Ago*. Princeton, NJ: Princeton University Press.

———. 2007. "Disparity: Morphological Pattern and Developmental Context." *Palaeontology* 50 (1): 57–73. doi:10.1111/j.1475-4983.2006.00614.x.

———. 2011. "Evolutionary Uniformitarianism." *Developmental Biology* 357 (1): 27–34. doi:10.1016/j.ydbio.2011.01.020.

———. 2015a. "Early Metazoan Life: Divergence, Environment and Ecology." *Philosophical Transactions of the Royal Society of London B* 370 (1684): 20150036. doi:10.1098/rstb.2015.0036.

———. 2015b. "Was the Ediacaran-Cambrian Radiation a Unique Evolutionary Event?" *Paleobiology* 41 (1): 1–15. doi:10.1017/pab.2014.2.

Erwin, Douglas H., and Eric H. Davidson. 2002. "The Last Common Bilaterian Ancestor." *Development* 129 (13): 3021–32.

———. 2009. "The Evolution of Hierarchical Gene Regulatory Networks." *Nature Reviews Genetics* 10 (2): 141–48. doi:10.1038/nrg2499.

Erwin, Douglas H., Marc Laflamme, Sarah M. Tweedt, Erik A. Sperling, Davide Pisani, and Kevin J. Peterson. 2011. "The Cambrian Conundrum: Early Divergence and Later Ecological Success in the Early History of Animals." *Science* 334 (6059): 1091–97. doi:10.1126/science.1206375.

Erwin, Douglas H., and James W. Valentine. 2013. *The Cambrian Explosion: The Construction of Animal Biodiversity*. Greenwood, CO: Roberts.

Erwin, Douglas H., James W. Valentine, and J. John Sepkoski Jr. 1987. "A Comparative Study of Diversification Events: The Early Paleozoic versus the Mesozoic." *Evolution* 41 (6): 1177–86. doi:10.2307/2409086.

Estes, Suzanne, and Stevan J. Arnold. 2007. "Resolving the Paradox of Stasis: Models with Stabilizing Selection Explain Evolutionary Divergence on All Timescales." *American Naturalist* 169 (2): 227–44. doi:10.1086/508302.

Falcon, Andrea. 2005. *Aristotle and the Science of Nature: Unity without Uniformity*. Cambridge: Cambridge University Press.

Fedonkin, Mikhail A., James G. Gehling, Kathleen Grey, Guy M. Narbonne, and Patricia Vickers-Rich. 2008. *The Rise of Animals: Evolution and Diversification of the Kingdom Animalia*. Baltimore: Johns Hopkins University Press.

Ferguson, Niall. 1997. *Virtual History: Alternatives and Counterfactuals*. London: Picador.

Field, Katharine G., Gary J. Olsen, David J. Lane, Stephen J. Giovannoni, Michael T. Ghiselin, Elizabeth C. Raff, Norman R. Pace, and Rudolf A. Raff. 1988. "Molecular Phylogeny of the Animal Kingdom." *Science* 239 (4841): 748–53. doi:10.1126/science.3277277.

Fisher, R. A. 1918. "The Correlation between Relatives on the Supposition of Mendelian Inheritance." *Philosophical Transactions of the Royal Society of Edinburgh* 52:399–433.

———. 1922. "On the Dominance Ratio." *Proceedings of the Royal Society of Edinburgh* 42:321–41. doi:10.1016/S0092-8240(05)80012-6.

———. 1927. "On Some Objections to Mimicry Theory; Statistical and Genetic." *Transactions of the Royal Entomological Society of London* 75 (2): 269–78. doi:10.1111/j.1365-2311.1927.tb00074.x.

———. 1930. *The Genetical Theory of Natural Selection*. Oxford: Clarendon Press.

———. 1932. *The Social Selection of Human Fertility*. The Herbert Spencer Lecture. Oxford: Oxford University Press.

———. 1934. "Indeterminism and Natural Selection." *Philosophy of Science* 1 (1): 99–117.

Flores-Moya, Antonio, Mónica Rouco, María Jesús García-Sánchez, Camino García-Balboa, Raquel González, Eduardo Costas, and Victoria López-Rodas. 2012. "Effects of Adaptation, Chance, and History on the Evolution of the Toxic Dinoflagellate *Alexandrium minutum* under Selection of Increased

Temperature and Acidification." *Ecology and Evolution* 2 (6): 1251–59. doi:10.1002/ece3.198.

Fong, Stephen S., Andrew R. Joyce, and Bernhard Ø. Palsson. 2005. "Parallel Adaptive Evolution Cultures of *Escherichia coli* Lead to Convergent Growth Phenotypes with Different Gene Expression States." *Genome Research* 15 (10): 1365–72. doi:10.1101/gr.3832305.

Fontana, Walter, and Leo W. Buss. 1994. "What Would Be Conserved If 'The Tape Were Played Twice'?" *Proceedings of the National Academy of Sciences* 91 (2): 757–61. doi:10.1073/pnas.91.2.757.

Foote, Mike. 1992. "Rarefaction Analysis of Morphological and Taxonomic Diversity." *Paleobiology* 18 (1): 1–16.

———. 1993. "Discordance and Concordance between Morphological and Taxonomic Diversity." *Paleobiology* 19 (2): 185–204.

———. 1997. "The Evolution of Morphological Diversity." *Annual Review of Ecology and Systematics* 28:129–52.

———. 1998. "Contingency and Convergence." *Science* 280 (5372): 2068–69. doi:10.1126/science.280.5372.2068.

Fortey, Richard A., Derek E. G. Briggs, and Matthew A. Wills. 1996. "The Cambrian Evolutionary 'Explosion': Decoupling Cladogenesis from Morphological Disparity." *Biological Journal of the Linnean Society* 57 (1): 13–33. doi:10.1006/bijl.1995.0002.

Fortuna, Miguel A., Luis Zaman, Aaron P. Wagner, and Charles Ofria. 2013. "Evolving Digital Ecological Networks." *PLoS Computational Biology* 9 (3): e1002928. doi:10.1371/journal.pcbi.1002928.

Foster, Patricia L. 2006. "Methods for Determining Spontaneous Mutation Rates." In *Methods in Enzymology*, edited by Judith L. Campbell and Paul Modrich, 409:195–213. DNA Repair, Part B. London: Academic Press.

Fraser, Hunter B. 2005. "Modularity and Evolutionary Constraint on Proteins." *Nature Genetics* 37 (4): 351–52. doi:10.1038/ng1530.

Frost, G. E., and H. Rosenberg. 1973. "The Inducible Citrate-Dependent Iron Transport System in *Escherichia coli* K12." *Biochimica et Biophysica Acta* 330 (1): 90–101. doi:10.1016/0005-2736(73)90287-3.

Fukami, Tadashi, Hubertus J. E. Beaumont, Xue-Xian Zhang, and Paul B. Rainey. 2007. "Immigration History Controls Diversification in Experimental Adaptive Radiation." *Nature* 446 (7134): 436–39. doi:10.1038/nature05629.

Galtier, Nicolas, Nicolas Tourasse, and Manolo Gouy. 1999. "A Nonhyperthermophilic Common Ancestor to Extant Life Forms." *Science* 283 (5399): 220–21. doi:10.1126/science.283.5399.220.

Garcia-Villada, Libertad, and John W. Drake. 2010. "Mutational Clusters Generated by Non-processive Polymerases: A Case Study Using DNA Polymerase β in Vitro." *DNA Repair* 9 (8): 871–78. doi:10.1016/j.dnarep.2010.05.002.

Garfinkel, Alan. 1981. *Forms of Explanation: Rethinking the Questions in Social Theory*. New Haven, CT: Yale University Press.

Garland, Theodore, and Michael R. Rose, eds. 2009. *Experimental Evolution: Concepts, Methods, and Applications of Selection Experiments*. Berkeley: University of California Press.

Gaucher, Eric A. 2007. "Ancestral Sequence Reconstruction as a Tool to Understand Natural History and Guide Synthetic Biology: Realizing and Extending the Vision of Zuckerkandl and Pauling." In *Ancestral Sequence Reconstruction*, edited by David A. Liberles, 20–33. Oxford: Oxford University Press.

Gaucher, Eric A., Sridhar Govindarajan, and Omjoy K. Ganesh. 2008. "Palaeotemperature Trend for Precambrian Life Inferred from Resurrected Proteins." *Nature* 451 (7179): 704–7. doi:10.1038/nature06510.

Gayon, Jean. 1998. *Darwinism's Struggle for Survival: Heredity and the Hypothesis of Natural Selection*. Cambridge: Cambridge University Press.

———. 2005. "Chance, Explanation, and Causation in Evolutionary Theory." *History and Philosophy of the Life Sciences* 27 (3–4): 395–405.

Gehling, James G. 1999. "Microbial Mats in Terminal Proterozoic Siliciclastics: Ediacaran Death Masks." *PALAIOS* 14 (1): 40–57. doi:10.2307/3515360.

Ghiselin, Michael T. 1994. "Darwin's Language May Seem Teleological, but His Thinking Is Another Matter." *Biology & Philosophy* 9:489–92.

Gigerenzer, Gerd, Zeno Swijtink, Theodore M. Porter, Lorraine Daston, John H. Beatty, and Lorenz Krüger. 1989. *The Empire of Chance: How Probability Changed Science and Everyday Life*. Cambridge: Cambridge University Press.

Gilbert, Scott F., John M. Opitz, and Rudolf A. Raff. 1996. "Resynthesizing Evolutionary and Developmental Biology." *Developmental Biology* 173 (2): 357–72. doi:10.1006/dbio.1996.0032.

Gildenhuys, Peter. 2009. "An Explication of the Causal Dimension of Drift." *British Journal for the Philosophy of Science* 60 (3): 521–55. doi:10.1093/bjps/axp019.

Gillespie, John H. 1983a. "A Simple Stochastic Gene Substitution Model." *Theoretical Population Biology* 23 (2): 202–15. doi:10.1016/0040-5809(83)90014-X.

———. 1983b. "Some Properties of Finite Populations Experiencing Strong Selection and Weak Mutation." *American Naturalist* 121 (5): 691–708.

———. 1984. "Molecular Evolution over the Mutational Landscape." *Evolution* 38 (5): 1116–29. doi:10.2307/2408444.

———. 1991. *The Causes of Molecular Evolution*. Oxford: Oxford University Press.

———. 2004. *Population Genetics: A Concise Guide*. Baltimore: Johns Hopkins University Press.

Gillespie, Rosemary. 2004. "Community Assembly through Adaptive Radiation in Hawaiian Spiders." *Science* 303 (5656): 356–59. doi:10.1126/science.1091875.

Gingerich, Philip D. 2009. "Rates of Evolution." *Annual Review of Ecology, Evolution, and Systematics* 40 (1): 657–75. doi:10.1146/annurev.ecolsys.39.110707.173457.

Glazier, Douglas S. 2010. "A Unifying Explanation for Diverse Metabolic Scaling in Animals and Plants." *Biological Reviews* 85 (1): 111–38. doi:10.1111/j.1469 -185X.2009.00095.x.

Glennan, Stuart S. 1997. "Probable Causes and the Distinction between Subjective and Objective Chance." *Noûs* 31 (4): 496–519.

Godfrey-Smith, Peter. 2009. *Darwinian Populations and Natural Selection*. Oxford: Oxford University Press.

Goldenfeld, Nigel, and Carl Woese. 2007. "Biology's Next Revolution." *Nature* 445 (7126): 369. doi:10.1038/445369a.

Goldschmidt, Richard. 1940. *The Material Basis of Evolution*. New Haven, CT: Yale University Press.

Gompel, Nicolas, and Benjamin Prud'homme. 2009. "The Causes of Repeated Genetic Evolution." *Developmental Biology* 332 (1): 36–47. doi:10.1016/j.ydbio.2009.04.040.

Gould, Stephen Jay. 1970. "Dollo on Dollo's Law: Irreversibility and the Status of Evolutionary Laws." *Journal of the History of Biology* 3 (2): 189–212.

———. 1977. "Eternal Metaphors of Palaeontology." In *Patterns of Evolution: As Illustrated by the Fossil Record*, edited by A. Hallam, 1–26. Amsterdam: Elsevier.

———. 1980a. *The Panda's Thumb: More Reflections in Natural History*. New York: Norton.

———. 1980b. "The Promise of Paleobiology as a Nomothetic, Evolutionary Discipline." *Paleobiology* 6 (1): 96–118.

———. 1981. *The Mismeasure of Man*. New York: Norton.

———. 1982. "The Uses of Heresy: An Introduction to Richard Goldschmidt's *Material Basis of Evolution*." In *The Material Basis of Evolution*, by Richard Goldschmidt, xiii—xliii. New Haven, CT: Yale University Press.

———. 1983. "The Hardening of the Modern Synthesis." In *Dimensions of Darwinism*, edited by Marjorie Grene, 71–93. Cambridge: Cambridge University Press.

———. 1985a. *The Flamingo's Smile: Reflections in Natural History*. New York: Norton.

———. 1985b. "The Paradox of the First Tier: An Agenda for Paleobiology." *Paleobiology* 11 (1): 2–12.

———. 1988. "On Replacing the Idea of Progress with an Operational Notion of Directionality." In *Evolutionary Progress*, edited by Matthew H. Nitecki, 319–38. Chicago: University of Chicago Press.

———. 1989. *Wonderful Life*. New York: Norton.

———. 1996. *Full House: The Spread of Excellence from Plato to Darwin*. New York: Harmony.

———. 1999. *Rocks of Ages: Science and Religion in the Fullness of Life*. New York: Ballantine Books.

———. 2002. *The Structure of Evolutionary Theory*. Cambridge, MA: Harvard University Press.

Gould, Stephen Jay, and Richard C. Lewontin. 1979. "The Spandrels of San Marco and the Panglossian Paradigm: A Critique of the Adaptationist Programme." *Proceedings of the Royal Society of London B* 205 (1161): 581–98.

Gould, Stephen Jay, David M. Raup, J. John Sepkoski Jr., Thomas J. M. Schopf, and Daniel S. Simberloff. 1977. "The Shape of Evolution: A Comparison of Real and Random Clades." *Paleobiology* 3 (1): 23–40.

Gray, Asa. 1862. "Letter 3489—Gray, A. to Darwin, C. R., 31 Mar. [1862]." https://www.darwinproject.ac.uk/entry-3489.

———. 1884. *Darwiniana: Essays and Reviews Pertaining to Darwinism*. New York: Appleton.

Griffiths, Paul E., and Russell D. Gray. 1994. "Developmental Systems and Evolutionary Explanation." *Journal of Philosophy* 91 (6): 277–304.

Gromiha, M. Michael, Motohisa Oobatake, and Akinori Sarai. 1999. "Important Amino Acid Properties for Enhanced Thermostability from Mesophilic to Thermophilic Proteins." *Biophysical Chemistry* 82 (1): 51–67. doi:10.1016/S0301-4622(99)00103-9.

Gros, Pierre-Alexis, Hervé Le Nagard, and Olivier Tenaillon. 2009. "The Evolution of Epistasis and Its Links with Genetic Robustness, Complexity and Drift in a Phenotypic Model of Adaptation." *Genetics* 182 (1): 277–93. doi:10.1534/genetics.108.099127.

Hacking, Ian. 1990. *The Taming of Chance*. Cambridge: Cambridge University Press.

Haig, David. 2007. "Weismann Rules! OK? Epigenetics and the Lamarckian Temptation." *Biology & Philosophy* 22 (3): 415–28. doi:10.1007/s10539-006-9033-y.

Hájek, Alan. 2007. "The Reference Class Problem Is Your Problem Too." *Synthese* 156 (3): 563–85. doi:10.1007/s11229-006-9138-5.

Halanych, Kenneth M. 2004. "The New View of Animal Phylogeny." *Annual Review of Ecology, Evolution, and Systematics* 35:229–56. doi:10.1146/annurev.ecolsys.35.112202.130124.

Haldane, J. B. S. 1924a. "A Mathematical Theory of Natural and Artificial Selection—I." *Transactions of the Cambridge Philosophical Society* 23:19–41. doi:10.1007/BF02459574.

———. 1924b. "A Mathematical Theory of Natural and Artificial Selection. Part II: The Influence of Partial Self-Fertilisation, Inbreeding, Assortative Mating, and Selective Fertilisation on the Composition of Mendelian Populations, and on Natural Selection." *Biological Reviews* 1 (3): 158–63. doi:10.1111/j.1469-185X.1924.tb00546.x.

———. 1932. *The Causes of Evolution*. London: Longmans, Green.

———. 1942. *New Paths in Genetics*. New York: Harper.

Hall, Barry G. 1982. "Chromosomal Mutation for Citrate Utilization by *Escherichia coli* K-12." *Journal of Bacteriology* 151 (1): 269–73.

Hall, Brian K. 2003. "Descent with Modification: The Unity Underlying Homology and Homoplasy as Seen through an Analysis of Development and Evolution." *Biological Reviews* 78 (3): 409–33. doi:10.1017/S1464793102006097.

Harding, David J. 2003. "Counterfactual Models of Neighborhood Effects: The Effect of Neighborhood Poverty on Dropping Out and Teenage Pregnancy." *American Journal of Sociology* 109 (3): 676–719. doi:10.1086/ajs.2003.109 .issue-3.

Harms, Michael J., and Joseph W. Thornton. 2013. "Evolutionary Biochemistry: Revealing the Historical and Physical Causes of Protein Properties." *Nature Reviews Genetics* 14 (8): 559–71. doi:10.1038/nrg3540.

Hartl, Daniel L., and Clifford H. Taubes. 1998. "Towards a Theory of Evolutionary Adaptation." *Genetica* 102–3:525–33. doi:10.1023/A:1017071901530.

Hegel, Georg Wilhelm Friedrich. 1953. *Reason in History*. Translated by Robert S. Hartman. New York: Liberal Arts Press.

Hegreness, Matthew, and Roy Kishony. 2007. "Analysis of Genetic Systems Using Experimental Evolution and Whole-Genome Sequencing." *Genome Biology* 8 (1): 201. doi:10.1186/gb-2007-8-1-201.

Hempel, Carl G. 1965. "Aspects of Scientific Explanation." In *Aspects of Scientific Explanation*, 331–496. New York: Free Press.

Herring, Christopher D., Anu Raghunathan, Christiane Honisch, Trina Patel, M. Kenyon Applebee, Andrew R. Joyce, Thomas J. Albert, et al. 2006. "Comparative Genome Sequencing of *Escherichia coli* Allows Observation of Bacterial Evolution on a Laboratory Timescale." *Nature Genetics* 38 (12): 1406–12. doi:10.1038/ng1906.

Hill, Kathleen A., Jicheng Wang, Kelly D. Farwell, William A. Scaringe, and Steve S. Sommer. 2004. "Spontaneous Multiple Mutations Show Both Proximal Spacing Consistent with Chronocoordinate Events and Alterations with p53-Deficiency." *Mutation Research* 554 (1–2): 223–40. doi:10.1016/j.mrfmmm.2004.05.005.

Hitchcock, Christopher, and Joel D. Velasco. 2014. "Evolutionary and Newtonian Forces." *Ergo* 1. doi:10.3998/ergo.12405314.0001.002.

Hodge, Charles. 1968. *Systematic Theology*. 3 vols. Grand Rapids, MI: Eerdmans.

———. 1994. *What Is Darwinism? And Other Writings on Science and Religion.* Edited by Mark A. Noll and David N. Livingstone. Grand Rapids, MI: Baker Books.

Hodge, M. J. S. 1972. "The Universal Gestation of Nature: Chambers' *Vestiges* and *Explanations*." *Journal of the History of Biology* 5 (1): 127–51. doi:10.1007/BF02113488.

———. 1985. "Darwin as a Lifelong Generation Theorist." In *The Darwinian Heritage: A Centennial Retrospect*, edited by David Kohn, 207–43. Princeton, NJ: Princeton University Press.

———. 1987. "Natural Selection as a Causal, Empirical, and Probabilistic Theory." In Krüger, Gigerenzer, and Morgan 1987, 233–70.

———. 1992a. "Biology and Philosophy (including Ideology): A Study of Fisher and Wright." In *The Founders of Evolutionary Genetics*, edited by Sahotra Sarkar, 231–93. Dordrecht: Kluwer Academic.

———. 1992b. "Natural Selection: Historical Perpsectives." In *Keywords in Evolutionary Biology*, edited by Evelyn Fox Keller and Elisabeth A. Lloyd, 212–19. Cambridge, MA: Harvard University Press.

———. 2009. "Capitalist Contexts for Darwinian Theory: Land, Finance, Industry and Empire." *Journal of the History of Biology* 42 (3): 399–416. doi:10.1007/s10739-009-9187-y.

———. 2011. "Darwinism after Mendelism: The Case of Sewall Wright's Intellectual Synthesis in His Shifting Balance Theory of Evolution (1931)." *Studies in History and Philosophy of Biological and Biomedical Sciences* 42 (1): 30–39. doi:10.1016/j.shpsc.2010.11.008.

Hodge, M. J. S., and David Kohn. 1985. "The Immediate Origins of Natural Selection." In *The Darwinian Heritage: A Centennial Retrospect*, edited by David Kohn, 185–206. Princeton, NJ: Princeton University Press.

Hodge, M. J. S., and Gregory Radick. 2009. "The Place of Darwin's Theories in the Intellectual Long Run." In *The Cambridge Companion to Darwin*, edited by M. J. S. Hodge and Gregory Radick, 246–73. 2nd ed. Cambridge: Cambridge University Press.

Hodgkinson, Alan, and Adam Eyre-Walker. 2011. "Variation in the Mutation Rate across Mammalian Genomes." *Nature Reviews Genetics* 12 (11): 756–66. doi:10.1038/nrg3098.

Hopf, Eberhard. 1934. "On Causality, Statistics and Probability." *Journal of Mathematics and Physics* 13:51–102.

Hou, Xian-guang, Richard Aldridge, Jan Bergström, David J. Siveter, Derek Siveter, and Xiang-Hong Feng. 2004. *The Cambrian Fossils of Chengjiang, China: The Flowering of Early Animal Life*. Oxford: Wiley-Blackwell.

Hou, Xian-guang, Lars Ramsköld, and Jan Bergström. 1991. "Composition and Preservation of the Chengjiang Fauna—a Lower Cambrian Soft-Bodied Biota." *Zoologica Scripta* 20 (4): 395–411. doi:10.1111/j.1463-6409.1991.tb00303.x.

Houle, David. 1998. "How Should We Explain Variation in the Genetic Variance of Traits?" *Genetica* 102-3:241–53. doi:10.1023/A:1017034925212.

Huey, Raymond B., George W. Gilchrist, Margen L. Carlson, David Berrigan, and Luís Serra. 2000. "Rapid Evolution of a Geographic Cline in Size in an Introduced Fly." *Science* 287 (5451): 308–9. doi:10.1126/science.287.5451.308.

Hughes, Austin L., and Robert Friedman. 2003. "Parallel Evolution by Gene Duplication in the Genomes of Two Unicellular Fungi." *Genome Research* 13 (5): 794–99. doi:10.1101/gr.714603.

Hull, David L. 1965. "The Effect of Essentialism on Taxonomy—Two Thousand Years of Stasis." *British Journal for the Philosophy of Science* 15 (60): 314–26.

———. 1973. *Darwin and His Critics: The Reception of Darwin's Theory of Evolution by the Scientific Community*. Cambridge, MA: Harvard University Press.

———. 1988. *Science as a Process: An Evolutionary Account of the Social and Conceptual Development of Science*. Chicago: University of Chicago Press.

Hussein, Saber, Klaus Hantke, and Volkmar Braun. 1981. "Citrate-Dependent Iron Transport System in *Escherichia coli* K-12." *European Journal of Biochemistry* 117 (2): 431–37. doi:10.1111/j.1432-1033.1981.tb06357.x.

Huxley, Julian S. 1912. *The Individual in the Animal Kingdom*. Cambridge: Cambridge University Press.

———. 1942. *Evolution: The Modern Synthesis*. London: Allen and Unwin.

Huxley, Thomas Henry. 1887. "On the Reception of the 'Origin of Species.'" In *The Life and Letters of Charles Darwin, Including an Autobiographical Chapter*, edited by Francis Darwin, 2:179–204. London: John Murray.

Inkpen, Rob, and Derek D. Turner. 2012. "The Topography of Historical Contingency." *Journal of the Philosophy of History* 6 (1): 1–19. doi:10.1163/187226312X625573.

International Theological Commission. 2004. "Communion and Stewardship: Human Persons Created in the Image of God." Vatican City. www.vatican.va/roman_curia/congregations/cfaith/cti_documents/rc_con_cfaith_doc_20040723_communion-stewardship_en.html.

Jablonski, David. 1986. "Background and Mass Extinctions: The Alternation of Macroevolutionary Regimes." *Science* 231 (4734): 129–33. doi:10.1126/science.231.4734.129.

———. 2007. "Scale and Hierarchy in Macroevolution." *Palaeontology* 50 (1): 87–109. doi:10.1111/j.1475-4983.2006.00615.x.

Jacob, François. 1977. "Evolution and Tinkering." *Science* 196 (4295): 1161–66. doi:10.1126/science.860134.

Jefferies, Richard P. S. 1979. "The Origin of Chordates—a Methodological Essay." In *The Origin of Major Invertebrate Groups*, edited by M. R. House, 443–47. London: Academic Press.

Jensen, Roy A. 1976. "Enzyme Recruitment in Evolution of New Function." *Annual Review of Microbiology* 30:409–25. doi:10.1146/annurev.mi.30.100176.002205.

Jensen, Sören, Mary L. Droser, and James G. Gehling. 2005. "Trace Fossil Preservation and the Early Evolution of Animals." *Palaeogeography, Palaeoclimatology, Palaeoecology* 220 (1–2): 19–29. doi:10.1016/j.palaeo.2003.09.035.

Jeong, Haeyoung, Valérie Barbe, Choong Hoon Lee, David Vallenet, Dong Su Yu, Sang-Haeng Choi, Arnaud Couloux, et al. 2009. "Genome Sequences of *Escherichia coli* B Strains REL606 and BL21(DE3)." *Journal of Molecular Biology* 394 (4): 644–52. doi:10.1016/j.jmb.2009.09.052.

Jerison, Harry J. 1973. *Evolution of the Brain and Intelligence*. New York: Academic Press.

John Paul II. 1996a. "Aux Membres de l'Académie pontificale des Sciences." Vatican City. http://w2.vatican.va/content/john-paul-ii/fr/messages/pont_messages/1996/documents/hf_jp-ii_mes_19961022_evoluzione.html.

———. 1996b. "Revised Translation of Pope's Message on Evolution to the Pontifical Academy of Sciences." *Origins* 26 (25).

Johnson, Curtis. 2015. *Darwin's Dice: The Idea of Chance in the Thought of Charles Darwin*. Oxford: Oxford University Press.

Johnson, Steven D., and Kim E. Steiner. 2000. "Generalization versus Specialization in Plant Pollination Systems." *Trends in Ecology & Evolution* 15 (4): 140–43. doi:10.1016/S0169-5347(99)01811-X.

Jones, Adam G., Stevan J. Arnold, and Reinhard Bürger. 2007. "The Mutation Matrix and the Evolution of Evolvability." *Evolution* 61 (4): 727–45. doi:10.1111/j.1558-5646.2007.00071.x.

Joyce, Paul, Darin R. Rokyta, Craig J. Beisel, and H. Allen Orr. 2008. "A General Extreme Value Theory Model for the Adaptation of DNA Sequences under Strong Selection and Weak Mutation." *Genetics* 180 (3): 1627–43. doi: 10.1534/genetics.108.088716.

Kacar, Betul, and Eric A. Gaucher. 2012. "Towards the Recapitulation of Ancient History in the Laboratory: Combining Synthetic Biology with Experimental Evolution." In *Artificial Life 13*, edited by Christoph Adami, David M. Bryson, Charles Ofria, and Robert T. Pennock, 11–18. Cambridge, MA: MIT Press.

———. 2013. "Experimental Evolution of Protein-Protein Interaction Networks." *Biochemical Journal* 453 (3): 311–19. doi:10.1042/BJ20130205.

Kahneman, Daniel. 2011. *Thinking, Fast and Slow*. New York: Farrar, Straus and Giroux.

Kant, Immanuel. 2000. *Critique of the Power of Judgment*. Edited by Paul Guyer and Eric Matthews. Cambridge: Cambridge University Press.

Kauffman, Stuart A. 1993. *The Origins of Order: Self-Organization and Selection in Evolution*. Oxford: Oxford University Press.

———. 1995. *At Home in the Universe: The Search for the Laws of Self-Organization and Complexity*. Oxford: Oxford University Press.

Kawecki, Tadeusz J., Richard E. Lenski, Dieter Ebert, Brian Hollis, Isabelle Olivieri, and Michael C. Whitlock. 2012. "Experimental Evolution." *Trends in Ecology & Evolution* 27 (10): 547–60. doi:10.1016/j.tree.2012.06.001.

Keller, Evelyn Fox. 2000. *The Century of the Gene*. Cambridge, MA: Harvard University Press.

Khersonsky, Olga, Cintia Roodveldt, and Dan S. Tawfik. 2006. "Enzyme Promiscuity: Evolutionary and Mechanistic Aspects." *Current Opinion in Chemical Biology* 10 (5): 498–508. doi:10.1016/j.cbpa.2006.08.011.

Kimura, Motoo. 1980. "Average Time until Fixation of a Mutant Allele in a Finite Population under Continued Mutation Pressure: Studies by Analytical, Numerical, and Pseudo-Sampling Methods." *Proceedings of the National Academy of Sciences* 77 (1): 522–26.

———. 1983. *The Neutral Theory of Molecular Evolution*. Cambridge: Cambridge University Press.

King, Jack Lester, and Thomas H. Jukes. 1969. "Non-Darwinian Evolution." *Science* 164 (881): 788–98.

Kinnison, Michael T., and Andrew P. Hendry. 2001. "The Pace of Modern Life II: From Rates of Contemporary Microevolution to Pattern and Process." *Genetica* 112–13 (1): 145–64. doi:10.1023/A:1013375419520.

Kirkpatrick, Mark. 1982. "Quantum Evolution and Punctuated Equilibria in Continuous Genetic Characters." *American Naturalist* 119 (6): 833–48.

Klingenberg, Christian Peter. 2005. "Developmental Constraints, Modules, and Evolvability." In *Variation: A Central Concept in Biology*, edited by Benedikt Hallgrimsson and Brian K. Hall, 219–47. Burlington, MA: Elsevier.

Knoll, Andrew H., and Sean B. Carroll. 1999. "Early Animal Evolution: Emerging Views from Comparative Biology and Geology." *Science* 284 (5423): 2129–37. doi:10.1126/science.284.5423.2129.

Kohn, David. 2009. "Darwin's Keystone: The Principle of Divergence." In Ruse and Richards 2009, 87–108.

Kolbe, Jason J., Manuel Leal, Thomas W. Schoener, David A. Spiller, and Jonathan B. Losos. 2012. "Founder Effects Persist despite Adaptive Differentiation: A Field Experiment with Lizards." *Science* 335 (6072): 1086–89. doi:10.1126/science.1209566.

Kolmogorov, A. N. 1968. "Three Approaches to the Quantitative Definition of Information." *International Journal of Computer Mathematics* 2 (1–4): 157–68. doi:10.1080/00207166808803030.

Kominek, Jacek, Jaroslaw Marszalek, Cécile Neuvéglise, Elizabeth A. Craig, and Barry L. Williams. 2013. "The Complex Evolutionary Dynamics of Hsp70s: A Genomic and Functional Perspective." *Genome Biology and Evolution* 5 (12): 2460–77. doi:10.1093/gbe/evt192.

Koonin, Eugene V. 2012. *The Logic of Chance: The Nature and Origin of Biological Evolution*. Upper Saddle River, NJ: FT Press Science.

Korona, Ryszard, Cindy H. Nakatsu, Larry J. Forney, and Richard E. Lenski. 1994. "Evidence for Multiple Adaptive Peaks from Populations of Bacteria Evolving in a Structured Habitat." *Proceedings of the National Academy of Sciences* 91 (19): 9037–41.

Kouchinsky, Artem, Stefan Bengtson, Bruce Runnegar, Christian Skovsted, Michael Steiner, and Michael Vendrasco. 2012. "Chronology of Early Cambrian Biomineralization." *Geological Magazine* 149 (02): 221–51. doi:10.1017/S0016756811000720.

Kozak, Kenneth H., Robert W. Mendyk, and John J. Wiens. 2009. "Can Parallel Diversification Occur in Sympatry? Repeated Patterns of Body-Size Evolution in

Coexisting Clades of North American Salamanders." *Evolution* 63 (7): 1769–84. doi:10.1111/j.1558-5646.2009.00680.x.

Kratzer, James T., Miguel A. Lanaspa, Michael N. Murphy, Christina Cicerchi, Christina L. Graves, Peter A. Tipton, Eric A. Ortlund, Richard J. Johnson, and Eric A. Gaucher. 2014. "Evolutionary History and Metabolic Insights of Ancient Mammalian Uricases." *Proceedings of the National Academy of Sciences* 111 (10): 3763–68. doi:10.1073/pnas.1320393111.

Krüger, Lorenz, Lorraine Daston, and Michael Heidelberger, eds. 1987. *The Probabilistic Revolution*. Vol. 1, *Ideas in History*. Cambridge, MA: Bradford Books.

Krüger, Lorenz, Gerd Gigerenzer, and Mary S. Morgan, eds. 1987. *The Probabilistic Revolution*. Vol. 2, *Ideas in the Sciences*. Cambridge, MA: Bradford Books.

Kumar, Sudhir, and Sankar Subramanian. 2002. "Mutation Rates in Mammalian Genomes." *Proceedings of the National Academy of Sciences* 99 (2): 803–8. doi:10.1073/pnas.022629899.

Laflamme, Marc. n.d. "Ediacaran Clades." Unpublished manuscript, to be submitted for publication in a special issue of *Canadian Journal of Zoology*. In Douglas H. Erwin's collection.

Laflamme, Marc, Simon A. F. Darroch, Sarah M. Tweedt, Kevin J. Peterson, and Douglas H. Erwin. 2013. "The End of the Ediacara Biota: Extinction, Biotic Replacement, or Cheshire Cat?" *Gondwana Research* 23 (2): 558–73. doi:10.1016/j.gr.2012.11.004.

Laflamme, Marc, James D. Schiffbauer, Guy M. Narbonne, and Derek E. G. Briggs. 2011. "Microbial Biofilms and the Preservation of the Ediacara Biota." *Lethaia* 44 (2): 203–13. doi:10.1111/j.1502-3931.2010.00235.x.

Lande, Russell. 1976. "Natural Selection and Random Genetic Drift in Phenotypic Evolution." *Evolution* 30 (2): 314–34. doi:10.2307/2407703.

———. 1979. "Quantitative Genetic Analysis of Multivariate Evolution, Applied to Brain:Body Size Allometry." *Evolution* 33 (1): 402–16. doi:10.2307/2407630.

———. 1985. "Expected Time for Random Genetic Drift of a Population between Stable Phenotypic States." *Proceedings of the National Academy of Sciences* 82 (22): 7641–45.

Lang, Gregory I., Daniel P. Rice, Mark J. Hickman, Erica Sodergren, George M. Weinstock, David Botstein, and Michael M. Desai. 2013. "Pervasive Genetic Hitchhiking and Clonal Interference in Forty Evolving Yeast Populations." *Nature* 500 (7464): 571–74. doi:10.1038/nature12344.

Lange, Marc. 2005. "A Counterfactual Analysis of the Concepts of Logical Truth and Necessity." *Philosophical Studies* 125 (3): 277–303. doi:10.1007/s11098-005-7774-0.

Lara, F. J. S., and J. L. Stokes. 1952. "Oxidation of Citrate by *Escherichia coli*." *Journal of Bacteriology* 63 (3): 415–20.

Largent, Mark A. 2009. "The So-Called Eclipse of Darwinism." In Cain and Ruse 2009, 3–21.

Lea, Douglas E., and Charles A. Coulson. 1949. "The Distribution of the Numbers of Mutants in Bacterial Populations." *Journal of Genetics* 49 (3): 264–85. doi:10.1007/BF02986080.

Le Gac, Mickaël, Tim F. Cooper, Stéphane Cruveiller, Claudine Médigue, and Dominique Schneider. 2013. "Evolutionary History and Genetic Parallelism Affect Correlated Responses to Evolution." *Molecular Ecology* 22 (12): 3292–3303. doi:10.1111/mec.12312.

Le Gac, Mickaël, Jessica Plucain, Thomas Hindré, Richard E. Lenski, and Dominique Schneider. 2012. "Ecological and Evolutionary Dynamics of Coexisting Lineages during a Long-Term Experiment with *Escherichia coli.*" *Proceedings of the National Academy of Sciences* 109 (24): 9487–92. doi:10.1073/pnas.1207091109.

Lennox, James G. 1993. "Darwin *was* a Teleologist." *Biology & Philosophy* 8:409–21.

———. 1994. "Teleology by Another Name: A Reply to Ghiselin." *Biology & Philosophy* 9:493–95.

———. 2000. *Aristotle's Philosophy of Biology.* Cambridge: Cambridge University Press.

———. 2010. "The Darwin/Gray Correspondence 1857–1869: An Intelligent Discussion about Chance and Design." *Perspectives on Science* 18 (4): 456–79.

———. 2013. "Darwin and Teleology." In *The Cambridge Encyclopedia of Darwin and Evolutionary Thought*, edited by Michael Ruse, 152–57. Cambridge: Cambridge University Press.

Lenormand, Thomas, Denis Roze, and François Rousset. 2009. "Stochasticity in Evolution." *Trends in Ecology & Evolution* 24 (3): 157–65. doi:10.1016/j.tree.2008.09.014.

Lenski, Richard E. 2004. "Phenotypic and Genomic Evolution during a 20,000-Generation Experiment with the Bacterium *Escherichia coli.*" Edited by Jules Janick. *Plant Breeding Reviews* 24 (2): 225–65.

Lenski, Richard E., Jeffrey E. Barrick, and Charles Ofria. 2006. "Balancing Robustness and Evolvability." *PLoS Biology* 4 (12): e428. doi:10.1371/journal.pbio.0040428.

Lenski, Richard E., and John E. Mittler. 1993. "The Directed Mutation Controversy and Neo-Darwinism." *Science* 259 (5092): 188–94. doi:10.1126/science.7678468.

Lenski, Richard E., Michael R. Rose, Suzanne C. Simpson, and Scott C. Tadler. 1991. "Long-Term Experimental Evolution in *Escherichia coli*. I. Adaptation and Divergence during 2,000 Generations." *American Naturalist* 138 (6): 1315–41.

Lenski, Richard E., and Michael Travisano. 1994. "Dynamics of Adaptation and Diversification: A 10,000-Generation Experiment with Bacterial Populations." *Proceedings of the National Academy of Sciences* 91 (15): 6808–14.

Levin, Bruce R., and Richard E. Lenski. 1985. "Bacteria and Phage: A Model System for the Study of the Ecology and Coevolution of Hosts and Parasites." In

Ecology and Genetics of Host-Parasite Interactions, edited by David Rollinson and R. M. Anderson, 227–42. London: Academic Press.

Levins, Richard. 1966. "The Strategy of Model Building in Population Biology." *American Scientist* 54 (4): 421–31.

Lewens, Tim. 2010. "Natural Selection Then and Now." *Biological Reviews* 85 (4): 829–35. doi:10.1111/j.1469-185X.2010.00128.x.

Lewis, David. 1986. "Causal Explanation." In *Philosophical Papers*, 2:214–40. Oxford: Oxford University Press.

Lewontin, Richard C. 1966. "Is Nature Probable or Capricious?" *BioScience* 16 (1): 25–27. doi:10.2307/1293548.

———. 1974. *The Genetic Basis of Evolutionary Change*. New York: Columbia University Press.

———. 2003. "Four Complications in Understanding the Evolutionary Process." *SFI Bulletin* 18 (1): center.

Li, Guoxiang, Michael Steiner, Xuejian Zhu, Aihua Yang, Haifeng Wang, and Bernd D. Erdtmann. 2007. "Early Cambrian Metazoan Fossil Record of South China: Generic Diversity and Radiation Patterns." *Palaeogeography, Palaeoclimatology, Palaeoecology* 254 (1–2): 229–49. doi:10.1016/j.palaeo.2007.03.017.

Liu, Jianni, Degan Shu, Jian Han, Zhifei Zhang, and Xingliang Zhang. 2008. "Origin, Diversification, and Relationships of Cambrian Lobopods." *Gondwana Research* 14 (1–2): 277–83. doi:10.1016/j.gr.2007.10.001.

Lobkovsky, Alexander E., and Eugene V. Koonin. 2012. "Replaying the Tape of Life: Quantification of the Predictability of Evolution." *Frontiers in Genetics* 3:246. doi:10.3389/fgene.2012.00246.

Look, Brandon C. 2006. "Blumenbach and Kant on Mechanism and Teleology in Nature: The Case of the Formative Drive." In Smith 2006, 355–74.

Losos, Jonathan B. 1994. "Historical Contingency and Lizard Community Ecology." In *Lizard Ecology: Historical and Experimental Perspectives*, edited by Laurie J. Vitt and Eric R. Planka, 319–33. Princeton, NJ: Princeton University Press.

———. 2010. "Adaptive Radiation, Ecological Opportunity, and Evolutionary Determinism." *American Naturalist* 175 (6): 623–39. doi:10.1086/648324.

———. 2011. "Convergence, Adaptation, and Constraint." *Evolution* 65 (7): 1827–40. doi:10.1111/j.1558-5646.2011.01289.x.

Losos, Jonathan B., Todd R. Jackman, Allan Larson, Kevin de Queiroz, and Lourdes Rodríguez-Schettino. 1998. "Contingency and Determinism in Replicated Adaptive Radiations of Island Lizards." *Science* 279 (5359): 2115–18. doi:10.1126/science.279.5359.2115.

Luria, Salvador E., and Max Delbrück. 1943. "Mutations of Bacteria from Virus Sensitivity to Virus Resistance." *Genetics* 28 (6): 491–511.

Lütgens, Malke, and Gerhard Gottschalk. 1980. "Why a Co-Substrate Is Required for Anaerobic Growth of *Escherichia coli* on Citrate." *Journal of General Microbiology* 119 (1): 63–70. doi:10.1099/00221287-119-1-63.

Mahler, D. Luke, Travis Ingram, Liam J. Revell, and Jonathan B. Losos. 2013. "Exceptional Convergence on the Macroevolutionary Landscape in Island Lizard Radiations." *Science* 341 (6143): 292–95. doi:10.1126/science.1232392.

Maloof, Adam C., Susannah M. Porter, John L. Moore, Frank Ö. Dudás, Samuel A. Bowring, John A. Higgins, David A. Fike, and Michael P. Eddy. 2010. "The Earliest Cambrian Record of Animals and Ocean Geochemical Change." *Geological Society of America Bulletin* 122 (11–12): 1731–74. doi:10.1130/B30346.1.

Mani, G. S., and B. C. Clarke. 1990. "Mutational Order: A Major Stochastic Process in Evolution." *Proceedings of the Royal Society of London B* 240 (1297): 29–37.

Markus, R. A. 1970. *Saeculum: History and Society in the Theology of St. Augustine.* Cambridge: Cambridge University Press.

Martin, Arnaud, and Virginie Orgogozo. 2013. "The Loci of Repeated Evolution: A Catalog of Genetic Hotspots of Phenotypic Variation." *Evolution* 67 (5): 1235–50. doi:10.1111/evo.12081.

Martin, Guillaume, and Thomas Lenormand. 2006. "A General Multivariate Extension of Fisher's Geometrical Model and the Distribution of Mutation Fitness Effects across Species." *Evolution* 60 (5): 893–907. doi:10.1111/j.0014-3820.2006.tb01169.x.

———. 2008. "The Distribution of Beneficial and Fixed Mutation Fitness Effects Close to an Optimum." *Genetics* 179 (2): 907–16. doi:10.1534/genetics.108.087122.

Martin-Löf, Per. 1966. "The Definition of Random Sequences." *Information and Control* 9:602–19. doi:10.1016/S0019-9958(66)80018-9.

Martí-Renom, Marc A., Ashley C. Stuart, András Fiser, Roberto Sánchez, Francisco Melo, and Andrej Šali. 2000. "Comparative Protein Structure Modeling of Genes and Genomes." *Annual Review of Biophysics and Biomolecular Structure* 29:291–325. doi:10.1146/annurev.biophys.29.1.291.

Matthen, Mohan. 2009. "Drift and 'Statistically Abstractive Explanation.'" *Philosophy of Science* 76 (4): 464–87. doi:10.1086/648063.

Matthen, Mohan, and André Ariew. 2002. "Two Ways of Thinking about Fitness and Natural Selection." *Journal of Philosophy* 99 (2): 55–83.

———. 2009. "Selection and Causation." *Philosophy of Science* 76 (2): 201–24. doi:10.1086/648102.

Maynard Smith, John. 2000. "The Concept of Information in Biology." *Philosophy of Science* 67 (2): 177–94.

Maynard Smith, John, Richard M. Burian, Stuart A. Kauffman, Pere Alberch, J. Campbell, Brian Goodwin, Russell Lande, David M. Raup, and Lewis Wolpert. 1985. "Developmental Constraints and Evolution: A Perspective from the Mountain Lake Conference on Development and Evolution." *Quarterly Review of Biology* 60 (3): 265–87.

Mayr, Ernst. 1942. *Systematics and the Origin of Species from the Viewpoint of a Zoologist.* New York: Columbia University Press.

———. 1963. *Animal Species and Evolution*. Cambridge, MA: Belknap.

———. 1970. *Populations, Species, and Evolution*. Cambridge, MA: Harvard University Press.

———. 1976. "Typological versus Population Thinking." In *Evolution and the Diversity of Life: Selected Essays*, 26–29. Cambridge, MA: Belknap.

———. 1982. *The Growth of Biological Thought: Diversity, Evolution, and Inheritance*. Cambridge, MA: Belknap.

———. 1988. *Toward a New Philosophy of Biology: Observations of an Evolutionist*. Cambridge, MA: Harvard University Press.

Mayr, Ernst, and William B. Provine, eds. 1980. *The Evolutionary Synthesis: Perspectives on the Unification of Biology*. Cambridge, MA: Harvard University Press.

McGee, Matthew D., and Peter C. Wainwright. 2013. "Convergent Evolution as a Generator of Phenotypic Diversity in Threespine Stickleback." *Evolution* 67 (4): 1204–8. doi:10.1111/j.1558-5646.2012.01839.x.

McGhee, George R. 2011. *Convergent Evolution: Endless Forms Most Beautiful*. Cambridge, MA: MIT Press.

McLaughlin, Peter. 2007. "On Selection Of, For, With and Against." In *Thinking about Causes: From Greek Philosophy to Modern Physics*, edited by Peter Machamer and Gereon Wolters, 265–83. Pittsburgh, PA: University of Pittsburgh Press.

McPherson, James M. 1988. *Battle Cry of Freedom: The Civil War Era*. New York: Oxford University Press.

McShea, Daniel W. 1993. "Arguments, Tests, and the Burgess Shale—a Commentary on the Debate." *Paleobiology* 19 (4): 399–402.

———. 2005. "The Evolution of Complexity without Natural Selection, a Possible Large-Scale Trend of the Fourth Kind." *Paleobiology* 31 (2): 146–56.

McShea, Daniel W., and Robert N. Brandon. 2010. *Biology's First Law: The Tendency for Diversity and Complexity to Increase in Evolutionary Systems*. Chicago: University of Chicago Press.

Melnyk, Anita H., and Rees Kassen. 2011. "Adaptive Landscapes in Evolving Populations of *Pseudomonas fluorescens*." *Evolution* 65 (11): 3048–59. doi:10.1111/j.1558-5646.2011.01333.x.

Merlin, Francesca. 2010. "Evolutionary Chance Mutation: A Defense of the Modern Synthesis' Consensus View." *Philosophy and Theory in Biology* 2: e103.

———. 2013. *Mutations et aléas, le hasard dans la théorie de l'évolution*. Paris: Editions Hermann.

Metcalf, Maynard M. 1913. "Adaptation through Natural Selection and Orthogenesis." *American Naturalist* 47 (554): 65–71.

Meyer, Justin R., Anurag A. Agrawal, Ryan T. Quick, Devin T. Dobias, Dominique Schneider, and Richard E. Lenski. 2010. "Parallel Changes in Host Resistance to Viral Infection during 45,000 Generations of Relaxed Selection." *Evolution* 64 (10): 3024–34. doi:10.1111/j.1558-5646.2010.01049.x.

Meyer, Justin R., Devin T. Dobias, Joshua S. Weitz, Jeffrey E. Barrick, Ryan T. Quick, and Richard E. Lenski. 2012. "Repeatability and Contingency in the Evolution of a Key Innovation in Phage Lambda." *Science* 335 (6067): 428–32. doi:10.1126/science.1214449.

Meyer, Justin R., and Rees Kassen. 2007. "The Effects of Competition and Predation on Diversification in a Model Adaptive Radiation." *Nature* 446 (7134): 432–35. doi:10.1038/nature05599.

Millstein, Roberta L. 2002. "Are Random Drift and Natural Selection Conceptually Distinct?" *Biology & Philosophy* 17:33–53. doi:10.1023/A:1012990800358.

———. 2005. "Selection vs. Drift: A Response to Brandon's Reply." *Biology & Philosophy* 20 (1): 171–75. doi:10.1007/s10539-004-6047-1.

———. 2006. "Discussion of 'Four Case Studies on Chance in Evolution': Philosophical Themes and Questions." *Philosophy of Science* 73 (5): 678–87. doi:10.1086/518522.

———. 2008. "Distinguishing Drift and Selection Empirically: 'The Great Snail Debate' of the 1950s." *Journal of the History of Biology* 41 (2): 339–67. doi:10.1007/s10739-007-9145-5.

———. 2011. "Chances and Causes in Evolutionary Biology: How Many Chances Become One Chance." In *Causality in the Sciences*, edited by Phyllis McKay Illari, Federica Russo, and Jon Williamson, 425–44. Oxford: Oxford University Press.

———. 2013. "Natural Selection and Causal Productivity." In *Mechanism and Causality in Biology and Economics*, edited by Hsiang-Ke Chao, Szu-Ting Chen, and Roberta L. Millstein, 147–63. New York: Springer.

Millstein, Roberta L., Robert A. Skipper, and Michael R. Dietrich. 2009. "(Mis)interpreting Mathematical Models: Drift as a Physical Process." *Philosophy and Theory in Biology* 1: e002. doi:10.3998/ptb.6959004.0001.002.

Mivart, St. George Jackson. 1871. *On the Genesis of Species*. 2nd ed. London: Macmillan.

Monod, Jacques. 1971. *Chance and Necessity: An Essay on the Natural Philosophy of Modern Biology*. New York: Alfred A. Knopf.

Moore, Francisco B.-G., and Robert J. Woods. 2006. "Tempo and Constraint of Adaptive Evolution in *Escherichia coli* (Enterobacteriaceae, Enterobacteriales)." *Biological Journal of the Linnean Society* 88 (3): 403–11. doi:10.1111/j.1095-8312.2006.00629.x.

Moore, James. 2010. "Darwin's Progress and the Problem of Slavery." *Progress in Human Geography* 34 (5): 555–82. doi:10.1177/0309132510362932.

Morgan, Mary S., and Margaret Morrison, eds. 1999. *Models as Mediators: Perspectives on Natural and Social Science*. Cambridge: Cambridge University Press.

Morgan, Stephen L., and Christopher Winship. 2007. *Counterfactuals and Causal*

Inference: Methods and Principles for Social Research. Cambridge: Cambridge University Press.

Morgan, Thomas Hunt. 1916. *A Critique of the Theory of Evolution*. Princeton, NJ: Princeton University Press.

———. 1923. "The Bearing of Mendelism on the Origin of Species." *Scientific Monthly* 16 (3): 237–47.

Moss, Lenny. 2003. *What Genes Can't Do*. Cambridge, MA: MIT Press.

Müller, Gerd B. 2007. "Evo–devo: Extending the Evolutionary Synthesis." *Nature Reviews Genetics* 8 (12): 943–49. doi:10.1038/nrg2219.

Muller, H. J. 1959. "One Hundred Years without Darwinism Are Enough." *School Science and Mathematics* 59 (4): 304–16. doi:10.1111/j.1949-8594.1959 .tb08235.x.

Müller-Wille, Staffan, and Hans-Jörg Rheinberger. 2012. *A Cultural History of Heredity*. Chicago: University of Chicago Press.

Nagel, Anna C., Paul Joyce, Holly A. Wichman, and Craig R. Miller. 2012. "Stick-breaking: A Novel Fitness Landscape Model That Harbors Epistasis and Is Consistent with Commonly Observed Patterns of Adaptive Evolution." *Genetics* 190 (2): 655–67. doi:10.1534/genetics.111.132134.

Neander, Karen. 1991. "Functions as Selected Effects: The Conceptual Analyst's Defense." *Philosophy of Science* 58 (2): 168–84.

Nei, Masatoshi. 2007. "The New Mutation Theory of Phenotypic Evolution." *Proceedings of the National Academy of Sciences* 104 (30): 12235–42. doi:10.1073/pnas.0703349104.

———. 2013. *Mutation-Driven Evolution*. Oxford: Oxford University Press.

Newton, Isaac. 1950. "A Short Scheme of the True Religion." In *Theological Manuscripts*, edited by H. McLachlan, 48–53. Liverpool: University Press. Originally published in 1710.

Ninio, Jacques. 1991. "Transient Mutators: A Semiquantitative Analysis of the Influence of Translation and Transcription Errors on Mutation Rates." *Genetics* 129 (3): 957–62.

———. 1996. "Gene Conversion as a Focusing Mechanism for Correlated Mutations: A Hypothesis." *Molecular and General Genetics* 251 (5): 503–8. doi:10.1007/BF02173638.

Noguera-Solano, Ricardo. 2013. "The Metaphor of the Architect in Darwin: Chance and Free Will." *Zygon* 48 (4): 859–74. doi:10.1111/zygo.12045.

Noll, Mark A., ed. 1983. *The Princeton Theology 1812–1921: Scripture, Science, and Theological Method from Archibald Alexander to Benjamin Breckinridge Warfield*. Grand Rapids, MI: Baker Books.

Nosil, Patrik, Bernard J. Crespi, and Cristina P. Sandoval. 2002. "Host-Plant Adaptation Drives the Parallel Evolution of Reproductive Isolation." *Nature* 417 (6887): 440–43. doi:10.1038/417440a.

Nyhart, Lynn K. 2009. "Embryology and Morphology." In Ruse and Richards 2009, 194–215.

Ogawa, Tomohisa, and Tsuyoshi Shirai. 2013. "Experimental Molecular Archeology: Reconstruction of Ancestral Mutants and Evolutionary History of Proteins as a New Approach in Protein Engineering." In *Protein Engineering—Technology and Application*, edited by Tomohisa Ogawa. Rijeka, Croatia: InTech.

Okasha, Samir. 2009. "Causation in Biology." In *The Oxford Handbook of Causation*, edited by Helen Beebee, Christopher Hitchcock, and Peter Menzies, 707–25. Oxford: Oxford University Press.

Orr, H. Allen. 1998. "The Population Genetics of Adaptation: The Distribution of Factors Fixed during Adaptive Evolution." *Evolution* 52 (4): 935–49. doi:10.2307/2411226.

———. 2000. "Adaptation and the Cost of Complexity." *Evolution* 54 (1): 13–20. doi:10.1111/j.0014-3820.2000.tb00002.x.

———. 2005a. "The Genetic Theory of Adaptation: A Brief History." *Nature Reviews Genetics* 6 (2): 119–27. doi:10.1038/nrg1523.

———. 2005b. "The Probability of Parallel Evolution." *Evolution* 59 (1): 216–20.

———. 2005c. "Theories of Adaptation: What They Do and Don't Say." *Genetica* 123 (1–2): 3–13. doi:10.1007/s10709-004-2702-3.

———. 2006. "The Distribution of Fitness Effects among Beneficial Mutations in Fisher's Geometric Model of Adaptation." *Journal of Theoretical Biology* 238 (2): 279–85. doi:10.1016/j.jtbi.2005.05.001.

Ospovat, Dov. 1981. *The Development of Darwin's Theory: Natural History, Natural Theology, and Natural Selection, 1838–1859*. Cambridge: Cambridge University Press.

Ostrowski, Elizabeth A., Robert J. Woods, and Richard E. Lenski. 2008. "The Genetic Basis of Parallel and Divergent Phenotypic Responses in Evolving Populations of *Escherichia coli*." *Proceedings of the Royal Society of London B* 275 (1632): 277–84. doi:10.1098/rspb.2007.1244.

Otsuka, Jun. 2014. "Causal Foundations of Evolutionary Genetics." *British Journal for the Philosophy of Science*. doi:10.1093/bjps/axu039.

Otsuka, Jun, Trin Turner, Colin Allen, and Elisabeth A. Lloyd. 2011. "Why the Causal View of Fitness Survives." *Philosophy of Science* 78 (2): 209–24. doi:10.1086/659219.

Paley, William. 1802. *Natural Theology; or, Evidences of the Existence and Attributes of the Deity*. London: R. Faulder.

Papadopoulos, Dimitri, Dominique Schneider, Jessica Meier-Eiss, Werner Arber, Richard E. Lenski, and Michel Blot. 1999. "Genomic Evolution during a 10,000-Generation Experiment with Bacteria." *Proceedings of the National Academy of Sciences* 96 (7): 3807–12. doi:10.1073/pnas.96.7.3807.

Papp, Balázs, Richard A. Notebaart, and Csaba Pál. 2011. "Systems-Biology Approaches for Predicting Genomic Evolution." *Nature Reviews Genetics* 12 (9): 591–602. doi:10.1038/nrg3033.

Parker, Joe, Georgia Tsagkogeorga, James A. Cotton, Yuan Liu, Paolo Provero, Elia Stupka, and Stephen J. Rossiter. 2013. "Genome-Wide Signatures of Convergent Evolution in Echolocating Mammals." *Nature* 502 (7470): 228–31. doi:10.1038/nature12511.

Pauling, Linus, and Emile Zuckerkandl. 1963. "Chemical Paleogenetics: Molecular 'Restoration Studies' of Extinct Forms of Life." *Acta Chemica Scandinavia* 17 (suppl 1): S9–S16.

Pearce, Trevor. 2012. "Convergence and Parallelism in Evolution: A Neo-Gouldian Account." *British Journal for the Philosophy of Science* 63 (2): 429–48. doi:10.1093/bjps/axr046.

Peirce, Charles S. 1893. "Evolutionary Love." *Monist* 3 (2): 176–200.

Pelosi, Ludovic, Lauriane Kühn, Dorian Guetta, Jérôme Garin, Johannes Geiselmann, Richard E. Lenski, and Dominique Schneider. 2006. "Parallel Changes in Global Protein Profiles during Long-Term Experimental Evolution in *Escherichia coli*." *Genetics* 173 (4): 1851–69. doi:10.1534/genetics.105.049619.

Pence, Charles H. 2011. "'Describing Our Whole Experience': The Statistical Philosophies of W. F. R. Weldon and Karl Pearson." *Studies in History and Philosophy of Biological and Biomedical Sciences* 42 (4): 475–85. doi:10.1016/j.shpsc.2011.07.011.

———. 2014. "Chance in Evolutionary Theory: Fitness, Selection, and Genetic Drift in Philosophical and Historical Perspective." PhD diss., University of Notre Dame. doi:10.6084/m9.figshare.988695.

———. 2015. "The Early History of Chance in Evolution." *Studies in History and Philosophy of Science* 50:48–58. doi:10.1016/j.shpsa.2014.09.006.

Pérez-Zaballos, F. J., L. M. Ortega-Mora, G. Álvarez-García, E. Collantes-Fernández, V. Navarro-Lozano, Libertad García-Villada, and Eduardo Costas. 2005. "Adaptation of *Neospora Caninum* Isolates to Cell-Culture Changes: An Argument in Favor of Its Clonal Population Structure." *Journal of Parasitology* 91 (3): 507–10. doi:10.1645/GE-381R1.

Philippe, Nadège, Estelle Crozat, Richard E. Lenski, and Dominique Schneider. 2007. "Evolution of Global Regulatory Networks during a Long-Term Experiment with *Escherichia coli*." *BioEssays* 29 (9): 846–60. doi:10.1002/bies.20629.

Philippe, Nadège, Ludovic Pelosi, Richard E. Lenski, and Dominique Schneider. 2009. "Evolution of Penicillin-Binding Protein 2 Concentration and Cell Shape during a Long-Term Experiment with *Escherichia coli*." *Journal of Bacteriology* 191 (3): 909–21. doi:10.1128/JB.01419-08.

Pieper, Rembert, Quanshun Zhang, David J. Clark, Shih-Ting Huang, Moo-Jin Suh, John C. Braisted, Samuel H. Payne, Robert D. Fleischmann, Scott N. Peterson, and Saul Tzipori. 2011. "Characterizing the *Escherichia coli* O157:H7 Proteome

Including Protein Associations with Higher Order Assemblies." *PLoS ONE* 6 (11): e26554. doi:10.1371/journal.pone.0026554.

Pigliucci, Massimo. 2007. "Do We Need an Extended Evolutionary Synthesis?" *Evolution* 61 (12): 2743–49. doi:10.1111/j.1558-5646.2007.00246.x.

———. 2008a. "The Borderlands between Science and Philosophy: An Introduction." *Quarterly Review of Biology* 83 (1): 7–15. doi:10.1086/529558.

———. 2008b. "Is Evolvability Evolvable?" *Nature Reviews Genetics* 9 (1): 75–82. doi:10.1038/nrg2278.

Pius XII. 1950. "Humani Generis." Encyclical letter. Vatican City. www.vatican.va/holy_father/pius_xii/encyclicals/documents/hf_p-xii_enc_12081950_humani-generis_en.html.

Poincaré, Henri. 1896. *Calcul des probabilités*. 1st ed. Paris: Gauthier-Villars.

Porter, Susannah M. 2004. "Closing the Phosphatization Window: Testing for the Influence of Taphonomic Megabias on the Pattern of Small Shelly Fossil Decline." *PALAIOS* 19 (2): 178–83. doi:10.1669/0883-1351(2004)019<0178:CTPWTF>2.0.CO;2.

Porter, Theodore M. 1986. *The Rise of Statistical Thinking, 1820–1900*. Princeton, NJ: Princeton University Press.

Pos, Klaas Martinus, Peter Dimroth, and Michael Bott. 1998. "The *Escherichia coli* Citrate Carrier CitT: A Member of a Novel Eubacterial Transporter Family Related to the 2-Oxoglutarate/malate Translocator from Spinach Chloroplasts." *Journal of Bacteriology* 180 (16): 4160–65.

Poulin, Robert. 2011. "The Many Roads to Parasitism: A Tale of Convergence." In *Advances in Parasitology*, edited by David Rollinson and Simon Iain Hay, 74: 1–40. London: Academic Press.

Powell, Russell. 2012. "Convergent Evolution and the Limits of Natural Selection." *European Journal for Philosophy of Science* 2 (3): 355–73. doi:10.1007/s13194-012-0047-9.

Price, Peter W. 2003. *Macroevolutionary Theory on Macroecological Patterns*. Cambridge: Cambridge University Press.

Price, Peter W., and Timothy G. Carr. 2000. "Comparative Ecology of Membracids and Tenthredinids in a Macroevolutionary Context." *Evolutionary Ecology Research* 2 (5): 645–65.

Prichard, James Cowles. 1826. *Researches into the Physical History of Man*. 2nd ed. 2 vols. London: John and Arthur Arch.

Prigogine, Ilya, and Isabelle Stengers. 1984. *Order out of Chaos: Man's New Dialogue with Nature*. New York: Bantam.

Provine, William B. 1971. *The Origins of Theoretical Population Genetics*. Princeton, NJ: Princeton University Press.

———. 1986. *Sewall Wright and Evolutionary Biology*. 1st ed. Chicago: University of Chicago Press.

————. 1989. *Sewall Wright and Evolutionary Biology*. 2nd ed. Chicago: University of Chicago Press.

Punnett, Reginald Crundall. 1915. *Mimicry in Butterflies*. Cambridge: Cambridge University Press.

Quandt, Erik M., Daniel E. Deatherage, Andrew D. Ellington, George Georgiou, and Jeffrey E. Barrick. 2014. "Recursive Genomewide Recombination and Sequencing Reveals a Key Refinement Step in the Evolution of a Metabolic Innovation in *Escherichia coli*." *Proceedings of the National Academy of Sciences* 111 (6): 2217–22. doi:10.1073/pnas.1314561111.

Quandt, Erik M., Jimmy Gollihar, Zachary D. Blount, Andrew D. Ellington, George Georgiou, and Jeffrey E. Barrick. Forthcoming. "Fine-Tuning Citrate Synthase Flux Potentiates and Refines Metabolic Innovation in the Lenski Evolution Experiment." *eLife*. doi:10.7554/eLife.09696.

Rainey, Paul B. 2009. "Evolutionary Biology: Arrhythmia of Tempo and Mode." *Nature* 461 (7268): 1219–21. doi:10.1038/4611219a.

Ralph, Peter, and Graham Coop. 2010. "Parallel Adaptation: One or Many Waves of Advance of an Advantageous Allele?" *Genetics* 186 (2): 647–68. doi:10.1534/genetics.110.119594.

Ramsey, Grant. 2013. "Can Fitness Differences Be a Cause of Evolution?" *Philosophy and Theory in Biology* 5:e401. doi:10.3998/ptb.6959004.0005.001.

Ramu, Avinash, Michiel J. Noordam, Rachel S. Schwartz, Arthur Wuster, Matthew E. Hurles, Reed A. Cartwright, and Donald F. Conrad. 2013. "DeNovoGear: De Novo Indel and Point Mutation Discovery and Phasing." *Nature Methods* 10 (10): 985–87. doi:10.1038/nmeth.2611.

Raup, David M. 1991. *Extinction: Bad Genes or Bad Luck?* New York: Norton.

Raup, David M., and Stephen Jay Gould. 1974. "Stochastic Simulation and Evolution of Morphology—Towards a Nomothetic Paleontology." *Systematic Biology* 23 (3): 305–22. doi:10.1093/sysbio/23.3.305.

Raup, David M., Stephen Jay Gould, Thomas J. M. Schopf, and Daniel S. Simberloff. 1973. "Stochastic Models of Phylogeny and the Evolution of Diversity." *Journal of Geology* 81 (5): 525–42.

Raymond, Michel, Christine Chevillon, Thomas Guillemaud, Thomas Lenormand, and Nicole Pasteur. 1998. "An Overview of the Evolution of Overproduced Esterases in the Mosquito *Culex pipiens*." *Philosophical Transactions of the Royal Society of London B* 353 (1376): 1707–11.

Reichenbach, Hans. 1948. *The Theory of Probability, an Inquiry into the Logical and Mathematical Foundations of the Calculus of Probability*. Berkeley: University of California Press.

————. 2008. *The Concept of Probability in the Mathematical Representation of Reality*. Edited by Frederick Eberhardt and Clark Glymour. Chicago: Open Court.

Re Manning, Russell, John Hedley Brooke, and Fraser Watts, eds. 2013. *The Oxford Handbook of Natural Theology*. Oxford: Oxford University Press.

Richards, Richard A. 2010. *The Species Problem: A Philosophical Analysis*. Cambridge: Cambridge University Press.

Richards, Robert J. 1992. *The Meaning of Evolution: The Morphological Construction and Ideological Reconstruction of Darwin's Theory*. Chicago: University of Chicago Press.

———. 2009. "The Descent of Man [review of *Darwin's Sacred Cause: How a Hatred of Slavery Shaped Darwin's Views on Human Evolution*]." *American Scientist*, September.

Ricoeur, Paul. 1985. *Time and Narrative*. Translated by Kathleen McLaughlin and David Pellauer. 3 vols. Chicago: University of Chicago Press.

Ridley, Mark. 2004. *Evolution*. 3rd ed. Malden, MA: Blackwell Science.

Riley, Merry S., Vaughn S. Cooper, Richard E. Lenski, Larry J. Forney, and Terence L. Marsh. 2001. "Rapid Phenotypic Change and Diversification of a Soil Bacterium during 1000 Generations of Experimental Evolution." *Microbiology* 147 (4): 995–1006.

Risso, Valeria A., Jose A. Gavira, Diego F. Mejia-Carmona, Eric A. Gaucher, and Jose M. Sanchez-Ruiz. 2013. "Hyperstability and Substrate Promiscuity in Laboratory Resurrections of Precambrian β-Lactamases." *Journal of the American Chemical Society* 135 (8): 2899–2902. doi:10.1021/ja311630a.

Roelants, Kim, Bryan G. Fry, Janette A. Norman, Elke Clynen, Liliane Schoofs, and Franky Bossuyt. 2010. "Identical Skin Toxins by Convergent Molecular Adaptation in Frogs." *Current Biology* 20 (2): 125–30. doi:10.1016/j.cub.2009.11.015.

Romero, Philip A., and Frances H. Arnold. 2009. "Exploring Protein Fitness Landscapes by Directed Evolution." *Nature Reviews Molecular Cell Biology* 10 (12): 866–76. doi:10.1038/nrm2805.

Rosche, William A., and Patricia L. Foster. 2000. "Determining Mutation Rates in Bacterial Populations." *Methods* 20 (1): 4–17. doi:10.1006/meth.1999.0901.

Rosenberg, Alex. 1994. *Instrumental Biology, or the Disunity of Science*. Chicago: University of Chicago Press.

Rosenberg, Susan M. 2010. "Spontaneous Mutation: Real-Time in Living Cells." *Current Biology* 20 (18): R810–R811. doi:10.1016/j.cub.2010.07.031.

Rosenblum, Erica Bree, Holger Römpler, Torsten Schöneberg, and Hopi E. Hoekstra. 2010. "Molecular and Functional Basis of Phenotypic Convergence in White Lizards at White Sands." *Proceedings of the National Academy of Sciences* 107 (5): 2113–17. doi:10.1073/pnas.0911042107.

Rosenthal, Jacob. 2010. "The Natural-Range Conception of Probability." In *Time, Chance, and Reduction: Philosophical Aspects of Statistical Mechanics*, edited by Gerhard Ernst and Andreas Hüttemann, 71–90. Cambridge: Cambridge University Press.

Ross-Ibarra, Jeffrey, Peter L. Morrell, and Brandon S. Gaut. 2007. "Plant Domestication, a Unique Opportunity to Identify the Genetic Basis of Adaptation."

Proceedings of the National Academy of Sciences 104 (suppl. 1): 8641–48. doi:10.1073/pnas.0700643104.

Rozen, Daniel E., and Richard E. Lenski. 2000. "Long-Term Experimental Evolution in *Escherichia coli*. VIII. Dynamics of a Balanced Polymorphism." *American Naturalist* 155 (1): 24–35. doi:10.1086/an.2000.155.issue-1.

Rozen, Daniel E., Nadège Philippe, J. Arjan G. M. de Visser, Richard E. Lenski, and Dominique Schneider. 2009. "Death and Cannibalism in a Seasonal Environment Facilitate Bacterial Coexistence." *Ecology Letters* 12 (1): 34–44. doi:10.1111/j.1461-0248.2008.01257.x.

Rozen, Daniel E., Dominique Schneider, and Richard E. Lenski. 2005. "Long-Term Experimental Evolution in *Escherichia coli*. XIII. Phylogenetic History of a Balanced Polymorphism." *Journal of Molecular Evolution* 61 (2): 171–80. doi:10.1007/s00239-004-0322-2.

Rundle, Howard D., Laura Nagel, Janette Wenrick Boughman, and Dolph Schluter. 2000. "Natural Selection and Parallel Speciation in Sympatric Sticklebacks." *Science* 287 (5451): 306–8. doi:10.1126/science.287.5451.306.

Runnegar, Bruce. 1982. "A Molecular-Clock Date for the Origin of the Animal Phyla." *Lethaia* 15 (3): 199–205. doi:10.1111/j.1502-3931.1982.tb00645.x.

Ruse, Michael. 1996. *Monad to Man: The Concept of Progress in Evolutionary Biology*. Cambridge, MA: Harvard University Press.

———. 1999. *The Darwinian Revolution: Science Red in Tooth and Claw*. 2nd ed. Chicago: University of Chicago Press.

———. 2003. *Darwin and Design: Does Evolution Have a Purpose?* Cambridge, MA: Harvard University Press.

———. 2005. *The Evolution-Creation Struggle*. Cambridge, MA: Harvard University Press.

———, ed. 2008. *The Oxford Handbook of Philosophy of Biology*. Oxford: Oxford University Press.

———. 2012. *The Philosophy of Human Evolution*. Cambridge: Cambridge University Press.

———. 2013. *The Gaia Hypothesis: Science on a Pagan Planet*. Chicago: University of Chicago Press.

Ruse, Michael, and Robert J. Richards, eds. 2009. *The Cambridge Companion to the "Origin of Species."* Cambridge: Cambridge University Press.

Salazar-Ciudad, Isaac, Jukka Jernvall, and Stuart A. Newman. 2003. "Mechanisms of Pattern Formation in Development and Evolution." *Development* 130 (10): 2027–37.

Salazar-Ciudad, Isaac, and Miquel Marín-Riera. 2013. "Adaptive Dynamics under Development-Based Genotype-Phenotype Maps." *Nature* 497 (7449): 361–64. doi:10.1038/nature12142.

Salmon, Wesley C. 1970. "Statistical Explanation." In *The Nature and Function of Scientific Theories: Essays in Contemporary Science and Philosophy*, edited by

Robert Garland Colodny and Grover Maxwell, 173–231. Pittsburgh, PA: University of Pittsburgh Press.

———. 1984. *Scientific Explanation and the Causal Structure of the World*. Princeton, NJ: Princeton University Press.

———. 1997. "Causality and Explanation: A Reply to Two Critiques." *Philosophy of Science* 64 (3): 461–77.

Saxer, Gerda, Michael Doebeli, and Michael Travisano. 2010. "The Repeatability of Adaptive Radiation during Long-Term Experimental Evolution of *Escherichia coli* in a Multiple Nutrient Environment." *PLoS ONE* 5 (12): e14184. doi:10.1371/journal.pone.0014184.

Schank, Jeffrey C., and William C. Wimsatt. 1986. "Generative Entrenchment and Evolution." *PSA: Proceedings of the Biennial Meeting of the Philosophy of Science Association* 1986:33–60.

Scheutz, Flemming, and Nancy A. Strockbine. 2005. "Genus I. *Escherichia.* Castellani and Chalmers 1919, 941T^AL." In *Bergey's Manual of Systematic Bacteriology*, vol. 2, *The Proteobacteria, Part B: The Gammaproteobacteria*, edited by George Garrity, Don J. Brenner, Noel R. Krieg, and James R. Staley, 607–24. 2nd ed. New York: Springer.

Schluter, Dolph. 1996. "Adaptive Radiation along Genetic Lines of Least Resistance." *Evolution* 50 (5): 1766–74. doi:10.2307/2410734.

———. 2000. *The Ecology of Adaptive Radiation*. Oxford: Oxford University Press.

Schluter, Dolph, Elizabeth A. Clifford, Maria Nemethy, and Jeffrey S. McKinnon. 2004. "Parallel Evolution and Inheritance of Quantitative Traits." *American Naturalist* 163 (6): 809–22. doi:10.1086/an.2004.163.issue-6.

Schönborn, Christoph. 2005. "Finding Design in Nature." *New York Times*, July 7. www.nytimes.com/2005/07/07/opinion/07schonborn.html.

———. 2007. *Chance or Purpose? Creation, Evolution, and a Rational Faith*. Edited by Hubert Weber. Translated by Henry Taylor. San Francisco: Ignatius Press.

Schrödinger, Erwin. 1944. *What Is Life? The Physical Aspect of the Living Cell*. Cambridge: Cambridge University Press.

Schwartz, Joel S. 1984. "Darwin, Wallace, and the *Descent of Man*." *Journal of the History of Biology* 17 (2): 271–89.

Sears, Stephen W. 1983. *Landscape Turned Red: The Battle of Antietam*. New York: Houghton Mifflin.

Sedley, David N. 2007. *Creationism and Its Critics in Antiquity*. Berkeley: University of California Press.

Seilacher, Adolf. 1984. "Late Precambrian and Early Cambrian Metazoa: Preservational or Real Extinctions?" In *Patterns of Change in Earth Evolution*, edited by H. D. Holland and A. F. Trendall, 159–68. Berlin: Springer.

Sepkoski, David. 2012. *Rereading the Fossil Record: The Growth of Paleobiology as an Evolutionary Discipline*. Chicago: University of Chicago Press.

Sepkoski, David, and Michael Ruse. 2009. *The Paleobiological Revolution: Essays on the Growth of Modern Paleontology*. Chicago: University of Chicago Press.

Shimizu, Yoshihiro, Akio Inoue, Yukihide Tomari, Tsutomu Suzuki, Takashi Yokogawa, Kazuya Nishikawa, and Takuya Ueda. 2001. "Cell-Free Translation Reconstituted with Purified Components." *Nature Biotechnology* 19 (8): 751–55. doi:10.1038/90802.

Shu, Degan, Simon Conway Morris, and Xingliang Zhang. 1996. "A *Pikaia*-like Chordate from the Lower Cambrian of China." *Nature* 384 (6605): 157–58. doi:10.1038/384157a0.

Shu, Degan, Simon Conway Morris, Zhifei Zhang, and Jian Han. 2010. "The Earliest History of the Deuterostomes: The Importance of the Chengjiang Fossil-Lagerstätte." *Proceedings of the Royal Society of London B* 277 (1679): 165–74. doi:10.1098/rspb.2009.0646.

Shull, A. Franklin. 1935. "Weismann and Haeckel: One Hundred Years." *Science* 81 (2106): 443–52. doi:10.1126/science.81.2106.443.

Simpson, George Gaylord. 1944. *Tempo and Mode in Evolution*. New York: Columbia University Press.

———. 1952. "How Many Species?" *Evolution* 6 (3): 342. doi:10.2307/2405419.

———. 1961. "One Hundred Years without Darwin Are Enough." *Teachers College Record* 62 (8): 617–26.

Sloan, Phillip R. 1987. "From Logical Universals to Historical Individuals: Buffon's Idea of Biological Species." In *Histoire du concept d'espèce dans les sciences de la vie*, 101–40. Paris: Editions de la Fondation Singer-Polignac.

———. 1992. "Organic Molecules Revisited." In *Buffon 88: Actes du Colloque international pour le bicentenaire de la mort de Buffon*, edited by Jean Gayon, 415–38. Paris: J. Vrin.

———. 2013. "The Species Problem and History." *Studies in History and Philosophy of Biological and Biomedical Sciences* 44 (2): 237–41. doi:10.1016/j.shpsc.2013.01.001.

Sloan, Phillip R., and Brandon Fogel. 2011. *Creating a Physical Biology: The Three-Man Paper and Early Molecular Biology*. Chicago: University of Chicago Press.

Slowinski, Joseph B., and Craig Guyer. 1989. "Testing the Stochasticity of Patterns of Organismal Diversity: An Improved Null Model." *American Naturalist* 134 (6): 907–21.

Smith, Justin E. H., ed. 2006. *The Problem of Animal Generation in Early Modern Philosophy*. Cambridge: Cambridge University Press.

Smith, Stacey D., and Mark D. Rausher. 2011. "Gene Loss and Parallel Evolution Contribute to Species Difference in Flower Color." *Molecular Biology and Evolution* 28 (10): 2799–2810. doi:10.1093/molbev/msr109.

Smocovitis, Vassiliki Betty. 1994a. "Disciplining Evolutionary Biology: Ernst Mayr and the Founding of the Society for the Study of Evolution and *Evolution* (1939–1950)." *Evolution* 48 (1): 1–8. doi:10.2307/2409996.

————. 1994b. "Organizing Evolution: Founding the Society for the Study of Evolution (1939–1950)." *Journal of the History of Biology* 27 (2): 241–309. doi:10.1007/BF01062564.

————. 1996. *Unifying Biology: The Evolutionary Synthesis and Evolutionary Biology*. Princeton, NJ: Princeton University Press.

Sniegowski, Paul D., Philip J. Gerrish, Toby Johnson, and Aaron Shaver. 2000. "The Evolution of Mutation Rates: Separating Causes from Consequences." *BioEssays* 22 (12): 1057–66. doi:10.1002/1521-1878(200012)22:12<1057 ::AID-BIES3>3.0.CO;2-W.

Sober, Elliott. 1984. *The Nature of Selection*. Cambridge, MA: MIT Press.

————. 2004. "The Design Argument." In *The Blackwell Guide to the Philosophy of Religion*, edited by William E. Mann, 117–47. Malden, MA: Blackwell.

————. 2009. "Metaphysical and Epistemological Issues in Modern Darwinian Theory." In *The Cambridge Companion to Darwin*, edited by M. J. S. Hodge and Gregory Radick, 302–22. 2nd ed. Cambridge: Cambridge University Press.

————. 2012. "Coincidences and How to Reason about Them." In *EPSA Philosophy of Science: Amsterdam 2009*, edited by Henk W. de Regt, Stephan Hartmann, and Samir Okasha, 355–74. The European Philosophy of Science Association Proceedings 1. Amsterdam: Springer Netherlands.

————. 2013. "Trait Fitness Is Not a Propensity, but Fitness Variation Is." *Studies in History and Philosophy of Biological and Biomedical Sciences*. doi:10.1016/ j.shpsc.2013.03.002.

Spencer, Herbert. 1857. "Progress: Its Law and Cause." *Westminster Review* 67 (2): 445–85.

Stackhouse, Joseph, Scott R. Presnell, Gerard M. McGeehan, Krishnan P. Nambiar, and Steven A. Benner. 1990. "The Ribonuclease from an Extinct Bovid Ruminant." *FEBS Letters* 262 (1): 104–6. doi:10.1016/0014-5793(90)80164-E.

Stanek, Mark T., Tim F. Cooper, and Richard E. Lenski. 2009. "Identification and Dynamics of a Beneficial Mutation in a Long-Term Evolution Experiment with *Escherichia coli*." *BMC Evolutionary Biology* 9 (1): 302. doi:10.1186/ 1471-2148-9-302.

Stanley, Steven M., Philip W. Signor III, Scott Lidgard, and Alan F. Karr. 1981. "Natural Clades Differ from 'Random' Clades: Simulations and Analyses." *Paleobiology* 7 (1): 115–27.

Stebbins, G. Ledyard. 1950. *Variation and Evolution in Plants*. New York: Columbia University Press.

Sterelny, Kim. 2005. "Another View of Life." *Studies in History and Philosophy of Biological and Biomedical Sciences* 36 (3): 585–93. doi:10.1016/j.shpsc.2005 .07.008.

Sterelny, Kim, and Paul E. Griffiths. 1999. *Sex and Death: An Introduction to Philosophy of Biology*. Chicago: University of Chicago Press.

Stern, David L. 2000. "Perspective: Evolutionary Developmental Biology and the

Problem of Variation." *Evolution* 54 (4): 1079–91. doi:10.1111/j.0014-3820 .2000.tb00544.x.

———. 2013. "The Genetic Causes of Convergent Evolution." *Nature Reviews Genetics* 14 (11): 751–64. doi:10.1038/nrg3483.

Sternberg, Meir. 1985. *The Poetics of Biblical Narrative: Ideological Literature and the Drama of Reading*. Bloomington: Indiana University Press.

Stewart, Frank M., David M. Gordon, and Bruce R. Levin. 1990. "Fluctuation Analysis: The Probability Distribution of the Number of Mutants under Different Conditions." *Genetics* 124 (1): 175–85.

Stoltzfus, Arlin. 2006. "Mutationism and the Dual Causation of Evolutionary Change." *Evolution and Development* 8 (3): 304–17. doi:10.1111/j.1525 -142X.2006.00101.x.

Stoltzfus, Arlin, and Lev Y. Yampolsky. 2009. "Climbing Mount Probable: Mutation as a Cause of Nonrandomness in Evolution." *Journal of Heredity* 100 (5): 637– 47. doi:10.1093/jhered/esp048.

Strecker, Ulrike, Bernhard Hausdorf, and Horst Wilkens. 2012. "Parallel Speciation in *Astyanax* Cave Fish (Teleostei) in Northern Mexico." *Molecular Phylogenetics and Evolution* 62 (1): 62–70. doi:10.1016/j.ympev.2011.09.005.

Streisfeld, Matthew A., and Mark D. Rausher. 2011. "Population Genetics, Pleiotropy, and the Preferential Fixation of Mutations during Adaptive Evolution." *Evolution* 65 (3): 629–42. doi:10.1111/j.1558-5646.2010.01165.x.

Strevens, Michael. 2003. *Bigger than Chaos: Understanding Complexity through Probability*. Cambridge, MA: Harvard University Press.

———. 2008. *Depth: An Account of Scientific Explanation*. Cambridge, MA: Harvard University Press.

———. 2011. "Probability out of Determinism." In *Probabilities in Physics*, edited by Claus Beisbart and Stephan Hartmann, 339–64. Oxford: Oxford University Press.

———. 2013. *Tychomancy: Inferring Probability from Causal Structure*. Cambridge, MA: Harvard University Press.

———. Forthcoming. "Special Science Autonomy and the Division of Labor." In *The Philosophy of Philip Kitcher*, edited by Mark Couch and Jessica Pfeifer. Oxford: Oxford University Press.

Studier, F. William, Patrick Daegelen, Richard E. Lenski, Sergei Maslov, and Jihyun F. Kim. 2009. "Understanding the Differences between Genome Sequences of *Escherichia coli* B Strains REL606 and BL21(DE3) and Comparison of the *E. Coli* B and K-12 Genomes." *Journal of Molecular Biology* 394 (4): 653–80. doi:10.1016/j.jmb.2009.09.021.

Takahata, Naoyuki. 2007. "Molecular Clock: An *anti*-Neo-Darwinian Legacy." *Genetics* 176 (1): 1–6. doi:10.1534/genetics.104.75135.

Taleb, Nassim Nicholas. 2007. *The Black Swan: The Impact of the Highly Improbable*. New York: Random House.

Tenaillon, Olivier, Alejandra Rodríguez-Verdugo, Rebecca L. Gaut, Pamela

McDonald, Albert F. Bennett, Anthony D. Long, and Brandon S. Gaut. 2012. "The Molecular Diversity of Adaptive Convergence." *Science* 335 (6067): 457–61. doi:10.1126/science.1212986.

Teotónio, Henrique, and Michael R. Rose. 2001. "Perspective: Reverse Evolution." *Evolution* 55 (4): 653–60. doi:10.1111/j.0014-3820.2001.tb00800.x.

Travisano, Michael, and Richard E. Lenski. 1996. "Long-Term Experimental Evolution in *Escherichia coli*. IV. Targets of Selection and the Specificity of Adaptation." *Genetics* 143 (1): 15–26.

Travisano, Michael, Judith A. Mongold, Albert F. Bennett, and Richard E. Lenski. 1995. "Experimental Tests of the Roles of Adaptation, Chance, and History in Evolution." *Science* 267 (5194): 87–90. doi:10.1126/science.7809610.

Travisano, Michael, Farida Vasi, and Richard E. Lenski. 1995. "Long-Term Experimental Evolution in *Escherichia coli*. III. Variation among Replicate Populations in Correlated Responses to Novel Environments." *Evolution* 49 (1): 189–200. doi:10.2307/2410304.

Treves, David S., Shannon Manning, and Julian Adams. 1998. "Repeated Evolution of an Acetate-Crossfeeding Polymorphism in Long-Term Populations of *Escherichia coli*." *Molecular Biology and Evolution* 15 (7): 789–97.

Treviranus, Gottfried Reinhold. 1802. *Biologie; oder die Philosophie der lebenden Natur für Naturforscher und Aerzte.* 6 vols. Göttingen: Johann Friedrich Röwer.

Turner, Caroline B., Zachary D. Blount, Daniel H. Mitchell, and Richard E. Lenski. Forthcoming. "Evolution and Coexistence in Response to a Key Innovation in a Long-Term Evolution Experiment with *Escherichia coli*." *Evolution*. Preprint available with doi:10.1101/020958.

Turner, Derek D. 2011. "Gould's Replay Revisited." *Biology & Philosophy* 26 (1): 65–79. doi:10.1007/s10539-010-9228-0.

Turner, J. R. G. 1987. "Random Genetic Drift, R. A. Fisher, and the Oxford School of Ecological Genetics." In Krüger, Gigerenzer, and Morgan, 1987, 313–54.

Tweedt, Sarah M., and Douglas H. Erwin. 2015. "Origin of Metazoan Developmental Toolkits and Their Expression in the Fossil Record." In *Evolutionary Transitions to Multicellular Life: Principles and Mechanisms*, edited by Iñaki Ruiz-Trillo and Aurora M. Nedelcu, 47–78. Advances in Marine Genetics 2. Dordrecht: Springer.

Uyeda, Josef C., Thomas F. Hansen, Stevan J. Arnold, and Jason Pienaar. 2011. "The Million-Year Wait for Macroevolutionary Bursts." *Proceedings of the National Academy of Sciences* 108 (38): 15908–13. doi:10.1073/pnas.1014503108.

Valentine, James W. 1980. "Determinants of Diversity in Higher Taxonomic Categories." *Paleobiology* 6 (4): 444–50.

Valentine, James W., and Douglas H. Erwin. 1987. "Interpreting Great Developmental Experiments: The Fossil Record." In *Development as an Evolutionary Process*, edited by Rudolf A. Raff and Elizabeth C. Raff, 71–107. New York: Alan R. Liss.

Van Roy, Peter, and Derek E. G. Briggs. 2011. "A Giant Ordovician Anomalocaridid." *Nature* 473 (7348): 510–13. doi:10.1038/nature09920.

Vasi, Farida, Michael Travisano, and Richard E. Lenski. 1994. "Long-Term Experimental Evolution in *Escherichia coli*. II. Changes in Life-History Traits during Adaptation to a Seasonal Environment." *American Naturalist* 144 (3): 432–56.

Vavilov, Nikolai I. 1922. "The Law of Homologous Series in Variation." *Journal of Genetics* 12 (1): 47–89. doi:10.1007/BF02983073.

Vermeij, Geerat J. 2006. "Historical Contingency and the Purported Uniqueness of Evolutionary Innovations." *Proceedings of the National Academy of Sciences* 103 (6): 1804–9. doi:10.1073/pnas.0508724103.

Voltaire. 1772. *Mélanges de philosophie*. Vol. 5. Collection Complete Des Oeuvres de Mr. de Voltaire 26. London.

von Kries, Johannes. 1886. *Die Principien von Warscheinlichkeitsrechnung: Eine logische Untersuchung*. Freiburg: J. C. B. Mohr.

von Plato, Jan. 1983. "The Method of Arbitrary Functions." *British Journal for the Philosophy of Science* 34 (1): 37–47.

Wagner, Günter P., and Jianzhi Zhang. 2011. "The Pleiotropic Structure of the Genotype–Phenotype Map: The Evolvability of Complex Organisms." *Nature Reviews Genetics* 12 (3): 204–13. doi:10.1038/nrg2949.

Wake, David B. 1991. "Homoplasy: The Result of Natural Selection, or Evidence of Design Limitations?" *American Naturalist* 138 (3): 543–67.

Wake, David B., Marvalee H. Wake, and Chelsea D. Specht. 2011. "Homoplasy: From Detecting Pattern to Determining Process and Mechanism of Evolution." *Science* 331 (6020): 1032–35. doi:10.1126/science.1188545.

Walsh, Denis M. 2007. "The Pomp of Superfluous Causes: The Interpretation of Evolutionary Theory." *Philosophy of Science* 74 (3): 281–303. doi:10.1086/520777.

Walsh, Denis M., Tim Lewens, and André Ariew. 2002. "The Trials of Life: Natural Selection and Random Drift." *Philosophy of Science* 69 (3): 429–46. doi:10.1086/342454.

Wang, Jicheng, Kelly D. Gonzalez, William A. Scaringe, Kimberly Tsai, Ning Liu, Dongqing Gu, Wenyan Li, Kathleen A. Hill, and Steve S. Sommer. 2007. "Evidence for Mutation Showers." *Proceedings of the National Academy of Sciences* 104 (20): 8403–8. doi:10.1073/pnas.0610902104.

Warfield, B. B. 2000. "Creation, Evolution and Mediate Creation." In *Evolution, Science, and Scripture: Selected Writings*, edited by Mark A. Noll and David N. Livingstone, 197–210. Grand Rapids, MI: Baker Books.

Waxman, D., and J. J. Welch. 2005. "Fisher's Microscope and Haldane's Ellipse." *American Naturalist* 166 (4): 447–57. doi:10.1086/444404.

Weill, Mylène, Georges Lutfalla, Knud Mogensen, Fabrice Chandre, Arnaud Berthomieu, Claire Berticat, Nicole Pasteur, Alexandre Philips, Philippe Fort, and

Michel Raymond. 2003. "Comparative Genomics: Insecticide Resistance in Mosquito Vectors." *Nature* 423 (6936): 136–37. doi:10.1038/423136b.

Weinberg, Steven. 1992. *Dreams of a Final Theory: The Search for the Fundamental Laws of Nature.* New York: Pantheon.

Weinreich, Daniel M., Nigel F. Delaney, Mark A. DePristo, and Daniel L. Hartl. 2006. "Darwinian Evolution Can Follow Only Very Few Mutational Paths to Fitter Proteins." *Science* 312 (5770): 111–14. doi:10.1126/science.1123539.

Weinreich, Daniel M., Richard A. Watson, and Lin Chao. 2005. "Perspective: Sign Epistasis and Genetic Constraint on Evolutionary Trajectories." *Evolution* 59 (6): 1165–74.

Wellner, Alon, Maria Raitses Gurevich, and Dan S. Tawfik. 2013. "Mechanisms of Protein Sequence Divergence and Incompatibility." *PLoS Genetics* 9 (7): e1003665. doi:10.1371/journal.pgen.1003665.

West, Geoffrey B., James H. Brown, and Brian J. Enquist. 1997. "A General Model for the Origin of Allometric Scaling Laws in Biology." *Science* 276 (5309): 122–26. doi:10.1126/science.276.5309.122.

West-Eberhard, Mary Jane. 2003. *Developmental Plasticity and Evolution.* Oxford: Oxford University Press.

Wichman, Holly A., Marty R. Badgett, L. A. Scott, Carla M. Boulianne, and James J. Bull. 1999. "Different Trajectories of Parallel Evolution during Viral Adaptation." *Science* 285 (5426): 422–24. doi:10.1126/science.285.5426.422.

Wichman, Holly A., and Celeste J. Brown. 2010. "Experimental Evolution of Viruses: Microviridae as a Model System." *Philosophical Transactions of the Royal Society of London B* 365 (1552): 2495–2501. doi:10.1098/rstb.2010.0053.

Wilkins, John S. 2009. *Species: A History of the Idea.* Berkeley: University of California Press.

Wills, Matthew A. 1998. "Cambrian and Recent Disparity: The Picture from Priapulids." *Paleobiology* 24 (2): 177–99.

Wills, Matthew A., Sylvain Gerber, Marcello Ruta, and Martin Hughes. 2012. "The Disparity of Priapulid, Archaeopriapulid and Palaeoscolecid Worms in the Light of New Data." *Journal of Evolutionary Biology* 25 (10): 2056–76. doi:10.1111/j.1420-9101.2012.02586.x.

Wilson, Jessica. 2007. "Newtonian Forces." *British Journal for the Philosophy of Science* 58:173–205. doi:10.1093/bjps/axm004.

Wimsatt, William C. 1972. "Complexity and Organization." *PSA: Proceedings of the Biennial Meeting of the Philosophy of Science Association* 1972:67–86.

———. 1986. "Developmental Constraints, Generative Entrenchment, and the Innate-Acquired Distinction." In *Integrating Scientific Disciplines*, edited by William Bechtel, 185–208. Amsterdam: Springer.

———. 2001. "Generative Entrenchment and the Developmental Systems Approach to Evolutionary Processes." In *Cycles of Contingency: Developmental Systems and Evolution*, edited by Susan Oyama, Paul E. Griffiths, and Russell D. Gray, 219–38. Cambridge, MA: MIT Press.

Winsor, Mary P. 2006. "The Creation of the Essentialism Story: An Exercise in Metahistory." *History and Philosophy of the Life Sciences* 28 (2): 149–74.

Wiser, Michael J., Noah Ribeck, and Richard E. Lenski. 2013. "Long-Term Dynamics of Adaptation in Asexual Populations." *Science* 342 (6164): 1364–67. doi:10.1126/science.1243357.

Witteveen, Joeri. 2013. "Rethinking 'Typological' vs. 'Population' Thinking: A Historical and Philosophical Reassessment of a Troubled Dichotomy." PhD diss., Cambridge: Trinity College, University of Cambridge.

Wood, Troy E., John M. Burke, and Loren H. Rieseberg. 2005. "Parallel Genotypic Adaptation: When Evolution Repeats Itself." *Genetica* 123 (1–2): 157–70. doi:10.1007/s10709-003-2738-9.

Woods, Robert J., Jeffrey E. Barrick, Tim F. Cooper, Utpala Shrestha, Mark R. Kauth, and Richard E. Lenski. 2011. "Second-Order Selection for Evolvability in a Large *Escherichia coli* Population." *Science* 331 (6023): 1433–36. doi:10.1126/science.1198914.

Woods, Robert J., Dominique Schneider, Cynthia L. Winkworth, Margaret A. Riley, and Richard E. Lenski. 2006. "Tests of Parallel Molecular Evolution in a Long-Term Experiment with *Escherichia coli*." *Proceedings of the National Academy of Sciences* 103 (24): 9107–12. doi:10.1073/pnas.0602917103.

Woodward, James. 2003. *Making Things Happen*. Oxford: Oxford University Press.

Wray, Gregory A., Jeffrey S. Levinton, and Leo H. Shapiro. 1996. "Molecular Evidence for Deep Precambrian Divergences among Metazoan Phyla." *Science* 274 (5287): 568–73. doi:10.1126/science.274.5287.568.

Wright, Larry. 1973. "Functions." *Philosophical Review* 82 (2): 139–68. doi:10.2307/2183766.

Wright, Sewall. 1930. "*The Genetical Theory of Natural Selection*: A Review." *Journal of Heredity* 21 (8): 349–56.

———. 1931. "Evolution in Mendelian Populations." *Genetics* 16 (2): 97–159.

———. 1932. "The Roles of Mutation, Inbreeding, Crossbreeding and Selection in Evolution." *Proceedings of the Sixth International Congress of Genetics* 1:356–66.

Xiao, Shuhai, and Marc Laflamme. 2009. "On the Eve of Animal Radiation: Phylogeny, Ecology and Evolution of the Ediacara Biota." *Trends in Ecology & Evolution* 24 (1): 31–40. doi:10.1016/j.tree.2008.07.015.

Yampolsky, Lev Y., and Arlin Stoltzfus. 2001. "Bias in the Introduction of Variation as an Orienting Factor in Evolution." *Evolution and Development* 3 (2): 73–83. doi:10.1046/j.1525-142x.2001.003002073.x.

Zahm, John. 1896. *Evolution and Dogma*. Chicago: D. H. McBride.

Zammito, John H. 2007. "Stealing Herder's Thunder: Kant's Debunking of Herder on History in 'Conjectural Beginning of the Human Race.'" In *Immanuel Kant:*

German Professor and World-Philosopher, edited by Günther Lottes and Uwe Steiner, 43–72. Hannover: Wehrhahn.

Zhou, Ying, Haruichi Asahara, Eric A. Gaucher, and Shaorong Chong. 2012. "Reconstitution of Translation from *Thermus thermophilus* Reveals a Minimal Set of Components Sufficient for Protein Synthesis at High Temperatures and Functional Conservation of Modern and Ancient Translation Components." *Nucleic Acids Research* 40 (16): 7932–45. doi:10.1093/nar/gks568.

Zuckerkandl, Emile. 1994. "Molecular Pathways to Parallel Evolution: I. Gene Nexuses and Their Morphological Correlates." *Journal of Molecular Evolution* 39 (6): 661–78. doi:10.1007/BF00160412.

Zuckerkandl, Emile, and Linus Pauling. 1965. "Molecules as Documents of Evolutionary History." *Journal of Theoretical Biology* 8 (2): 357–66. doi:10.1016/0022-5193(65)90083-4.

J. Matthew Ashley
 Department of Theology
 University of Notre Dame
 Notre Dame, IN 46556
 USA

Thomas Bataillon
 Bioinformatics Research Center
 Aarhus University
 8000 Aarhus C
 Denmark

Zachary D. Blount
 BEACON Center for the Study of Evolution in Action
 East Lansing, MI 48824
 USA
 and
 Department of Microbiology and Molecular Genetics
 Michigan State University
 East Lansing, MI 48824
 USA

Luis-Miguel Chevin
 Centre d'Ecologie Fonctionnelle & Evolutive, UMR 5175
 34293 Montpellier 5
 France

David J. Depew

  Professor of Communication Studies and Rhetoric of Inquiry, Emeritus
  University of Iowa
  Iowa City, IA 52242
  USA

Eric Desjardins

  Rotman Institute of Philosophy
  University of Western Ontario
  London, Ontario N6A 5B8
  Canada

Douglas H. Erwin

  Department of Paleobiology
  National Museum of Natural History
  Washington, DC 20013
  USA

Jonathan Hodge

  School of Philosophy, Religion, and History of Science
  University of Leeds
  Leeds, LS2 9JT
  United Kingdom

Betul Kacar

  Organismic and Evolutionary Biology
  Harvard University
  Cambridge, MA 02144
  USA

Thomas Lenormand

  Centre d'Ecologie Fonctionnelle & Evolutive, UMR 5175
  34293 Montpellier 5
  France

Lucas John Matthews

  Department of Philosophy
  University of Utah
  Salt Lake City, UT 84112
  USA

Francesca Merlin
Institut d'histoire et de philosophie des sciences et des techniques
CNRS / Univ. Paris 1 / ENS
Paris
France

Daniel Molter
Department of Philosophy
University of Utah
Salt Lake City, UT 84112
USA

Charles H. Pence
Department of Philosophy and Religious Studies
Louisiana State University
Baton Rouge, LA 70803
USA

Anya Plutynski
Department of Philosophy
Washington University in St. Louis
St. Louis, MO 63130
USA

Grant Ramsey
Institute of Philosophy
KU Leuven
BE-3000 Leuven
Belgium

Michael Ruse
Program in the History and Philosophy of Science
Florida State University
Tallahassee, FL 32306
USA

Michael Strevens
Department of Philosophy
New York University
New York, NY 10003
USA

Kenneth Blake Vernon
Department of Philosophy
University of Utah
Salt Lake City, UT 84112
USA

INDEX